Many-Particle Theory
and Its Application to Plasma

RUSSIAN MONOGRAPHS AND TEXTS
ON ADVANCED MATHEMATICS AND PHYSICS

Volume I

A. I. KHINCHIN, A Course of Mathematical Analysis

Volume II

S. G. MIKHLIN, Linear Integral Equations

Volume III

P. P. KOROVKIN, Linear Operators and Approximation Theory

Volume IV

L. E. EL'SGOL'TS, Differential Equations

Volume V

L. A. LUSTERNIK and V. J. SOBOLEV, Elements of Functional Analysis

Volume VI

V. L. GINZBURG, Propagation of Electromagnetic Waves in Plasma

Volume VII

A. A. VLASOV, Many-Particle Theory and Its Application to Plasma

Additional volumes in preparation

Many-Particle Theory and Its Application to Plasma

by A. A. VLASOV

GORDON AND BREACH
Science Publishers, Inc., New York

Originally published by

State Publishing House for Technical-Theoretical Literature
Moscow and Leningrad, 1950.

Library of Congress Catalog Card Number 61-18831

GORDON AND BREACH, Science Publishers, Inc.
150 Fifth Avenue, New York 11, New York

PRINTED IN THE UNITED STATES OF AMERICA

FOREWORD

The present monograph is an exposition of a new theory of many particles, developed by the author and his students. The book generaiizes work of the author previously published in the periodical press. The monograph is designed for theoretical and experimental physicists.

The basic problems of the theory expounded here arose in the course of an investigation into the property of many-electron systems, an important place among which is taken by what is called electron plasma. Such systems have special properties, which required the elaboration of a method that took into account not only "near" interactions (at distances smaller than the mean distance between particles), but also "distant" interactions, operating at distances greater than the magnitude in question. A particular method for calculating these interactions proved to be extremely accurate in expressing the fundamental properties of many-electron systems. On the basis of the proposed method, it was found possible to explain a large number of the observed phenomena. The principal superiority of this method consisted in its comparative simplicity and effectiveness.

In the course of working these questions out, the problem arose as to the interrelationship between the distant and near interactions. The distant interactions express the connection of each particle with the entire ensemble as a whole. On the contrary, the near interactions are usually calculated with the help of the method of "collisions," employed in classical kinetic theory of gases. The distinguishing feature of this method is that only the interactions are taken into account that arise between two colliding particles.

Is it possible, by means of a systematic calculation of double, triple, quadruple, etc. collisions, finally to obtain the distant interactions in an ensemble of particles? Does the method based on calculating the distant interactions include the ordinary method of collisions as a particular or limiting case? This problem proved to be a complicated one. In the first place, there arose the question of the presence, the role and the correct calculation of the distant interactions for neutral particles as well. Application of this method to systems of neutral particles led to a definite success, since the problem of the nature of the crystallization process was solved. Continuing along this path, it proved to be possible to discover a number of characteristic properties of the propagation of waves in a system of many particles.

Introducing the distant interactions for an ensemble of neutral particles made the problem raised above sharper. In point of fact, collision conceptions proved to be thoroughly applicable in the case of neutral particles, and the method proposed was based exclusively on the equation of continuity and on the integral method of calculating the interactions. This method did not seem clearly to contain impact conceptions. Its greater generality only emerged gradually.

The second stage in the development of the theory consisted in attempts to supply a basis for this method, starting with a precise formulation of the n-body problem in Gibbs' statistical mechanics and quantum mechanics. Physical and mathematical analysis of this problem showed that a complete derivation of the proposed method from the theories mentioned above is impossible. This was connected with the presence of a number of difficulties in the theories mentioned, and likewise with the physical elements contained in the method in question. For example, attempts to include electrodynamic interactions in the statistical mechanics of many particles, on the basis of Liouville's theorem, lead to difficulties connected with the localization of charged particles. The method set forth, on the other hand, arose in the course of a study of an ensemble of electrons, and does not contain difficulties of this nature. Another difficulty of classical and statistical mechanics consists in the impossibility of obtaining from those theories the mere fact of the formation of the crystal structure from a liquid upon continuous decrease of temperature. This circumstance clearly indicates that the apparatus of statistical mechanics fails to take fully into account the effects of collective interaction. Quantum mechanics is likewise not in a position to answer this question. From the mathematical point of view, this follows from the fact that the equations of quantum mechanics do not lead to "branching" solutions, which play a great part in the theory set forth. This relates equally to the approximate quantum-mechanics method, known as the method of the self-consistent field.

These considerations led the author to feel that the method set forth is qualitatively new.

In the construction of the theory, three factors seemed to be the guiding ones from the point of view of methodology.

1) *Abandonment of a strictly localized description of microparticles.* The conception of a particle is a point conserving this property independently of any connection with the medium and other particles is only an approximate reflection of the reality. At the present time, both the experimental facts and the internal difficulties of existing theories make it essential to create a theory based on a new treatment of the concept of the particle. In particular, the theory should express independence of the extension of the particle on the physical conditions encountered by the particle.

2) *A new approach to the concept of the closure of a physical system.* It seems to us that in essence there is even in the existing theories a departure from the concept of closure in the sense of classical mechanics (the apparatus of statistical and quantum mechanics). The particle dynamics developed in this book is characterized by a special method of calculating the connection of the particle with the surrounding medium. Thanks to this, the concept of closure acquires a new sense. The transition to classical theories here is linked with the introduction of certain limitations to this connection.

3) *The attempt to construct a theory in which motion would be an inseparable property of the object, and would not be the result of the action of any "sources" (forces — in classical mechanics, the heat reservoir — in statistics, charges and currents — in electrodynamics).* The effect of sources of this nature is that in essence the motion is introduced into the physical system from without. The idea in question is realized in the present theory by having each particle described by an extended function of the space distribution of coordinates and velocities.

It should be stressed that the present theory is not in contradiction to the classic theories, but makes it possible to set more precisely the boundaries within which they apply. For example, it is explained in what cases the classical concepts of the localized particle, the closed system, etc. are applicable, and in what cases they lose their meaning and should be rendered more general. We shall see that the equations of classical mechanics, and in a certain sense, those of statistics as well, are obtained from the equations of the theory advanced here as particular cases under definite conditions.

The applications of the theory are still far from having been exhausted. Up to the present, the following results have been obtained:

1. The vibration properties of many-electron ensembles have been studied in detail, and a number of phenomena considered for which these properties may be considered as decisive.

2. Various means of exciting characteristic oscillations in the plasma have been proposed and analyzed.

3. An explanation of the layers has been given.

4. An explanation has been given of the nature of crystallization and some new properties of the crystal indicated.

5. The existence has been shown of an effect of sudden appearance of sound, if the medium is sufficiently dense and the temperature sufficiently low.

6. The existence has been shown of certain velocities of sound in sets of interacting particles, if the forces of repulsion dominate over the forces of attraction.

7. An analysis was given of the role of interactions between electrons in electron-discharge tubes.

In conclusion, I feel it my duty to express my thanks to my students A. G. Pereleshin, V. A. Yakovlev, A. A. Luchina, I. P. Bazarov, N. I. Goldman, K. S. Tokareva, and G. M. Myashkev, who took part in working out particular questions of theory, and likewise to V. T. Khozyairov, who gave great help in preparing the book for the press.

A. VLASOV

TABLE OF CONTENTS

PART II
APPLICATIONS OF THE THEORY

PART I

FOUNDATIONS OF THE THEORY

CHAPTER I

NEED FOR THE PROPOSED THEORY

Section 1. Difficulties Arising Upon Application of Boltzmann's Method to Sets of Charged Particles, and Their Elimination

Is Boltzmann's equation (and the gas-kinetic method connected with it) correct for systems of particles interacting according to Coulomb's law? This question was considered in an article by the author*, in which it was clearly shown that the gas-kinetic method cannot be applied in this case, wherefore a new method was proposed to replace it.

On the other hand, present theories of such ensembles are based on the analogy with ordinary gas consisting of neutral particles. The assumption is here that the motion of each particle takes place by inertia, with the exception of brief periods of proximity (impact). Systematic elaboration of this view was based on the basic concept of kinetic gas theory (diameter of cross-section, length and time of free path, etc.), and on Boltzmann's basic equation for this theory.

As we know, if $f(x, y, z, \xi, \eta, \zeta, t)$ is the distribution function determining the probable number of particles in the element of phase space dr dv, then Boltzmann's equation takes the form

$$+\frac{\partial f}{\partial t}+\mathrm{div}_r\, vf=\left[\frac{\partial f}{\partial t}\right]^{st}, \tag{1.1}$$

where the term $\left[\frac{\partial f}{\partial t}\right]^{st}$ expresses the change in distribution function due to impact and is defined by the well-known relation

$$\left[\frac{\partial f}{\partial t}\right]^{st}=\int_{-\infty}^{\infty}\int_{0}^{\infty}\int_{0}^{2\pi}(ff_1-f'f'_1)\, Vb\, db\, d\omega_1\, d\varepsilon. \tag{1.2}$$

Here the integration extends over all the velocities ξ_1, η_1, ζ_1 and for all values of the parameters of the impact b and ε. It is assumed that Equations (1.1) and (1.2) express the dynamics of particles with any law of the forces of interaction. That is, the law of interaction between the particles determines the dependence of the range b on the angle of deflection θ and angle ε. Accordingly, the specialization of the system under discussion reduces merely

* A. A. Vlasov, Zh. E.T.F. 8, 291 (1938).

to fixing the law of interaction among the particles. In the case under consideration, that of charged particles, this interaction is a Coulomb interaction.

The apparatus of the kinetic theory of gases is applied to systems of electrons in the investigations of Thomas, Langmuir, Landau, Druveystein, Davydov, Gvozdover, et al. Work in this field is continuously appearing down to the present time (a part of it having gone into monographs and textbooks)*.

It is clear, however, that the scheme of the kinetic equation, and the ordinary conception of a gas on which that sheme is based, are a particular approximation in the solution of the problem of many bodies, an approximation based on a special method for calculating the interaction, that is, the calculation is made only for paired interactions between the particles. The assumption is made that each particle interacts at a given moment only with a single particle of the rest of the ensemble, and that that interaction has the character of an "impact." However, for such an approximation to be meaningful, it is necessary for the following three conditions to be satisfied.:

1. The forces of interaction between the particles must be of such a nature that it is possible to introduce the concept of sphere of action. This means that we may ignore the changes of the distribution function due to the influence of particles going off a long distance (as compared to the radius of the sphere of action).

2. The gas must be sufficiently rarefied, so that the average distance between the particles is large compared to the radius of action of the forces.

3. We must be able to ignore the intervals of time in the course of which the process of "collision" takes place. This condition is employed implicitly in the apparatus of kinetic theory, as appears quite explicitly in the apparatus of stochastic processes, which is essentially equivalent to the first apparatus.

Any deviation from these three postulates takes us outside the ordinary framework of kinetic theory. For example, in the case of interaction according to Coulomb's law, we are concerned with forces that decrease sufficiently slowly with the distance. In this case, condition 1 is not satisfied; each particle interacts at once with a whole system of other particles, and therefore the problem of many bodies can no longer be reduced to the ordinary pattern of the kinetic equation.

The inadequacy of calculating merely the paired interaction is visible,

---

* L. Thomas, Proc. Roy. Soc. a. 124, 464 (1928); L. Langmuir, Proc. Nat. Acad. 14, 627 (1928); L. Landau, Phys. Z. Sow. Union 10, H. 2 (1936); M. Druveystein, Physica 562 (1938); B. Davydov, Phys. Z. Sow. Union 9, 433 (1936); S. Gvozdover, Phys. Z. Sow. Union 12, 164 (1937); L. Gurevich, Osnovy fizicheskoi kinetiki /Foundations of Physical Kinetics/ (1940); I. H. Cahn, Phys. Rev. 75, 293, 346, 838 (1949).

for example, from the fact that a variation of the relative density of positive and negative particles at any point of space will be connected with the appearance of a space charge, which will act on the motion of the charged particles located enormously more distant than the average distance between particles. In that case, calculating only the successive paired interactions is by no means an expression of the reality. An essential role must be played here by forces of interaction at distances greater than the mean distance between particles (we shall call them distant forces), the action of which cannot be calculated by the ordinary pattern of the kinetic equation.

This circumstance also appears, for example, in the fact that solving the problem of dispersion at the Coulomb center leads to a divergent expression for the entire diameter, and in the specific as a result of the action of forces at great distances. For, as we know, the effective diameter of elastic dispersion at the Coulomb center for small angles of deflection is given by the expression

$$\underset{\theta \to 0}{d\sigma} = 4\left(\frac{e^2}{mv^2}\right)^2 \frac{d\omega}{\theta^4},$$

in which v, m and e are the initial velocity, mass and charge, respectively, of the dispersed particles. For $\theta \to 0$ (distant paths), the expression above approaches infinity, so that the entire diameter of dispersion $\left(\int d\sigma\right)$ is divergent.

The physical cause for the diversion of this integral consists in the fact that, although for long paths the particles change their trajectories only slightly, still the probability of such paths increases rapidly as the range increases. As a result, the radius of the sphere of action becomes infinite, and the corresponding integrals in the kinetic equation have no meaning.

A similar state of affairs holds true as well for nonelastic collisions. According to the conceptions of the theory of paired collisions, the physical cause of the losses of energy of a charge moving in a system of free electrons, is the transfer of the kinetic energy of the moving particle to other electrons in a series of paired interactions of the type of impacts. In this way, the problem of the complex interaction of a particle colliding with an entire ensemble of electrons is replaced at each moment by a problem of only two bodies.

In the case where the velocity of the moving particle considerably exceeds the velocity of the electrons of the ensemble, the relationships are the simplest, and are well known+. If N is the concentration of electrons and p the range, then the mean value for the number of collisions along a path dz, for which the magnitude of the range lies between p and p + dp, will be $2\pi Np\,dp\,dz$. The energy loss of a moving particle with charge e

+ N. Bohr, Phil. Mg. 30, 581 (1915); 25, 10 (1913).

and mass M in a single act of interaction with a free electron (charge e and mass m) is equal to:

$$Q = \frac{2\varepsilon^2 e^2}{mv^2} \frac{1}{p^2 + a^2}, \quad \text{where} \quad a = \frac{\varepsilon e (M+m)}{Mmv^2}.$$

Hence, for complete loss of energy over unit path, we have:

$$-\frac{dE}{dz} = \frac{4\pi\varepsilon^2 e^2}{mv^2} \int \frac{p\,dp}{p^2 + a^2},$$

where integration should be performed for all values of p from 0 to ∞. **The integral obtained diverges.** Once again, the divergence is due to interactions at great distances (greater than p). The cause is the same as in the case of elastic dispersion. The fact of the divergence of the integral clearly indicates the role of distant interactions and, strictly speaking, makes it impossible to describe a system of electrons by the ordinary methods of the kinetic theory of gases, since a fundamental condition for the existence of an entire diameter of dispersion is not satisfied.

Instead of calculating the distant forces in the theory and evaluating their action, all the above-mentioned works are characterized by an effort to retain the ideas and conceptions of the theory of paired collisions, and along with them, the entire apparatus of the kinetic theory of gases. To this end, in integrating the range is cut off at some definite value. Langmuir and Landau limited the range to Debye's length, Druveystein to a magnitude greater than the Debye distance, Davydov and Gvozdover to the average distance between particles. It is obvious that by acting in this way, the authors in question avoid calculating the distant interactions, failing to appreciate their role. These theories cannot be regarded as of full value.

On the basis of the foregoing, it may be said that the method of the kinetic equation, taking into account only paired interaction (interaction by means of impact) for a system of charged particles is an unsatisfactory approximation, and that in the theory of such ensembles a comparable role must be played by the forces of interaction both at near and at great distances, and that consequently a system of charged particles is essentially not a gas but some kind of system with a nature entirely its own, and held together by distant forces.

It cannot be held that we are making progress in the solution of the problem in question if we proceed by calculating sufficiently great ranges (as is done in the above-mentioned works of Langmuir, Landau, Davydov, et al.), since this means retaining the fundamental defect of the gas-kinetic method: the reduction of particle interaction to the interaction of only two bodies moving in isolation. It is of little use to introduce triple, quadruple, etc. collisions, since even the problem of three bodies has not yet been solved in classical mechanics.

A new step must be taken in this complex question.

We divide the Coulomb field into two parts: first, interaction and distances less than the average distance between particles, and secondly, distant interactions. To distinguish the near interactions, we cut off the Coulomb field, for example, at half the mean interparticle distance. This makes it possible to retain the idea of collisions, and in fact we can calculate this part of the interaction merely according to the pattern of the kinetic equation.

As we have already shown, in the statistical terms (1.2), the law of interaction between particles appears in the definition of the function $b = b(\theta, \varepsilon)$. For example, for the impact of elastic spheres

$$b = \sqrt{\sigma} \cos \frac{\theta}{2}.$$

$$\left[\frac{\partial f}{\partial t}\right]_{11}^{st} = \int (ff_1 - f'f'_1) v\sigma \cos \frac{\theta}{2} \sin \frac{\theta}{2} d \frac{\theta}{2} d\varepsilon.$$

For Coulomb interaction

$$b = \frac{2e^2}{mv^2} \operatorname{ctg} \frac{\theta}{2}.$$

For angles of dispersion θ (θ = obtuse angle) lying in the interval $\frac{\pi}{2}$ to π, the functions cosine $\frac{\theta}{2}$ and ctg $\frac{\theta}{2}$ and their first derivatives are almost identical. For $\frac{\theta}{2} = \frac{\pi}{2}$ the terms cosine $\frac{\theta}{2}$ sin $\frac{\theta}{2}$ at ctg $\frac{\theta}{2}$ $\sin^2 \frac{\theta}{2}$ coincide, for $\frac{\theta}{2}$ = 80° the ratio of the second to the first is approximately 1.08 for 60°, 1.7 and for 45° 3. Hence, the formulas for these angles have a similar appearance, but only in the case of Coulomb interaction instead of the constant cross-section diameter σ we have the variable quantity $\frac{2e^2}{mv^2}$. If we designate by v the constant mean velocity of the particles (which is quite permissible in evaluating the orders of magnitude), the corresponding integrals will actually be almost identical, if only the magnitudes of the cross-sections are identical. In this way, the formula

$$\sigma = 4\left(\frac{e^2}{mv^2}\right)^2 \qquad (\alpha)$$

should define the order of magnitude of the sphere of action for large angles of deviation, considering as v the mean velocity of particles.

For angles less than $\frac{\pi}{2}$, there is no longer an analogy with the case of elastic spheres, and an evaluation of the role of $\left[\frac{\partial f}{\partial t}\right]_{11}^{st}$ must be made in a different manner. Calculation of only the distant paths $\left(0 < \theta < \frac{\pi}{2}\right)$ in the framework of the usual pattern of the kinetic equation was conducted by Langmuir (1928) and Landau (1936). They give as the formula for the length of the free path $l \simeq \frac{(kT)^2}{e^4 LN}$.

whence for cross-section σ we have:

$$\sigma = \frac{1}{\pi l N} \cong \frac{1}{\pi}\left(\frac{e^2}{kT}\right)^2 L,$$

or, since $3kT/2 = mv^2/2$:

$$\sigma = \frac{9}{\pi}\left(\frac{e^2}{mv^2}\right)^2 L,$$

$$(\beta)$$

which differs from formula (α) only by the logarithmic term $L = \ln \frac{b_2}{b_1}$, where b_1 and b_2 are the minimum and maximum values of the range. Consequently, calculating merely the near or merely the distant paths leads, apart from the logarithmic term, to the same sort of expressions for the sphere of action. The minimum distance should be taken as that corresponding to $\theta = \pi/2$, namely,

$$b_1 = \frac{2e^2}{mv^2}.$$

The maximum value b_2 should lie between b_1 and half the mean interparticle distance+. Setting $b = \frac{1}{2}N^{-\frac{1}{3}}$, we shall have L:

$$L = \ln \frac{3kT}{4N^{1/3}e^2}.$$

Considering the relationship of densities and temperatures, L may be greater or less than unity, and correspondingly the size of the sphere of action may be defined essentially both for distance ($b_2 > b_1$) and for near ($b_2 < b_1$) paths. For ordinary conditions, in plasma $L \sim 10$. Hence, the order of magnitude of the sphere of action (taking into account both near and distant paths) may be evaluated by the formula

$$\sigma = \alpha\left(\frac{e^2}{mv^2}\right)^2,$$

where $\alpha \sim 10$.

+ It was not quite correct to say that b_2 is a Debye distance, since it may be greater than the mean interparticle difference, and since in this case several other particles will be within the sphere of action of a single particle, the concept of "impact" loses its meaning, and with it the original customary scheme of the kinetic equation, as well.

In the conditions of the ionosphere N $\sim 10^6$ el/cm^3, we assume T equal to 300° K, and we then obtain the following values for sphere of action σ, the length of the free path 1 and the frequency of collisions ν_{st}:

$$\sqrt{\sigma} \cong \sqrt{3\alpha}10^{-6}\ см;$$

$$l = \frac{1}{\pi\sigma^2 N} \cong \frac{1}{3\pi\alpha}10^6 \cong 10^4\ см;$$

$$\nu_{st} = \frac{v}{l} \cong 10^8\ сек^{-1}.$$

For electron plasma in discharge tubes, assuming N = 10^{10} el/cm^3, T = 10^4°K, we have:

$$\sqrt{\sigma} \cong \sqrt{27\alpha}10^{-8} \sim 10^{-7}\ см, \qquad l \sim 10^8\ см, \qquad \nu_{st} \sim 10^4\ сек^{-1}.$$

For the interaction by means of impact between electrons and ions, the expression for cross-section σ has, as will easily be seen, the same expression (A), where m will be the effective mass of these particles and, consequently, the valuation of σ, l, ν_{st} will remain in fact identical.

Calculation of the distant interactions may be conducted in the following manner. On the basis of considerations concerning the general structure of the equation continuity, we may add to equation (1.1) a term taking into account the action of forces. It should have the form

$$\mathrm{div}_v\, gf,$$

where the divergence is taken in the velocity space, and g is the acceleration vector. We calculate the effect of particles as distant as we please on the variation of the distribution function at a given point of the space if we require that the acceleration be determined by the strains in the electric and magnetic fields, which in turn are conditioned by the values of f at all points of the space:

$$\left.\begin{array}{c} g = \frac{e}{m}\left(e + \frac{1}{c}[vh]\right), \\ \mathrm{div}\, e = 4\pi\rho; \quad \mathrm{rot}\, h - \frac{1}{c}\frac{\partial e}{\partial t} = \frac{4\pi}{c}j, \\ \rho = e\int f\,dv; \quad j = e\int vf\,dv. \end{array}\right\} \qquad (1.3)$$

The expressions for the charge and current densities are defined on the basis of ideas as to the mean values. If ψ is any magnitude connected with the particle, then

$$\bar{\psi} = \frac{\int \psi f\,dv}{\int f\,dv}.$$

Normalizing the distribution function for the density of particles n

$$\int f\,dv = n$$

the expression for the density ρ_ψ of any quantity ψ will be:

$$\rho_\psi = n\bar{\psi} = \int \psi f\,dv.$$

Assuming $\psi = 1,\ v,$, we obtain an expression for the density of the particles and the density of the current.

Thus, in calculating the distant interactions, we must write down the initial equation for f in the form

$$\left.\begin{aligned}
&\frac{\partial f}{\partial t} + \operatorname{div}_r vf + \operatorname{div}_v \frac{e}{m}\left(e + \frac{1}{c}[vh]\right)f = \left[\frac{\partial f}{\partial t}\right]^{st},\\
&\operatorname{div} e = 4\pi e\int f\,dv,\\
&\operatorname{rot} h - \frac{1}{c}\frac{\partial e}{\partial t} = 4\pi e\int vf\,dv\\
&\quad\left(\operatorname{div} h = 0;\quad \operatorname{rot} e + \frac{1}{c}\frac{\partial h}{\partial t} = 0\right);
\end{aligned}\right\} \quad (1.4)$$

The generalization to the case of the presence of several kinds of particles is elementary. Thus, for instance, for a mixture of three kinds of particles (electrons, ions and neutral atoms) we have:

$$\left.\begin{aligned}
&\frac{\partial f_1}{\partial t} + \operatorname{div}_r vf_1 + \operatorname{div}_v \frac{e_1}{m_1}\left(e + \frac{1}{c}[vh]\right)f_1 =\\
&\qquad = \left[\frac{\partial f_1}{\partial t}\right]^{st}_{11} + \left[\frac{\partial f_1}{\partial t}\right]^{st}_{12} + \left[\frac{\partial f_1}{\partial t}\right]^{st}_{13},\\
&\frac{\partial f_2}{\partial t} + \operatorname{div}_r vf_2 + \operatorname{div}_v \frac{e_2}{m_2}\left(e + \frac{1}{c}[vh]\right)f_2 =\\
&\qquad = \left[\frac{\partial f_2}{\partial t}\right]^{st}_{21} + \left[\frac{\partial f_2}{\partial t}\right]^{st}_{22} + \left[\frac{\partial f_2}{\partial t}\right]^{st}_{23},\\
&\frac{\partial f_3}{\partial t} + \operatorname{div}_r vf_3 = \left[\frac{\partial f_3}{\partial t}\right]^{st}_{31} + \left[\frac{\partial f_3}{\partial t}\right]^{st}_{32} + \left[\frac{\partial f_3}{\partial t}\right]^{st}_{33},\\
&\operatorname{div} e = 4\pi\rho,\quad \operatorname{rot} h - \frac{1}{c}\frac{\partial e}{\partial t} = \frac{4\pi}{c}j,\\
&\operatorname{div} h = 0,\quad \operatorname{rot} e = -\frac{1}{c}\frac{\partial h}{\partial t},\\
&\rho = e_1\int f_1\,dv + e_2\int f_2\,dv;\ j = e_1\int vf_1\,dv + e_2\int vf_2\,dv,
\end{aligned}\right\} \quad (1.5)$$

where $\left[\frac{\partial f_1}{\partial t}\right]_{11}^{nt}$, $\left[\frac{\partial f_2}{\partial t}\right]_{21}^{nt}$ etc. contain the interactions by means of impact between particles of the same and different kinds.

The proposed method for describing a system of electrodynamically interacting particles raises the following fundamental question.

Up to the present, in introducing the distribution function f into Boltzmann's equation, it has always been assumed that there exists a physically infinitely small volume, which, despite the discreteness of the medium, ensures the continuity of the function f and its first derivatives. The particles are here understood in the sense of classical mechanics, that is, as point masses. which are centers of force.

Does the new method of calculating interaction require the previous conception of the f function, linked up with ideas of points moving along trajectories of classical mechanics?

For what follows we introduce the concept of a <u>localized</u> particle as a point, conserving its localization and individuality independent of the nature of its physical connections with the surrounding medium. On the other hand, <u>abandonment of localization</u> will denote the assumption that the particle has properties that make the point conception unacceptable and require the introduction of a space-extension continuous model for the particle. The abandonment of localization, as we shall see below, would be linked up with calculation of the interaction of the particle with the surrounding medium.

This interaction may in a particular case lead to strict localization of the particles in the sense given above. The theory advanced may therefore be called a theory of fully localizable particles, which is different in principle from quantum mechanics.

To answer the question asked above, we observe:

1. The continuity equation (one of the modifications of which is Boltzmann's equation) in the general form

$$-\frac{\partial f}{\partial t} = \mathrm{div}_r\, \boldsymbol{v} f + \mathrm{div}_v\, \boldsymbol{g} f + \mathrm{div}_g\, \boldsymbol{w} f + \ldots \qquad (1.6)$$

(where the values of <u>g</u>, <u>w</u>, etc. have not yet been discovered) does not require any "Boltzmannian" understanding of function f. This equation retains its meaning even for a single particle, whose motion is described by extended function f. The continuity equation constitutes merely a conservation law.

2. The calculation made of the interaction between particles by means of currents and charges, expressed through f, is likewise free of the condition of compulsory localization of particles in the sense of classical mechanics.

In the case of point particles, we must write:

$$\left.\begin{aligned} \rho &= \sum e_i\, \delta(\boldsymbol{r} - \boldsymbol{r}_i(t)), \\ \boldsymbol{j} &= \sum_i e_i \boldsymbol{v}_i(t)\, \delta(\boldsymbol{r} - \boldsymbol{r}_i(t)). \end{aligned}\right\} \qquad (a)$$

where δ is Dirac's delta function. But equations (1.4) contain the expressions

$$\rho = e \int f\, dv; \quad j = e \int vf\, dv, \qquad (\beta)$$

and for the case of different particles:

$$\rho = \sum_i e_i \int f_i\, dv, \quad j = \sum_i e_i \int vf_i\, dv. \qquad (\gamma)$$

Expressions (α) and (γ) are different: equation (γ) includes (α) as a particular case, when f_i - the distribution function for each particle - differs significantly from zero only at points $\underline{r}_i$ and $\underline{v}_i$. In the general case, strict localization of the particles in phase space is not assumed in expressions (γ) and (β).

3. Finally, the passage from the f function to mean values is a calculatory procedure, which likewise does not require necessary satisfaction of the laws of classical mechanics.

Accordingly, the entire pattern of the proposed method is free from the necessity of point fixation of the particle in the coordinate and velocity space. At the same time, the conception of classical mechanics as to the localization of particles is essentially applied to obtain the Boltzmann term considered above $\left[\frac{\partial f}{\partial t}\right]^{st}$. Consequently, this term is not in direct correspondence with the other terms of the equation.

The presence of the Boltzmann term decreases the field of application of the equation.

Function f may describe the behavior of a single arbitrarily chosen particle of the ensemble. And it is only in particular cases, when the region in which f is different from zero, may be approximately replaced by a point, that is it meaningful to speak of collisions. But if we do not wish to confine ourselves in this particular case, then we must omit $\left[\frac{\partial f}{\partial t}\right]^{st}$ in equation (1.4). With this (1.4) requires a new form and a new content. In this equation (without collisions), the number of particles of the system under consideration plays no part, and there is as yet no reason for considering it inapplicable to any number of particles, including the case of the single particle.

It will be seen from the above that the difference in principle of the method proposed for describing systems of charged particles from Boltzmann's method is comprised in the following:

1. The principle of strict point localization of particles in the sense of classical mechanics is abandoned.

2. The behavior of each particle of the system is described by means of the f-function extended in phase space.

3. Calculation is made not only of the near interactions but also of interactions as distant as desired of each particular particle with the entire collective of particles as a whole.

It is essential to note here that the mathematical apparatus of the method proposed does not as yet require a strict definition of the meaning of the f-function; it may express the statistical dispersion in the position of the point particle, or may be a generalization of the very concept of particle. In the second case, the f-function gives us a new model of a particle, extended in six-dimensional phase space.

As we shall see below, the present theory is basically developed on the basis of the second conception, although the possibility of another interpretation of the f-function is also taken into consideration. This question is also discussed in Sections 6 and 7.

Section 2. Collective Interactions in the Case of Arbitrary Central Forces

A particularly characteristic feature of the method set forth (see Sec. 1) is the dynamic connection of every particle with the ensemble as a whole. The question arises: is there, in other real systems with a non-Coulomb law of forces, any such type of interaction, and if there is, how is it to be calculated?

If we conceive of an ensemble of particles with forces acting at a short distance, but with the ensemble so dense that several particles are present at once within the sphere of action of the forces, then the interactions connecting many particles at once must be calculated. Accordingly, from our point of view, it is not only the strictly Coulomb forces that can act as distant forces. For the necessity of calculating the distant collective interactions in the system to arise, all that is required is the presence of forces applying to many particles at once, which is in fact usually the case.

Actually, we know that there are in general no forces in nature that become precisely zero at a certain distance. Our intuitive tendency to ignore such weak, but really existing interactions at great distances (distances exceeding the mean distance between particles) is unjustified, since this means neglecting the collectivizing effect, and at the same time, it must be expected, that we will have to abandon the explanation of a number of phenomena.

In a 1944 article[+] on "Generalization of the Conception of Electron Plasma," the means of calculating collective interactions was extended to arbitrary central forces.

For the Coulomb law of forces calculation of the distant interactions was effected by means of introducing into the continuity equation for the distribution function forces that in turn depended on the values of the distribution function at all the points of the space, for example, as follows:

[+] A. Vlasov, Isvestiya Akademiya Nauk, seriya fizich. VIII, No. 5, 248 (1944).

or

$$
\left.\begin{aligned}
F &= -\operatorname{grad}_r e\varphi, \quad \Delta\varphi = -4\pi e \int f\, dv \\
F &= -\operatorname{grad}_r \int \frac{e^2}{|r-r'|} \rho(r_1', t)\, dr', \\
\rho(r,t) &= \int f(r, v, t)\, dv,
\end{aligned}\right\} \tag{2.1}
$$

For any arbitrary central forces we can generalize this expression by the following source-like conception:

$$
\left.\begin{aligned}
F &= -\operatorname{grad}_r \int K(|r-r'|)\, \rho(r', t)\, dr', \\
\rho(r, t) &= \int f(r, v, t)\, dv.
\end{aligned}\right\} \tag{2.2}
$$

Hence, we shall have, for the equation with distant-acting forces:

$$
\left.\begin{aligned}
&\frac{\partial f}{\partial t} + \operatorname{div}_r vf - \operatorname{div}_r \frac{1}{m} \operatorname{grad} V \cdot f = \left[\frac{\partial f}{\partial t}\right]^{st}, \\
&V(r, t) = \int K(|r-r'|) \int f(r', v' t)\, dv'\, dr'.
\end{aligned}\right\} \tag{2.3}
$$

This equation corresponds to equation (1.4) for the case of charged particles. In studying it we therefore follow the same path as Sec. 1.

What is the specific nature of the problem arising in connection with equation (2.3)?

Equation (2.3) includes simultaneously interaction of two different types: impact and integral interactions. Undoubtedly both of these types must occur in real systems. However, equation (2.3) does not answer the question as to under what circumstances one type occurs, and under what circumstances the other.

Further, it is impossible to know the simultaneous presence in the equation of both interaction terms. In the integral term, calculation is made of interactions at all possible distances, both large and as small as desired. This is connected with the fact that nucleus K (r, r') expresses the complete and accurate interaction energy, and the integration is over all distances. Consequently, keeping both interaction terms in equation (2.3) signifies simultaneous and independent calculation of one and the same physical fact in different terms of one and the same equation, which seems unreasonable.

How can we hope to obtain a synthesis of the near paired consecutive interactions with the distant collective ones? This problem arose in the author's mind after the articles of 1938 and 1944 to which reference has been made.

On the one hand, there is reason for completely abandoning the statistical member $\left[\frac{\partial f}{\partial t}\right]^{st}$, in equation (2.3), since first, introducing it involves many limitations (on the character of the action of the forces, on the density of the gas, on the time interval), while the integral calculation of interaction is free from such limitations; secondly, as has been said, the integral members should include the near interactions as well.

On the other hand, the following too clearly hold good. Despite the desirability, already referred to, of introducing integral interactions, it appears at first glance that this method could not pretend to generality and precision. The fact is that in the integral terms the interaction is calculated only summarily, by means of the distribution functions. It therefore appears that the change of state of any definite fixed particle due to near interactions giving rise to particularly strong fluctuations in the motion of the particles, is not taken into account by this means of description. The problem therefore appears to be a complicated one.

A publication of the author in 1946[+] gave a new treatment of this problem; essentially, it was a solution of it.

As we know, the development of the idea of strict spatial localization of particles historically preceded the introduction of the distribution functions. But the elaboration of the theories based on this idea leads not only to practical difficulties (see Secs. 3, 4, 5) but also to difficulties of a general physical nature (see Sec. 7). Inasmuch as motion is an inseparable property of matter, organically inherent in it, the possibility seemed an obvious one of constructing a theory such that in it the very fact of the presence of motion as a form of the existence of matter would be organically connected with the form of the material object, in this case the particle.

In our theory, the idea of the presence of motion, the fact of the existence of which does not depend on the concrete physical conditions, is expressed in the very initial model of the particle, extended not only in coordinate space but also in velocity space. Here, the entire ensemble of the other particles, and likewise the external physical conditions affect only the magnitude of the six-dimensional region occupied by the particle.

Accordingly, the extension of function f in six-dimension space is not the result of any compromises, but expresses a more general point of view on the dynamics of the behavior of particles in general. This point of view should include the previous one within it, since function f may in particular cases differ from zero only in extremely small regions of space, which here may be replaced by points.

From the point of view that is being set forth, equation (2.3) acquires a radically different meaning. First of all, the statistical member $\left[\frac{\partial f}{\partial t}\right]^{st}$, in obtaining which the assumption of localization of particles was essentially employed, must be abandoned, as referring to a completely special

[+] A. Vlasov, Vestnik Moskovskovo universiteta, Nos. 3-4, 63 (1946).

case. The rest of the equation is an equation of continuity, which does not require particular derivation. The integral interaction generalizes the force interactions of classical mechanics, since it takes into account the fact of the extension of the distribution function in six-dimensional space. Thus, omitting the collision term in (2.3), we already arrive at a new equation.

The method proposed solves the problem, posed above, of synthesizing the distant and the near interactions on the basis of advancing strict localization of particles, which is given once for all. This method generalizes the classical models of point particles (introducing the extension of particles by means of the distribution function), and in a natural manner includes the picture of collisions, but only in the particular case.

Section 3. Elimination of the Difficulties Arising in the Attempt to Unite Classical Electrodynamics and Mechanics

As we know, attempts to unite electrodynamics and mechanics have presented a number of difficulties.

1. The strains in the fields ($\underline{e}$ and $\underline{h}$) created by point charges are defined by the system of electrodynamic equations

$$\left.\begin{aligned} &\operatorname{div} \boldsymbol{e} = 4\pi \sum_i e_i \delta(\boldsymbol{r} - \boldsymbol{r}_i(t)), \\ &\operatorname{rot} \boldsymbol{h} - \frac{1}{c}\frac{\partial \boldsymbol{e}}{\partial t} = \frac{4\pi}{c} \sum_i e_i \boldsymbol{v}_i(t)\, \delta(\boldsymbol{r} - \boldsymbol{r}_i(t)), \\ &\left(\operatorname{div} \boldsymbol{h} = 0, \quad \operatorname{rot} \boldsymbol{e} + \frac{1}{c}\frac{\partial \boldsymbol{h}}{\partial t} = 0\right), \end{aligned}\right\} \tag{3.1}$$

(where the δ denotes Dirac's delta function), and become infinite at the point at which each particle is located. Accordingly, the precise value of the ponderomotor force (including the characteristic field of the particle) at the point occupied by the point particle itself loses its meaning.

More precisely, the expression for the ponderomotor force at the place at which the i-th particle is located has an indeterminate value (includes infinite values as well), depending on the law according to which the transition to the limit takes place in the formula

$$\lim_{\substack{\Delta r \to 0 \\ \Delta t \to 0}} \boldsymbol{e}(\boldsymbol{r}_i + \Delta \boldsymbol{r}, t + \Delta t) + \frac{1}{c} [\boldsymbol{v}_i \boldsymbol{h}(\boldsymbol{r}_i + \Delta \boldsymbol{r}, t + \Delta t)]^{*}).$$

Since the values of the fields at the point at which the point particle is located are infinite, the characteristic electrostatic energy likewise diverges. This circumstance makes it difficult to construct a rational dynamics of charged particles.

---

* See, e.g., Wentzel, Zs. f. Phys. 86, 489 (1933); 87, 726 (1934).

2. According to classical electron theory, the dynamics of charged particles qualitatively differs from the dynamics of neutral particles. The most important property of the dynamics of charged particles is the presence of a force of spontaneous action by means of which the process of radiation as an experimental fact is explained. This property makes it difficult to make a synthesis of electrodynamics and mechanics.

3. The difficulties that arise make it impossible consistently to include heat phenomena as well in the pattern of electron theory. For the thermal phenomena, the Gibbs' apparatus of statistical mechanics was developed, which essentially includes the idea of localized particles. If we should try to extend Gibbs' method of ensembles to systems of electrodynamically interacting particles as well, we should at once encounter the same difficulties as we find in unifying the mechanics of point particles with electrodynamics (see Sec. 4 of this chapter). The impossibility of synthesizing two parts of theoretical physics represents an essential defect from the point of view of general many-particle theory.

In the theory developed here, our starting point in defining the strains of the electric and magnetic fields should not be the delta problem of the positions and velocities of the charges, but only the distribution function. This likewise eliminates the basic difficulties referred to above.

We combine the results of Secs. 1 and 2. An ensemble of uniform particles interacting not only with electrodynamic forces, but also with forces of some other arbitrary nature, should be described by the following system of equations:

$$\left.\begin{aligned} &-\frac{\partial f}{\partial t} = \operatorname{div}_r \boldsymbol{v} f + \operatorname{div}_v \frac{1}{m}\left\{-\operatorname{grad} V + \frac{e}{m}\left(\boldsymbol{e} + \frac{1}{c}[\boldsymbol{v}\boldsymbol{h}]\right)\right\} f, \\ &V(\boldsymbol{r}, t) = \int K(|\boldsymbol{r} - \boldsymbol{r}'|) \int f(\boldsymbol{r}', \boldsymbol{v}', t)\, d\boldsymbol{r}'\, d\boldsymbol{v}', \\ &\operatorname{div} \boldsymbol{e} = 4\pi e \int f\, d\boldsymbol{v}; \quad \operatorname{rot} \boldsymbol{h} - \frac{1}{c}\frac{\partial \boldsymbol{e}}{\partial t} = 4\pi e \int \boldsymbol{v} f\, d\boldsymbol{v} \\ &\left(\operatorname{div} \boldsymbol{h} = 0, \ \operatorname{rot} \boldsymbol{e} + \frac{1}{c}\frac{\partial \boldsymbol{h}}{\partial t} = 0\right). \end{aligned}\right\} \quad (3.2)$$

We indicate the fundamental properties of these equations.

1. Equations (3.2) are valid for any number of similar particles and apply even for a single particle.

2. The method of description does not comprise any kind of singular functions (of the type of the delta-functions), and it may therefore be asserted that it guarantees freedom from the difficulties to which reference has been made.

3. It was an essential feature of the attempts hitherto made to introduce standard models of elementary charged particles that these models were introduced into the theory from without, and arbitrarily, while here the region in which function f exists is defined by the solution of equations, which include the interaction of the fixed particle with all the remaining particles, and, in addition, depends on the boundary conditions, determination

of which is defined by the physical meaning of the problem.

4. By virtue of the uniformity of inclusion of both the electrodynamic forces and the forces of any other nature, equations (2.3) are free from the second difficulty. It is therefore possible to speak of the unification of electrodynamics and mechanics on a new basis.

5. It will be shown later that in contradistinction to classic electron theory as well as to the theories in which attempts have been made to eliminate the well-known difficulties, the theory set forth comprises the thermal form of the movement of particles. Equations (3.2) have solutions (see Chapter III) containing a parameter adequate to the temperature.

The work of Born (nonlinear electrodynamics), Bopp-Podolsky (electrodynamics with higher derivatives) and Dirac (the classical theory of the point electron without electromagnetic mass)*, in addition to other difficulties, are characterized by a complete divorce from statistical methods in physics. If we recognize that the mere fact of the motion of particles (including the thermal form) should be organically comprised in the theory, we must recognize the point indicated in point 5 as an essential feature of the proposed method.

In the equation proposed, the effect of spontaneous action is clearly included even for a single particle. However, another approach is also possible, in which the effect of spontaneous action does not occur for an isolated particle, and only appears as a result of the presence of the collective. Both methods may have a meaning, depending on what kind of particles we are dealing with.

For the case of N different particles, we have

$$\left.\begin{aligned} -\frac{\partial f_i}{\partial t} &= \operatorname{div}_r \boldsymbol{v} f_i + \operatorname{div}_r \frac{1}{m_i}\left\{-\operatorname{grad} V + e_i\left(\boldsymbol{e} + \frac{1}{c}\,[\boldsymbol{vh}]\right)\right\} f_i, \\ \operatorname{div} \boldsymbol{e} &= 4\pi \sum_{j\neq i} e_j \int f_j \, d\boldsymbol{v}, \\ \operatorname{rot} \boldsymbol{h}_i - \frac{1}{c}\frac{\partial \boldsymbol{e}_i}{\partial t} &= 4\pi \sum_{j\neq i} e_j \int \boldsymbol{v} f_j \, d\boldsymbol{v}, \\ V_i(\boldsymbol{r}, t) &= \sum_{j\neq i} \int K_{ij} \int f_j \, d\boldsymbol{v}' \, d\boldsymbol{r}', \\ \left(\operatorname{div} \boldsymbol{h}_i = 0;\right. & \left.\ \operatorname{rot} \boldsymbol{e}_i + \frac{1}{c}\frac{\partial \boldsymbol{h}_i}{\partial t} = 0\right) \quad i = 1, 2, \ldots, N. \end{aligned}\right\} \qquad (3.3)$$

* M. Born, Proc. Roy. Soc. (A) 143, 410 (1934); F. Bopp, Ann. d. Phys. 38, 345 (1940); B. Podolsky, Phys. Rev. 62, 60 (1941); P. A. M. Direc, Proc. Roy. Soc. (A) 167, 148 (1938).

Here the spontaneous action is omitted in all the members: the summation extends over all the j, excluding the i-th value.

For an ensemble of similar particles, described by similar distribution functions for each particle ($f_1 = f_2 = \dots = f_N = f$), the system of N equations reduces to a single one:

$$\left.\begin{aligned}
&-\frac{\partial f}{\partial t} = \operatorname{div}_r vf + \operatorname{div}_v \left\{-\frac{1}{m}\operatorname{grad} V + \frac{e}{m}\left(e + \frac{1}{c}[vh]\right)\right\} f,\\
&\operatorname{div} e = 4\pi (N-1) e \int f\, dv,\\
&\operatorname{rot} h - \frac{1}{c}\frac{\partial e}{\partial t} = \frac{4\pi}{c}(N-1) e \int vf\, dv,\\
&\left(\operatorname{div} h = 0;\quad \operatorname{rot} e + \frac{1}{c}\frac{\partial h}{\partial t} = 0\right),\\
&V(r, t) = (N-1) \int K(|r-r'|) \int f\, dv.
\end{aligned}\right\} \tag{3.4}$$

For a single particle, isolated from the collective, setting N = 1 in (3.4), we obtain:

$$-\frac{\partial f}{\partial t} = \operatorname{div}\ vf, \tag{3.5}$$

that is, merely the continuity equation for function f. For N = 1 (replacing the factor N - 1 by N and introducing, after multiplying the continuity equation N, merely f instead of Nf), we have:

$$\left.\begin{aligned}
&-\frac{\partial f}{\partial t} = \operatorname{div}_r vf + \operatorname{div}_v \left\{-\frac{1}{m}\operatorname{grad} V + \frac{e}{m}\left(e + \frac{1}{c}[vh]\right)\right\} f,\\
&\operatorname{div} e = 4\pi e \int f\, dv,\\
&\operatorname{rot} h - \frac{1}{c}\frac{\partial e}{\partial t} = 4\pi e \int vf\, dv,\\
&\left(\operatorname{div} h = 0;\quad \operatorname{rot} e + \frac{1}{c}\frac{\partial h}{\partial t} = 0\right),\\
&V(r, t) = \int K(|r-r'|) \int f(r', v', t)\, dr'\, dv'.
\end{aligned}\right\} \tag{3.6}$$

Here the spontaneous action effect arises as the result of the presence of the collective of particles.

It is essential to note the following:

1. The last case (particles without spontaneous action) brings out the radical difference of the dynamics of charged particles that has been set forth from the ordinary dynamics. A particle isolated from the collective loses its individual properties (charge and mass) and there is left merely the fact of the space-time existence of the particle and the presence of motion in it $(f(r, v, t) \neq 0)$. The charge and mass appear only as quantities characterizing the connections of the particles in the collective. The

fundamental problem of the characteristic mass of the particles thus is converted into the problem of defining the interactions in the collective.

2. For a sufficiently large number of particles ($N \quad 1$), the equations for both cases (3.2) and (3.6) are identical. The properties of the equations depend only to a slight extent from the total number of particles.

3. The spontaneous action effect appears only if the number of particles exceeds one. Hence, it is possible to solve the problem of radiation without ascribing to the particle a charge as an individual characteristic.

Section 4. Abandonment of the Principle of Point Localization, and Connection With Gibbs' Statistical Mechanics

Inasmuch as the method proposed is based on new physical assumptions, it is clear in advance that it cannot be deduced from existing theories. It might rather be expected that these theories would be obtained as a particular (limiting) case. We have seen that abandonment of the principle of localization of particles (in the sense given above) constitutes a generalization of previous conceptions. Thus, not eliminating, but generalizing the conception of classical mechanics and electron theory, the method set forth makes it possible to synthesize these theories on a new basis. A similar situation exists with relationship to Gibbs' statistical mechanics as well.

Gibbs' statistical method is based on two principal hypotheses:

1. Since it is assumed that dynamic systems are subject to the laws of classical mechanics, the assumption is also made of strict localization of particles.

2. The theory considers an infinite number of noninteracting systems (ensembles) of the same nature, with differing initial conditions; in this connection, the second assumption is made, namely, that the averages for such systems correspond to the observable macroscopic quantities.

It would appear that the method of noninteracting ensembles in statistical mechanics, determining the dispersion of states of the system under consideration, as a whole, and the method of f-functions in the theory now set forth, have much in common. Actually, this community is merely external.

If we abandon the first assumption, we must start out from a different dynamic theory of the action of systems, one that differs from classical mechanics. This dynamics does not assume the existence of any set of noninteracting ensembles, that is, it does not require that the second condition be fulfilled. At the same time, if, in the theory being set forth, we pass to the case of localized particles, which is quite possible, as we shall see, and then introduce ensembles of independent systems, we shall then go over into the apparatus of Gibbs' method. In this sense, accordingly, Gibbs' statistics is a <u>particular case</u> of the theory being set forth. The explanation of this fact was not obtained at once. The apparent closeness to Gibbs' theory originally suggested the conclusion that the method of calculating the collective interactions and the method of paired collisions are approximative methods for solving the problem of many particles moving according to the laws of classical mechanics. Both these methods eliminate, from different sides, the complicated problem of many particles, which appears in Gibbs' statistics.

We consider the initial equation of statistical mechanics

$$-\frac{\partial D}{\partial t}=[HD], \tag{4.1}$$

where $H(q_i, p_i)$ is the Hamiltonian function for a system of N particles

$$H(q_i, p_i)=\sum_i \frac{p_i^2}{2m_i}+\sum_{(1\leqslant i<j\leqslant N)} V(|q_i-q_j|), \tag{4.2}$$

and $[HD]$ is Poisson's parenthesis:

$$\frac{\partial H}{\partial p}\frac{\partial D}{\partial q}-\frac{\partial H}{\partial q}\frac{\partial D}{\partial p}.$$

Here we may introduce into the potential energy of interaction among particles not only the paired but also triple, quadruple, etc. types of interaction, which do not reduce merely to the sum of paired connections. We may take as an example of such interactions the triple type (for hydrogen atoms in the approximation of Van der Walls forces):

$$\frac{\text{const.}}{r_{12}^3 r_{13}^3 r_{23}^3}.$$

In the general case, we may assume

$$V=\sum K_{ij}+\sum K_{ijk}+\ldots+\sum K_{ijk\ldots l}, \tag{4.3}$$

where the form of the functions K_{ij}, K_{ijk}, etc. is not specialized.

For the case of three particles, for example, the initial equation of statistical mechanics is:

$$\begin{aligned}-\frac{\partial D(1,2,3)}{\partial t}&=\operatorname{div}_{r_1}v_1D(1,2,3)+\operatorname{div}_{r_2}v_2D(1,2,3)+\\&+\operatorname{div}_{r_3}v_3D(1,2,3)-\operatorname{div}_{v_1}\frac{1}{m_1}\nabla_{r_1}(K_{12}+K_{13}+K_{123})D(1,2,3)-\\&-\operatorname{div}_{v_2}\frac{1}{m_2}\nabla_{r_2}(K_{21}+K_{23}+K_{123})D(1,2,3)-\\&-\operatorname{div}_{v_3}\frac{1}{m_3}\nabla_{r_3}(K_{31}+K_{32}+K_{123})D(1,2,3).\end{aligned} \tag{4.4}$$

We now to pass from the function D (1, 2, ..., N) in a space of 6N variables to the function f(1) in a space of only six variables.

Integration (4.4) consecutively for the space and velocity variables of each of all the particles except one, we have:

$$-\frac{\partial f(1)}{\partial t}=\operatorname{div}_{r_1}v_1 f(1)-\operatorname{div}_{v_1}\frac{1}{m_1}\Big\{\nabla_{r_1}\int K(1,2)f(1,2)\,dr_2\,dv_2+$$
$$+\nabla_{r_1}\int K(1,3)f(1,3)\,dr_3\,dv_3+$$
$$+\nabla_{r_1}\int K(1,2,3)f(1,2,3)\,dr_2\,dr_3\,dv_2\,dv_3\Big\}, \qquad (4.5)$$

where

$$f(1)=\int f(1,2,3)\,d\tau_2\,d\tau_3;\quad f(1,2)=\int f(1,2,3)\,d\tau_3;\quad f(1,3)=$$
$$=\int f(1,2,3)\,d\tau_2,\quad f(2,3)=\int f(1,2,3)\,d\tau_1,\quad (d\tau=dr\,dv).$$

The last members reduce to surface integrals. For f(2) or f(3), the equations will be analogous.

We see that if we introduce into Gibbs' statistics the following conditions, which are not inherent in it intrinsically:

1) The vanishing of the surface integrals of the type

$$\int\operatorname{div}_{r_2}v_2 f(1,2,3)\,d\tau_2\,d\tau_3;\qquad \int\operatorname{div}_{v_2}\frac{1}{m_2}\nabla_{r_2}(K_{21}+K_{23}+K_{123})\,d\tau_2\,d\tau_3,$$
$$\int\operatorname{div}_{r_3}v_3 f(1,2,3)\,d\tau_2\,d\tau_3;\qquad \int\operatorname{div}_{v_3}\frac{1}{m_3}\nabla_{r_3}(K_{31}+K_{32}+K_{123})\,d\tau_2\,d\tau_3;$$

2) The condition of the multiplicative nature of the distribution function

$$f(1,2,3)=f(1)f(2)f(3),$$

3) The normalization condition

$$\int f_i(i)\,d\tau_i=1,$$

we then obtain:

$$\left.\begin{aligned}
-\frac{\partial f(1)}{\partial t}&=\operatorname{div}_{r_1}v_1 f(1)-\\
&\quad-\operatorname{div}_{v_1}\frac{1}{m_1}(\nabla_{r_1}V_{12}+\nabla_{r_1}V_{13}+\nabla_{r_1}V_{123})f(1),\\
V_{12}(r_1,t)&=\int K(|r_1-r_2|)\int f(r_2,v_2,t)\,dr_2\,dv_2,\\
V_{13}(r_1,t)&=\int K(|r_1-r_3|)\int f(r_3,v_3,t)\,dr_3\,dv_3)\\
V_{123}(r_1,t)&=\int\int K(|r_1-r_2|,\ |r_1-r_3|)\int f(r_2,v_2,t)\,dv_2\times\\
&\quad\times\int f(r_3,v_3,t)\,dv_3\,dr_2\,dr_3,
\end{aligned}\right| \qquad (4.6)$$

that is, equations of the type (3.2), which are at the basis of our theory (without spontaneous action).

If we remain within the framework of Gibbs' statistical mechanics, we may make the following deductions from the results that have been obtained.

1. The solution of the problem of many particles in six-dimensional space cannot pretend to precision insofar as we do not obtain an equation describing a change only of f(1). In equations (4.5) f(1) is connected with f (1, 2), with f(1, 3) and even with f (1, 2, 3).

2. The requirement of **multiplicativity** is a foreign requirement introduced into Gibbs' theory from outside, and cannot by any means be taken as an accurate relationship. Furthermore, it is precisely the difference of the distribution function of the system from the mere product of the functions of the particles making up the system that conditions a number of effects that are in correspondence with the experimental facts.

Thus, for instance: a) if we assume that the property of multiplacativity of distribution function is correct, we are unable to solve certain problems of real gases. For example, for the fluctuation of the number of particles in a region G, we have, as is well known[+]:

$$(N_G - \bar{N}_G)^2 =$$

$$= N_G\left(1 - \frac{\bar{N}_G}{N}\right) + \frac{N(N-1)}{V^2}\int\limits_G\int\limits_G \{F_2(q_1, q_2) - F_1(q_1)\,F_2(q_2)\}\,dq_1\,dq_2,$$

where N denotes the number of particles in the region V $(V \gg G)$ and

$$F_1(q_1) = \int f\,d\mathbf{v}_1; \qquad F_2(q_2) = \int f\,d\mathbf{v}_2; \qquad F_{12}(q_1 q_2) = \int\int f(1, 2)\,d\mathbf{v}_1\,d\mathbf{v}_2,$$

F being the space function of distribution, depending exclusively on the coordinates of the particles $(q \to x, y, z)$.

The expression $N_G\left(1 - \frac{\bar{N}_G}{N}\right)$ holds true for noninteracting molecules as well, so that the influence of molecular interaction should include a second term. If function F_2 is multiplicative, it vanishes, and hence, we must abandon explanation of the varying amount of the fluctuations of density in condensed and rarefied gases.

b) It is experimentally known, for gases and liquids, from studies of the dispersion of X-rays, that there is a connection among the fluctuations in

[+] See e.g. N. Bogoliubov, Problemy dinamicheskoi teorii v statisticheskoi fizike /Problems of Dynamic Theory in Statistical Physics/, 1946.

spatially dissociated regions. This dependence may be explained[++] by introducing a radial distribution function for f(1, 2)

$$F_2(q_1, q_2) = \mu(|q_1 - q_2|)$$

with $F_1(q_1) = F_2(q_2) =$ const. (in the case of spatially uniform distribution), where function $F_{12}(q_1, q_2)$ is obviously not equal to the product $F_1(q_1)F_2(q_2)$.

c) Finally, $F_2(q_1, q_2)$ defines certain properties of the deviation of the equation of state of real gases from that for ideal gases.

All this speaks for the correctness of the relation

$$F_{12}(q_1, q_2) \neq F_1(q_1) F_2(q_2)$$

as essentially an experimental facts.

N. N. Bogoliubov tries in one of the chapters of his monograph to derive an equation of the type (4.6) on the basis of Liouville's theorem (4.1). However, he was not able entirely to derive equation (4.6). In the first place, as shown in this chapter, it is impossible without difficulty to include the electrodynamic interactions in the pattern of the classical equation of Liouville. In addition, particular boundary conditions must be introduced from without, conditions which are not intrinsic to equation (4.6), in connection with which the calculation can be carried out only under the condition that the law of the interaction forces is expressed by a monotonic function, and on the smallness of these forces, as a necessary condition. The existence of these powerful limitations indicates the need of a consistent and thorough derivation of equation (4.6) from Gibbs' statistical mechanics.

Abandonment of the problem in this form is dictated not only by the inadequacy of the apparatus of statistical mechanics, which is discussed in this and subsequent chapters, but also by the presence of properties in equations (4.6) that are unknown in the equations of statistical mechanics (see Chapters II, IV, V) as well as by physical considerations (see Sec. 7 of this chapter).

Does the existence of multiplicativity of the distribution function f indicate that the method of calculating collective interactions developed in the preceding paragraph is clearly approximative? We consider the equation of continuity in a space of 6N variables

$$-\frac{\partial D}{\partial t} = \sum_{i=1}^{N} \mathrm{div}_{r_i} v_i D + \sum_{i=1}^{N} \mathrm{div}_{v_i} g_i D. \qquad (4.7)$$

++ Zernicke, Prins. Zeits. f. Phys. 41, 184 (1927).

If we introduce the condition of strict point localization of particles, we must assume for the vector of acceleration:

$$g_i = -\frac{1}{m_i}\sum_{j=1}^{N} \nabla_{r_i} K_{ij}(|r_i - r_j|), \tag{4.8}$$

then the continuity equation contains along with (4.8) Gibbs' statistics.

On the other hand, from the point of view of unlocalized particles, we must write:

$$g_i = -\frac{1}{m_i}\sum_{j\neq i}^{N} \nabla_{r_i} \int K_{ij}(|r_i - r_j|) \times$$
$$\times \underbrace{\int \cdots \int}_{(N-1)} D(1, 2, \ldots, N) \prod_{s\neq j} d\tau_s \cdot dv_j; \tag{4.9}$$

in which $d\tau_s$ is an element of phase space in the six-dimensional space of the s-th particle. Then, the equation of continuity together with (4.9) constitutes an equation for the definition of function D, having no direct relationship to Gibbs' theory, inasmuch as (4.8) and (4.9) are intrinsically different.

In the equation obtained, vector $\underline{g}_i$ depends only on the coordinates of the i-th particle (integration having been performed for the other variables) and the time by means of function D. This ensures the presence in the new equation of precise solution in the form of the product

$$D(1, 2, \ldots, N) = f(1) f(2) \ldots f(N). \tag{4.10}$$

In point of fact, inserting (4.10) in the continuity equation and making use of (4.9), we obtain:

$$\sum_i \left(\frac{\partial f_i}{\partial t} + \mathrm{div}_{r_i} v_i f_i + \mathrm{div}_{v_i} g_i f_i\right) \prod_{j\neq i} f_j = 0, \tag{4.11}$$

$$g_i = -\frac{1}{m_i}\sum_j \nabla_{r_i} \int K_{ij}(|r_i - r_j|) \int f(r_j, v_j, t)\, dr_j\, dv_j$$

$$\left(f(j) = \int D(1, 2, \ldots, N) \prod_{s\neq j} d\tau_s\right).$$

Since the equation in parentheses under the summation sign now depends only on the variables for the i-th particle, we may satisfy the equation obtained by equating each member of the sum to zero. In this way, we have:

$$-\frac{\partial f_i}{\partial t} = \operatorname{div}_{r_i} v_i f_i + \operatorname{div}_{v_j} g_i f_i, \tag{4.12}$$

$$g_i = -\frac{1}{m_i} \sum \nabla_{r_i} \int K_{ij}(|r_i - r_j|) \int f_j(r_j, v_j, t)\, dr_j\, dv_j,$$

i.e., the previous equation (without spontaneous action). In this way, it has been shown that the multiplicativity of the distribution function appears as a result of the abandonment of the condition of strict localization of the particles and cannot be considered as a defect or as approximativity in the theory.

Quite analogous considerations could be applied for the case of the electrodynamic forces as well. As we have pointed out, it is impossible to introduce electrodynamic interactions into the ordinary apparatus of Gibbs' theory.

The fact is that, although Gibbs' method is based on a statistical interpretation of function D(1, 2, ..., N), at the same time it involves the necessity of calculating the forces of interaction among point particles at two points in space that are not statistically dispersed but precisely determined. This also leads to the appearance of the difficulties considered above (Sec. 3) in generalizing this method to electrodynamic systems.

In the equation of continuity (4.7), the effect of powerful interaction among particles is defined by expressions of the form

$$\operatorname{div}_{r_i} g_i D(1, 2, \ldots, N).$$

In the spirit of electron theory, it is natural to write, instead of $\underline{g_i}$:

$$g_i = \frac{e_i}{m_i}\left(e + \frac{1}{c}[v_i h]\right),$$

where $\underline{e}$ and $\underline{h}$ are the strains in the electric and magnetic fields, which are determined by the motion and position of all N particles. From the point of view of the old theories, we must determine $\underline{e}$ and $\underline{h}$ from the equations of the electromagnetic field, where the density of currents and charges is defined as

$$\left.\begin{aligned} \rho &= \sum_i e_i \delta(r - r_i(t)), \\ j &= \sum_i e_i v_i(t)\, \delta(r - r_i(t)). \end{aligned}\right\} \tag{4.13}$$

We formulate our abandonment of the principle of localization by a new expression for these quantities

$$\left.\begin{aligned}\rho &= \sum_i \varepsilon_i \int D(1, 2, \ldots, N) \prod_{s \neq i} d\tau_s \, dv_i, \\ j &= \sum_i \varepsilon_i \int v_i D(1, 2, \ldots, N) \prod_{s \neq i} d\tau_s \, dv_i,\end{aligned}\right\} \quad (4.14)$$

or

$$\left.\begin{aligned}\rho &= \sum_i \varepsilon_i \int f_i \, dv_i, \\ j &= \sum_i \varepsilon_i \int v_i f_i \, dv_i.\end{aligned}\right\} \quad (4.15)$$

Combining these expressions with the equation of continuity in 6N-dimensional space and with the field equations, we realize that the requirement of multiplicativity (4.10) here too is fulfilled as a precise relationship.

It gives the previous equation of Sec. 3 (with spontaneous action).

In this way, both for any central forces and for electrodynamic forces, the generalized point of view as to the behavior of particles may be expressed in terms of 6N-dimensionsal space by introducing the condition of multiplicativity of the distribution function.

We consider a particular example, in which the need of interpreting the multiplicativity property as a precise relationship is especially marked. Instead of the function D(1, 2, 3, ..., N) many authors have introduced the sequence of distribution functions

$$\left.\begin{aligned}f(1) &= \int D(1, 2, \ldots, N) \, d\tau_2 \ldots d\tau_N, \\ f(1, 2) &= \int D(1, 2, \ldots, N) \, d\tau_3 \ldots d\tau_N, \\ &\cdots\cdots\cdots\cdots\cdots\cdots \\ &\cdots\cdots\cdots\cdots\cdots\cdots \\ f(1, 2, \ldots, k) &= \int D(1, 2, \ldots, N) \, d\tau_{k+1} \ldots d\tau_N, \\ &\cdots\cdots\cdots\cdots\cdots\cdots\end{aligned}\right\} \quad (4.16)$$

In order to determine for what physical magnitudes this sequence is applicable, we consider for simplicity's sake the case of similar particles, for which the initial function D may be considered as independent of permutation of the arguments. Then, for the mean values of the quantities of additive type

$$A = \sum_{i=1}^{N} A(i)$$

we have:

$$\bar{A} = \int D(1, 2, \ldots, N) \sum_i A(i)\, d\tau_1 \ldots d\tau_N = N \int f(1)\, A(1)\, d\tau_1. \quad (4.17)$$

Here we have made use of the symmetry of function D with respect to permutation of the arguments. For quantities depending only on pairs of arguments A(i, j) (binary type), we have:

$$A = \sum_i \sum_j A(i, j).$$

The mean value of this magnitude will be defined as:

$$\bar{A} = \frac{N(N-1)}{2} \int f(1, 2)\, A(1, 2)\, d\tau_1\, d\tau_2. \quad (4.18)$$

Among the binary magnitudes is the potential energy of the interaction between two particles $K(|r_1 - r_2|)$. The mean value of this magnitude for a cloud consisting of N similar particles must be, on the basis of (4.18):

$$\bar{K} = \frac{N(N-1)}{2} \int K(|r_1 - r_2|) \int f(1, 2)\, d\tau_1\, d\tau_2 = U_{макр}, \quad (4.19)$$

Where integration is over the six-dimensional space of each particle $(d\tau = dr\, dv)$

Accordingly, the mean potential energy of interaction (according to the ideas of Gibbs' statistics, the macroscopically observable energy) is expressed by the binary function f(1, 2).

But, on the other hand, formal electrodynamics and the theory of gravitation of distributed masses gives the following expression for the potential energy:

$$U_{\text{макр}} = \int K(|r_1 - r_2|)\,\rho(r_1)\,\rho(r_2)\,dr_1\,dr_2. \tag{4.20}$$

formulas (4.19) and (4.20) constitute different assertions. Only for

$$f(1, 2) = f(1)\,f(2)$$

and for a sufficiently large number of particles, when the factor N - 1 may be replaced by N, we may assume:

$$N\int f(1)\,dv_1 = \rho(r_1);\quad N\int f(2)\,dv_2 = \rho(r_2)$$
$$(dv = d\xi\,d\eta\,d\zeta)$$

and reduce (4.19) to (4.20). But the requirement of multiplicativity is justified only for noninteracting particles (from the point of view of Gibbs' statistics) and therefore cannot be precise.

In this way, Gibbs' statistics denies the correctness of formula (4.20). At the same time, there is no doubt that the theory of electrostatic and gravitational interactions, not including correlation functions, has at least as good a foundation as the entire apparatus of Gibbs' statistical mechanics. Expression (4.20) demands the condition of multiplicativity as a precise property.

The resolution of this contradiction cannot be sought in the special properties of the electrostatic or gravitational interactions. The apparatus of statistical mechanics by its very nature employs solely the laws of classical mechanics, in the system of which electrostatic and gravitational laws enter without any difficulty.

The solution consists in the fact that the formula (4.19) and formula (4.20) depart from different physical assumptions, and that they therefore relate to different physical cases. Formula (4.19) assumes that the particles are strictly localized and remain so upon passing into an ensemble of states. Formula (4.20) is based on the transition from the interaction energy of point particles

$$\frac{1}{2}\sum_{i\neq j}\sum \frac{e_i e_j}{|r_i - r_j|}$$

to space-distributed charges (or masses). This transition leads to the abandonment of localization, i.e., it includes in itself the conception of unlocalized particles.

Which relationship is more general, (4.19) or (4.20)? Contrary to our first impression, connected with the presence of multiplicativity, we must admit that formula (4.20) is the more general one, since it makes it possible in a particular case to go over to point particles, and hence, after introducing statistical ensembles, to expression (4.19). Formula (4.19) does not permit us to make the inverse transition, since from the point of view of localized particles, the multiplicativity of the distribution function cannot be assumed as a precise property.

Section 5. Theory of Crystallization and Gibbs' Statistics

By considering the problem of the crystalline state, we can convince ourselves more clearly as to the limitedness of the apparatus of statistical mechanics as a theory which employs the conception of point localization of particles.

Since Gibbs' statistics includes factors connected with the temperature, it would seem that it should give an explanation of the formation of the crystalline phase from the spatially uniform distribution. This question of the theory of the crystalline state is the clearest and simplest in its setting both from the physical and mathematical sides. A rational many-particle theory should explain in the first place the qualitative fact known from observation: the sudden appearance of a spatial-periodic distribution at a certain critical value of the temperature.

However, remaining within the framework of statistical mechanics, we come across difficulties of an essential nature in investigation the problem of crystals.

We consider these difficulties.

1. The canonical distribution of Gibbs

$$D_N = Q_N^{-1} \exp\left(-\frac{U_N}{\theta}\right), \tag{5.1}$$

in which U_N is the potential energy of a system of N similar particles

$$U_N = \sum_{(1 < i < j < N)} V(|q_i - q_j|) \tag{5.2}$$

(Q_N^{-1} is a normalizing factor), includes within it the classical theorem as to the uniform distribution of the mean kinetic energy for all the degrees of freedom of a system. This is of importance for answering the question of whether a periodic structure and the distribution of the particle density, which is a characteristic for crystals, can be obtained from (5.1).

It is precisely because the mean kinetic energy is uniformly distributed among all the degrees of freedom of the system (including likewise the degrees of freedom present in the motion of the system's center of mass, and likewise the changes in space of the entire system) that we see the fruitlessness of the attempts to obtain a periodic structure by Gibbs' method.

2. The usual picture of the crystalline state, in which each atom has a sharply expressed maximum probability of occurring around a corresponding intersection of the space lattice, may be obtained, generally speaking, by taking as the starting point the following distribution function:

$$D_{N\to\infty} = \sum_n \Delta(0, q_n - na), \tag{5.3}$$

where a is the period of the lattice and Δ the corresponding distribution function for a single atom (its exact form is not essential for us just now).

But such a form for function D contradicts the initial formula by (5.1), since formula (5.3) is not symmetrical with respect to permutation of the arguments, while formula (5.1) is symmetrical. It is therefore impossible, starting from the canonical expression (5.1) to obtain the classical picture of the lattice structure with atoms that can be denumerated.

3. The attempt can be made tp construct a theory of the crystalline state on the basis of Gibbs' statistical mechanics, applying multiplicative distribution functions. Here, multiplicativity must be regarded as an approximation. One of the fundamental results of our theory leads naturally to such an attempt. In point of fact, as we shall see below, the solution of the many-particle problem in the theory set forth leads directly to the occurrence of a periodic distribution of particles for a definite value of the temperature parameter (see Part II, Chapter I). At first sight, both methods have much in common. This will be seen if only from comparing the fundamental equations. Thus, for instance, the equation for distant interactions has the form ($\partial/\partial t = 0$):

$$\operatorname{div}_r vf - \frac{1}{m}\operatorname{div}_v \nabla_r \int K(|r-r'|) f(r', v', t)\, dr'\, dv \cdot f = 0. \tag{5.4}$$

The analogous equation in Gibbs' statistics is

$$\operatorname{div}_{r_1} v_1 f - \frac{1}{m}\operatorname{div}_{v_1} \nabla_{r_1} \int_{-\infty}^{\infty} K(|r_1 - r_2|) f(1,2)\, dr_2\, dv_2 = 0. \tag{5.5}$$

Hence it is obvious that if (5.4) has periodic solutions, (5.5) also will have them on the condition that

$$f(1,2) = f(1) f(2), \tag{5.6}$$

* A. Vlasov, Uch. Zpiski MGU (fizika) 75, No. 1, 30 (1945).

since in this case the equations coincide. We note that the periodic solution for function f that follows from equation (5.4) can be obtained on a more general assumption than simple multiplicativity. Since the concrete form of the interaction energy $K(|r - r'|)$ does not affect the result, then any form of f(1, 2) differing from (5.6) will likewise lead to a periodic solution if only the variation in function f(1, 2) could be converted into a variation of kernel K, which can be attained in various ways, For example, we may assume[+],

$$f(1,2) = f(1)f(2)G(|r_1 - r_2|), \tag{5.7}$$

where $G(|r_1 - r_2|) = e^{-\frac{K(|r_1 - r_2|)}{a}}$, a an arbitrary constant.

Then, substitution of (5.7) in (5.5) gives:

$$\nabla_{r_1} \int K(|r_1 - r_2|) f(r_2, v_2, t) e^{-\frac{K(|r_1 - r_2|)}{a}} dr_2 dv_2 =$$

$$= \nabla_{r_1} \int \{a(1 - e^{-\frac{K(|r_1 - r_2|)}{a}})\} f(2) dr_2 dv_2,$$

since

$$\nabla_r \{a(1 - e^{\frac{K(|r_1 - r_2|)}{a}})\} = \nabla_r K \cdot e^{-\frac{K}{a}}. \tag{5.8}$$

Consequently, combing (5.7) and (5.8) we have once again an equation of the form (5.4), only with a different kernel K^+:

$$K^*(|r_1 - r_2|) = a(1 - e^{-\frac{K(|r_1 - r_2|)}{a}}). \tag{5.9}$$

In 1947, Tyablikov published an article[++], in which he proposed a more general form instead of (5.9), assuming

$$\nabla_r K(|r - r'|)\, G(|r_1 - r_2|) = \nabla_r \int_{-\infty}^{|r - r'|} \frac{\partial K}{\partial r} G(r)\, dr,$$

from which the kernel

$$K^*(|r - r'|) = \int_{-\infty}^{|r - r'|} \frac{\partial K}{\partial r} G(r)\, dr,$$

+ A. Vlasov, Journ. of Phys. (Moscow) 9, 130 (1945).
++ S. Tyablikov, Zh. E. T. F. (1947).

He proposed to derive G(r) from observation, combining this expression with the correlation function for liquids. He asserted erroneously in this article that the crystallization effect can be obtained taking as the initial point Gibbs' statistics in the form given to it by Bogoliubov. The crystallization criterion advanced by Tyablikov coincides with mine, although the author does not mention this.

It would therefore seem at first sight that it is possible to abandon the requirement of strict multiplicativity, and consider it as merely an approximation, conserving the usual apparatus of Gibbs' method. But this conclusion is incorrect.

a. The fact is that if we remain on the positions of Gibbs' method, the periodicity of the function

$$f(1) = \int D(1,2, \ldots, N)\, d\tau_2 \ldots d\tau_N,$$

cannot be explained from the physical point of view. The periodicity of function f(1) would signify adopting radical assumptions that are unknown in Gibbs' statistical classical mechanics. Actually, $f(1)\,dr_1$ is the probability that the particle is in the space element of volume $d\underline{r}_1$. The periodicity of function f(1) would signify the possibility that the atom could be simultaneously at all the intersections of the lattice. Such a result already takes us outside the pattern of classical mechanics, which lies at the basis of Gibbs' statistics. Consequently, the periodicity of f(1) is physically inconsistent with the point of view of Gibbs' statistics.

b. Basically, description of the fact of the sudden appearance of a periodic structure out of a uniform phase as the temperature decreases continuously does not require the introduction of independent ensembles. The appearance of the periodic distribution is sudden, and, hence, may be effected in as small an element of time as required.

c. Gibbs' statistics cannot lead to strict ordering, since fluctuations of the center of mass of the system always take place. The derivation, however, did not assume any kind of limitations on the amplitude of fluctuations of the entire system as a whole.

4. It is essential to note that Born and Green+, in their investigations into the kinetic theory of liquids, do not even raise the problem of the proof of the periodicity of f(1) for explaining the appearance of the crystal from the liquid phase. They seek the periodicity of f(1, 2), assuming that f(1) = const. (is independent of the coordinates x, y, z). They assume here that

$$\int \int f(1,\ 2)\, d\boldsymbol{v}_1 d\boldsymbol{v}_2 = G(|\boldsymbol{r}_1 - \boldsymbol{r}_2|).$$

However, it is not at all clear what relationship the periodicity of the correlation function G(r) has to the fact of the triply periodic structure of the crystal. This is so especially in view of the fact that f(1) is

+ M. Born and H. Green, Proc. Roy. Soc. A. 188, 10 (1946); 189, 455 (1947).

spatially uniform. It would seem, on the contrary, that to obtain periodicity in a crystal, it would be impossible to assume that f(1) = const., since it is required to obtain properties of periodicity, for example, the dialectric constant or an expression for the density, which belong to average values of additive type, i.e., are expressed by means of f(1). It should be expressed particularly that first, Born and Green have not proved the periodicity of f(1, 2) (the mathematical difficulties are too great along this path), and secondly, the introduction of triple, quadruple, etc. interactions (see Sec. 4, likewise Part II, Chapter I) complicates the discussion of Born and Green to such an extent that it is extremely hard to say anything at all as to the conservation under these circumstances of the periodicity of f(1, 2), even if such periodicity had been demonstrated for paired interactions.

Section 6. Theory of Nonlocalized Particles and Quantum Mechanics

Here attention should be paid to the following questions:

1. Does the proposed method of describing systems of many particles contain elements of such a nature that it is known they cannot be included in the quantum mechanics treatment of this problem? To answer this question, we consider the simplest case, the problem of two particles, from the point of view of both theories. We have, on the one hand:

$$\left.\begin{aligned} &-\frac{\partial f_i}{\partial t}=[H_i f_i], \\ &H_i(r,p)=\frac{p_i^2}{2\mu_i}+\int K(|r_i-r_j|)\int f_j(r_j,v_j,t)\,dr_j\,dv_j \\ &(i=1,\ 2;\ j=2,\ 1), \\ &\left([Hf]=\frac{\partial H}{\partial p}\frac{\partial f}{\partial x}-\frac{\partial H}{\partial x}\frac{\partial f}{\partial p}\right), \end{aligned}\right\} \qquad (6.1)$$

and on the other, Schroedinger's equation

$$\left.\begin{aligned} &i\hbar\frac{\partial\psi}{\partial t}=H\psi, \\ &H=-\frac{\hbar^2}{2\mu_1}\Delta_1-\frac{\hbar^2}{2\mu_2}\Delta_2+K(|r_1-r_2|). \end{aligned}\right\} \qquad (6.2)$$

It will be shown below (Chapters III, IV and VI) that the system of equations (6.1) has various solutions, including solutions corresponding to a uniform distribution for any $K(|r_1-r_2|)$, spatial-periodic solutions, solutions having a temperature parameter, "branching" solutions, "self-accelerating" solutions. This fact definitely indicates that these equations contain properties of a system of many particles that are known not to be present in equations (6.2). The quantum mechanics problem of two bodies is precisely solved and contains nothing of the sort.

2. Does equation (6.1) contain solutions for equation (6.2)? It is hard to answer this question. Although the possibility is not excluded that

among the infinite number of solutions of equation (6.1), there are some that may coincide in sense with the corresponding quantum solutions, still our basic purpose is to prove the existence of a new complex of physical phenomena, not included in the pattern of quantum and statistical mechanics.

This naturally raises the question as to the field of application of the two methods. A precise determination of the boundaries of the theory can only be made from the point of view of some more general theory, that would include both quantum mechanics and our theory as particular cases. Up to the present, there is no such theory.

We now consider what we should regard as the inadequacies of equations of the second type. These consist in the fact that the theory based on these equations does not consistently carry through the idea of nonlocalization of particles. Abandonment of the principle of localization is included in quantum mechanics only in the very method of description by means of extended ψ-functions. The expression of the potential energy of interaction $K(|r_1-r_2|)$ requires precise fixation of the condition of both particles (one at a point with the coordinate $\underline{r}_1$, the other at a point with coordinate $\underline{r}_2$), so that the expression of this energy in quantum theory is taken over unchanged from classical mechanics.

The conclusion may be drawn from this that a theory should be constructed that is more general than quantum mechanics and in which the abandonment of the principle of localization would be completely embodied. This more general theory should include (6.1) and (6.2) as particular cases.

3. In connection with what has been said, a special discussion should be made of the Hartree-Fock self-consistent field method. It would appear at first sight that this method consistently carries through the principle of nonlocalization of particles, on the basis of the ψ-function apparatus. For, describing the behavior of each particle by means of a ψ-function, we have:

$$i\hbar\frac{\partial\psi_i}{\partial t}=H_i\psi_i, \tag{6.3}$$

where

$$H_i=\frac{p_i^2}{2\mu_i}+\int K(|r_i-r_j|)\,\psi_j^*(r_j)\,\psi_j(r_j)\,dr_j$$

$$(p_i=-i\hbar\,\mathrm{grad}_{ri})\ (i=1,\ 2;\ j=2,\ 1).$$

This is the so-called Hartree equation. It is clear that although the functional representation of interaction by means of the f-function generalized the treatment of the problem of N bodies in the sense of classical mechanics, still the introduction of functional interactions by means of a ψ-function does not generalize quantum theory, since it is impossible to derive (6.2) from (6.3).

On the other hand, multiplying each of the equations of system (6.3) by the corresponding function for the second particle and combining, we arrive at an equation of the type of (6.2), but with a multiplicative function

$$\psi(r_1 r_2) = \psi_1(r_1)\psi_2(r_2),$$

which, however, is not a solution of (6.2).

Hence, Hartree's equations cannot be considered as a rational expression of the abandonment of the principle of point localization as a generalizing assumption.

4. We must deal with the widespread opinion that Hartree's method is based on quantum mechanics. From this point of view, it might be natural to hold that in general abandonment of the principle of localizations in the expression for the potential energy is only an approximation, and not at all the expression of a new theory.

It is important to elucidate first of all whether Hartree's equation is actually founded, that is, whether the conditions of its applicability are determined.

To this end, we consider the work of Fock[+].

The equation of quantum mechanics (for example, equation (6.2) may be derived from the variation principle

$$\delta \int\int \psi^*(r_1, r_2)\{H - E\}\psi(r_1, r_2)\, dr_1\, dr_2 = 0, \qquad (6.4)$$

where

$$H = H_1 + H_2 + K(|r_1 - r_2|)$$
$$\left(H_i = -\frac{h^2}{2\mu_i}\Delta_i\right), \quad i = 1, 2.$$

it is easy to prove that equation (6.4) is equivalent to

$$\int\int \delta\psi^*(H - E)\psi\, dr_1\, dr_2 = 0 \qquad (6.5)$$

(the transition from formula (6.4) to formula (6.5) requires only that operator H be self-conjugate.)

To obtain the Hartree equation, Fock inserts in equation (6.5) the multiplicative function

$$\psi(r_1, r_2) = \psi_1(r_1)\psi_2(r_2).$$

[+] V. Fock, Zs. f. Phys. 61, 126 (1930).

This is a radical step, since what was to be proved is postulated from the outset. It is important to keep still another matter in mind: neither the variation form nor the precise weight equation itself do not in any way limit the assumption of multiplicativity. This assumption is equally correct (or incorrect) either for two or many particles, either for short-range acting or for Coulomb interactions. The insertion of the multiplicative function leads to

$$\iint [\delta\psi_1^*\psi_2^* + \psi_1^*\delta\psi_2^*]\{H_1 + H_2 + K(|r_1 - r_2|) - E\}\psi_1\psi_2\, dr_1\, dr_2 = 0$$

or

$$\int dr_1 \delta\psi_1^*(r_1) \int \psi_2^*(r_2)\{H_1 + H_2 + K(|r_1 - r_2|) - E\}\psi_1(r_1)\psi_2(r_2)\, dr_2 +$$
$$+ \int dr_2 \delta\psi_2^*(r_2) \int \psi_1^*(r_1)\{H_1 + H_2 +$$
$$+ K(|r_1 - r_2|) - E\}\psi_1(r_1)\psi_2(r_2)\, dr_1 = 0.$$

The internal integrals may be written as:

$$H_1\psi_1 + G_{12}(r_1)\psi_1 - E_1\psi_1; \quad H_2\psi_2 + G_{21}(r_2)\psi_2 - E_2\psi_2.$$

where

$$G_{12}(r_1) = \int K(|r_1 - r_2|)\psi_2^*(r_2)\psi_2(r_2)\, dr_2,$$
$$G_{21}(r_2) = \int K(|r_2 - r_1|)\psi_1^*(r_1)\psi_1(r_1)\, dr_1,$$
$$E_1 = E - \int \psi_2^*(r_2) H_2\psi_2(r_2)\, dr_2; \quad E_2 = E - \int \psi_1^*(r_1) H_1\psi_1(r_1)\, dr_1.$$

In this way, we have:

$$\int dr_1 \delta\psi_1^*(r_1)[H_1 + G_{12}(r_1) - E_1]\psi_1(r_1)\, dr_1 +$$
$$+ \int dr_2 \delta\psi_2^*(r_2)[H_2 + G_{21} - E_2]\psi_2\, dr_2 = 0. \tag{6.6}$$

The last step in Fock's discussion consists in the postulation of the independence of $\delta\psi_1^*$ and $\delta\psi_2^*$, which gives:

$$\left.\begin{array}{l}\{H_1 + G_{12} - E_1\}\psi_1 = 0,\\ \{H_2 + G_{21} - E_2\}\psi_2 = 0.\end{array}\right\} \tag{6.7}$$

These equations coincide with equations (6.3) (for the stationary case).

Does this discussion give any criteria for evaluating the degree of approximation of the Hartree equation? The answer must be negative. For obviously, introducing the condition of multiplicativity of the ψ - functions, in fact predetermines the result. The precise equation (6.2), inserting the multiplicative function, leads to Hartree's equation, and in this point, the variation method does not furnish anything more. (Here we do not deal with the important results with symmetrized wave functions). Consequently, the question of the foundation of Hartree's method remains an open one even after Fock's work.

5. We discuss the possible connection of equation (6.1) with equation (6.3). This question, of course, arises because of the fact that (6.1) and (6.3) have the same structure. We write equation (6.3) by means of the density matrix.

Let

$$\left.\begin{aligned} (x|\rho_1|x') &= \sum_\alpha P_{1\alpha}\psi_{1\alpha}(x, t)\psi^*_{1\alpha}(x', t), \\ (x|\rho_2|x') &= \sum_\alpha P_{2\alpha}\psi_{2\alpha}(x, t)\psi^*_{2\alpha}(x', t), \\ (\sum_\alpha P_{1\alpha} &= \sum_\alpha P_{2\alpha} = 1) \end{aligned}\right\} \tag{6.8}$$

be the elements of two density matrices for the first and second particle, respectively. Differentiating (6.8) with respect to the time (the $P_{1\alpha}$ do not depend on t) and making use of equation (6.3) in the x form

$$i\hbar\frac{\partial\psi_{1\alpha}(x, t)}{\partial t} = \int (x|H_1|x'')\psi_{1\alpha}(x'', t)\,dx'',$$

$$-i\hbar\frac{\partial\psi^*_{1\alpha}(x', t)}{\partial t} = \int (x'|H_1|x'')^*\psi^*_{1\alpha}(x'', t)\,dx''$$

and corresponding expressions for the second particle, we have:

$$\frac{\partial}{\partial t}(x|\rho_1|x') = \\ = \frac{1}{i\hbar}\int (x|H_1|x'')(x''|\rho_1|x')\,dx'' - \frac{1}{i\hbar}\int (x|\rho_1|x'')(x''|H_1|x')\,dx'' \tag{6.9}$$

and similarly for the second particle where

$$H_1 = -\frac{\hbar^2}{2\mu_1}\Delta_1 + \int K(|x_1 - x_3'|) \int (x_2'|\rho_2|x_3'')\,\delta(x_2'' - x_3')\,dx_2'' \cdot dx_2'. \tag{6.9a}$$

The expression

$$\int (x_2'|\rho_2|x_3'')\,\delta(x_2' - x_3'')\,dx_3'' = w(x_3') \tag{6.9b}$$

is the analogue of the space density probability. Equations (6.9) and (6.9a) are an expression of Hartree's method.

To pass from equation (6.9) to equation (6.1), we employ Blokhintsev's matrices* and his method for passing from quantum to classical ensembles.

The matrix elements of these matrices R(x, p), H(x, p) are defined as Fourier components in expressions of the following form:

$$\left.\begin{aligned} (x|\rho_1|x') &= \int R_1(x, p)\frac{e^{-\frac{ip(x'-x)}{\hbar}}}{2\pi\hbar}\,dp, \\ (x|H_1|x') &= \int H_1(x, p)\frac{e^{\frac{-ip(x'-x)}{\hbar}}}{2\pi\hbar}\,dp. \end{aligned}\right\} \tag{6.10}$$

Inserting these expression in (6.9) and (6.9a), we obtain:

$$i\hbar\int\frac{\partial R_1(x, p')}{\partial t}\cdot\frac{e^{\frac{-ip'(x'-x)}{\hbar}}}{2\pi\hbar}\,dp' =$$

$$= \int\int\int\left\{H_1(x, p'')\frac{e^{\frac{-ip''(x''-x)}{\hbar}}}{2\pi\hbar}R_1(x'', p''')\frac{e^{\frac{-ip'''(x'-x'')}{\hbar}}}{2\pi\hbar}\,dp''\,dp'''\,dx'' -\right.$$

$$\left. - \text{analogous expression}\right\} \tag{6.11}$$

and similarly for the second particle.
Expression (6.9b) goes over into

$$\int (x'|\rho_2|x'')\,\delta(x' - x'')\,dx'' = \int R_2(x', p)\frac{dp}{2\pi\hbar} = w(x'). \tag{6.12}$$

---

* D. Blokhintsev, Journ. of Phys. USSR 2, 71 (1940).

Multiplying (6.11) by $e^{\frac{ip(x'-x)}{\hbar}}$, and integrating for dx', we have:

$$i\hbar\,\frac{\partial R_1(x,p)}{\partial t}=\iint\left\{H_1(x,p'')\frac{e^{\frac{-ip''(x''-x)}{\hbar}}}{2\pi\hbar}R_1(x'',p)\times\right.$$

$$\times e^{\frac{+ip(x''-x)}{\hbar}}\,dx''\,dp''-\text{anal. expr.}$$

Or, setting

$$x''-x=\xi,$$
$$p''-p=\eta,$$

we obtain, finally:

$$\left.\begin{aligned}i\hbar\frac{\partial R_1}{\partial t}=-\iint\frac{d\xi d\eta}{2\pi\hbar}e^{-i\frac{\xi\eta}{\hbar}}\{H_1(x+\xi,p)R_1(x,p+\eta)-\\-R_1(x+\xi,p)H_1(x,p+\eta)\},\\H_1(x,p)=\frac{p^2}{2\mu_1}+\int K(|x-x'|)\int R_2(x',p)\frac{dp\,dx'}{2\pi\hbar}\end{aligned}\right\}(6.13)$$

and similarly for the second particle with $R_2(x, p)$).

Following Blokhintsev, we expand the expression under the integral sign in (6.13) into a series by powers of ξ and η, and obtain:

$$\left.\begin{aligned}&\frac{\partial R_1}{\partial t}=-[H_1,R_1]-\sum_{n=2}^{\infty}\frac{(-i\hbar)^{n-1}}{n!}[H_1,R_1]_n,\\&H_1(x,p)=\frac{p^2}{2\mu_1}+\int K(|x-x'|)\int R_2(x',p)\frac{dp\,dx'}{2\pi\hbar},\\&[H_1,R_1]_n=\frac{\partial^nH_1}{\partial p^n}\cdot\frac{\partial^nR_1}{\partial q^n}-\frac{\partial^nH_1}{\partial q^n}\cdot\frac{\partial^nR_1}{\partial p^n}\end{aligned}\right\}\quad(6.14)$$

(and similarly for the second particle with R_2 and H_2). If we omit all the terms containing $\hbar$, and postulate:

$$\lim_{\hbar\to 0}R_i(x,p)\backsim 2\pi\hbar f_i(x,p)\quad(i=1,2),$$

then we obtain equation (6.1).

Accordingly, the fundamental equation of our theory has been obtained taking Hartree's method as our starting point. This result applies both for two and for many particles, since it has been emphasized that the variation principle does not impose any limitations on the number of particles. It is also essential to remark that what has been said retains its validity not only for Coulomb, but also for short-acting forces, since the derivation did not entail any limitations in this direction either.

The result in question is important, since the quantum-mechanics method of the self-consistent field is experimentally confirmed. However, is it possible to consider this result as a derivation of equation (6.1) from equation (6.3)? This question must be answered in the negative, for the following reasons. First of all, for $\hbar = 0$, R_1 likewise vanishes, while the introduction of the f-function is not connected with the quantity $\hbar$. Further, in the derivation, we assumed that the functions $R_1(x, p)$ and $R_2(x, p)$, which play the roles of functions $f_1(x, p)$ and $f_2(x, p)$ are subject to the condition of being capable of expansion in a Fourier integral (that is, the problem is solved in what is called the L^2 class of functions). But we shall see later (see Chapters II, IV, V) that the most important solutions of the equation for f do not belong to this class of functions.

Another factor is likewise of importance. The requirement that the f-function vanish at some boundary (or sufficiently decrease at infinity) in space of coordinates and velocities is an essential factor in introducing localization in the sense of classical mechanics (see Chapter III), since this limiting condition excludes the possibility of a fluctuating expulsion of particles from the volume under consideration.

This circumstance, like equation (6.1), in principle includes the temperature effect, and possibility of such an expulsion is avoided. Hence, in addition to the tendency condition $\hbar \to 0$, we must introduce the fundamental conditions of localization of particles, whereas it is precisely abandonment of localization that is characteristic of our theory. Since the radius of decreasing decomposition is not evaluated according to $\hbar$, it is not entirely sure that the particles can be covered by their f-functions. But even if there were such covering, we should obtain new solutions, not entering into the pattern of classical mechanics, which would again raise the question as to the causes for this fact. In the case that the radius of decrease is sufficiently small as compared with the sphere of action of the particles, we simply go over into classical mechanics, which raises the question as to how we obtain precise equations from an approximated method. As a result, we must come to the conclusion that the equation obtained is not equation (6.1).

There is no reason to believe that another more adequate derivation can be found, by means of which it would be possible to obtain all the properties of equation (6.1), since the precise equation (6.2) simply does not have them.

In view of the fact that the equations of quantum mechanics do not contain a parameter adequate to the temperature, it is impossible within the framework of this apparatus even to pose the question of crystallization, since the temperature of crystallization is an essential condition in defining the criteria of crystallization. For this reason alone, it can be stated that the equations of quantum theory in themselves are inadequate to solve the problem of the crystal, Only in the limiting case $\hbar \longrightarrow 0$, does a quantum ensemble of different particles behave like a classical Gibbs' ensemble, therewith losing its quantum characteristics. Consequently, it is only possible in this limiting case to speak of the introduction of temperatures in the usual sense of the term. We do not here have in mind the mixed methods in which Gibbs' theory and quantum mechanics are unsystematically combined. Blokhintsev*, using the apparatus of density matrices, showed that in the case of uniform particles, the transition to the classical ensemble is quite impossible without an essential "smoothing" of all the quantities in phase space.

Born++ attempted to explain periodicity in crystals by means of the method ordinarily employed for explaining the formation of arbitrary molecules. He considered a system of N (a large number) of different nuclei and electrons and solved Schroedinger's equation, expanding the wave function in powers of

$$\varkappa = \sqrt[4]{\frac{m}{M}}$$

(m being the mass of the electron and M the average mass of the nuclei). The motion of such a system is described to an adequate degree of accuracy by a function (x_e, X_μ), which plays the part of the potential energy of the electrons (x_e) for the given configuration of nuclei (X_μ). Born proved a theorem to the effect that the ψ-function of the electrons differ from zero only for values of X_μ differing very little X_μ^0, which are the roots of the equation $\frac{\partial \Phi}{\partial x_\mu} = 0$. But it is not proved, merely postulated, that these last equations give a periodic configuration of the nuclei. The impossibility of proving this, by virtue even of the inadequacy of the number of equations, leads Born to a position where he must adduce additional thermodynamic considerations in order to find parameters for the lattice.

This method, in addition to the defect in principle that has been pointed out, is not a general one, so that in the case of crystals built up of uniform particles, considered as undeformed centers of force (inert gases in the solid state), there does not exist in general a small parameter analogous to κ. For this case, it might be possible as a starting point to take the equation of secondary quantization:

$$\left. \begin{array}{l} i\hbar \frac{\partial \psi}{\partial t} = H\psi, H = -\frac{\hbar^2}{2\mu} \Delta\psi + \int K(|r - r'|)\psi^*(r')\psi(r')dr' \\ \text{or} \\ \Delta\psi + \frac{2m}{\hbar^2}\left(E - \int K(|r - r'|)\psi^*(r')\psi(r')dr'\right)\psi(r) = 0. \end{array} \right\} \quad (6.15)$$

* Op. cit.

++ M. Born, Rev. Mod. Phys. 17, 245 (1945).

Here ψ, ψ^* are the quantized wave functions satisfying the permutative relations (in the case of Bose-Einstein particles)

$$\psi(r)\psi^*(r') - \psi^*(r')\psi(r) = \delta(r-r'),$$
$$\psi(r)\psi(r') - \psi(r')\psi(r) = 0.$$

To investigate the problem interesting us, we may pass from this precise equation to an equation with Hartree's self-consistent field, which will have the same form, but with ordinary unquantized functions $\psi(r), \psi^*(r)$, , and try to find periodic solutions for $|\psi(r)|^2$. This attempt was made by us but did not lead to any success.

We rewrite equation (6.15) in the integral form, introducing Green's function of the differential equation $\Delta\psi = 0$, and then

$$\psi(r) + \lambda \int_{-\infty}^{\infty} G(|r-r'|)\psi(r')\,dr' =$$
$$= \int_{-\infty}^{\infty}\int G(|r-r'|)\psi(r')K(|r'-r''|)\psi^*(r')\psi(r)\,dr'dr'', \quad (6.16)$$
$$\lambda = \frac{2m}{\hbar^2}E.$$

This equation is of the type of nonlinear integral equations. As is known from the mathematical theory of such equations*, branching solutions for them exist only on the condition that the linear integral equation at the left has singular points in the range of values of λ . Inasmuch as the equation $\Delta\psi + \lambda\psi = 0$ over an <u>infinite</u> interval has no singular points λ , then the linear part of equation (6.16) cannot have any either. It may therefore be asserted that nonlinear equation (6.16) does not have any "branching" solutions, as required for the description of the crystal
It might be hoped that consideration of quantum ensembles might change this result. We shall show, however, that introducing quantum ensembles my means of the density matrix will not lead to periodic solutions by means of which it would be possible to construct a crystalline phase.

For this purpose, we shall start from the Hartree-Fock equation for the density matrix

* See literature for Chapter IV.

where

$$\left.\begin{aligned} i\hbar\frac{\partial\rho_{xx'}}{\partial t} &= \int H_{xx''}\rho_{x''x'}\,dx'' - \int \rho_{xx''}H_{x''x'}\,dx'', \\ H_{xx'} &= -\frac{\hbar^2}{2\mu}\frac{\partial^2}{\partial x'^2}\delta(x'-x) + V(x')\,\delta(x'-x) \end{aligned}\right\} \tag{6.17}$$

and

$$V(x) = \int K(|x-x'|)\,\rho_{x'x'}\,dx'.$$

We write this equation in the following form:

$$i\hbar\frac{\partial\rho_{xx'}}{\partial t} = -\frac{\hbar^2}{2\mu}\left(\frac{\partial^2}{\partial x^2} - \frac{\partial^2}{\partial x'^2}\right)\rho_{xx'} + \\ + \left(\int K(|x-x'|)\,\rho_{x'x'}\,dx' - \int K(|x'-x|)\,\rho_{xx}\,dx\right)\rho_{xx'}. \tag{6.18}$$

Equation (6.18) has a trivial solution, corresponding to a uniform distribution of the particles

$$\rho_{xx'} = \rho_0 = \text{const.}, \tag{6.19}$$

since

$$\int_{-\infty}^{\infty} K(|x-x'|)\,dx' = \int_{-\infty}^{\infty} K(|x'-x|)\,dx = \text{const.} \tag{6.20}$$

We shall seek a solution of equation (6.18) in the form

$$\rho_{xx'} = \rho_0 + \sigma_{xx'}, \quad \sigma_{xx'} \ll \rho_0.$$

In a linear approximation, we have:

$$i\hbar\frac{\partial\sigma_{xx'}}{\partial t} = -\frac{\hbar^2}{2\mu}\left(\frac{\partial^2}{\partial x^2} - \frac{\partial^2}{\partial x'^2}\right)\sigma_{xx'} + \\ + \left(\int_{-\infty}^{\infty} K(|x-x'|)\,\sigma_{x'x'}\,dx' - \int_{-\infty}^{\infty} K(|x'-x|)\,\sigma_{xx}\,dx\right)\rho_0. \tag{6.21}$$

This equation has a solution of the form exp $(ik(x - x')$ with <u>any</u> value for k. It is evident that periodicity of this kind has nothing in common with crystals, and is an expression of the presence of free translation of particles, not subject to the influence of the field of forces. Such solutions are always present in the Hartree-Fock equation, as will easily be seen from equation (6.15).

This example shows that the Hartree-Fock method does not exhaust the multiplicity of solutions of our equation.

6. We now show that it is possible to generalize the Hartree-Fock equation in such a way as to obtain a result approaching our periodic solutions. (See Part II, Chapter I).

To this end, we observe, in the first place, that Hartree's equations (6.18) include a functional interaction of the form:

$$V(x) = \int K(|x - x'|)\, \rho_{x'x'}\, dx'.$$

In this way, the potential energy V(x) is connected solely with the diagonal elements of the matrix of density ρ. We take a generalizing step, including in our expressions the nondiagonal element as well, introducing instead of integrals

$$\int_{-\infty}^{+\infty} K(|x - x'|)\, \rho_{x'x'}\, dx' \qquad \text{and} \qquad \int_{-\infty}^{+\infty} K(|x' - x|)\, \rho_{xx}\, dx$$

the new integrals:

$$\int_{-\infty}^{+\infty} K(|x - x'|)_{x''x'}\, dx' \qquad \text{and} \qquad \int_{-\infty}^{+\infty} K(|x' - x''|)\, \rho_{xx''}\, dx''.$$

in which case we have:

$$i\hbar \frac{\partial \rho_{xx'}}{\partial t} = -\frac{\hbar^2}{2\mu}\left(\frac{\partial^2}{\partial x^2} - \frac{\partial^2}{\partial x'^2}\right)\rho_{xx'} +$$

$$+\left(\int_{-\infty}^{\infty} K(|x - x''|)\, \rho_{x''x'}\, dx'' - \int_{-\infty}^{\infty} K(|x' - x''|)\, \rho_{xx''}\, dx''\right)\rho_{xx'}. \quad (6.23)$$

Linearizing this equation $(\rho_{xx'} = \rho_0 + \sigma_{xx'},\ \sigma_{xx'} \ll \rho_0)$ gives:

$$i\hbar \frac{\partial \sigma_{xx'}}{\partial t} = -\frac{\hbar^2}{2\mu}\left(\frac{\partial^2}{\partial x^2} - \frac{\partial^2}{\partial x'^2}\right)\sigma_{xx'} + \left(\int_{-\infty}^{\infty} K(|x-x''|)\,\sigma_{x''x'}\,dx'' - \int_{-\infty}^{\infty} K(|x'-x''|)\,\sigma_{xx''}\,dx''\right)\rho_0. \tag{6.24}$$

Equation (6.24) has a solution of the form:

$$\sigma(x)\,\sigma^*(x') = \sigma_{xx'},$$

and for $\sigma(x)$ we obtain, for $\partial/\partial t = 0$:

$$C\sigma(x) - \frac{\hbar^2}{2\mu}\frac{\partial^2 \sigma(x)}{\partial x^2} + \rho_0 \int_{-\infty}^{\infty} K(|x-x'|)\,\sigma(x')\,dx' = 0, \tag{6.25}$$

where C is an arbitrary constant, appearing as the result of the separation of variables. A completely analogous equation applies for $\sigma^+(x')$ as well. The equation obtained has a solution of the type exp(ikx), where k is subject to a definite condition. In point of fact, if we substitute function e^{ikx} in equation (6.25) instead of $\sigma(x)$, we have:

$$Ce^{ikx} + \frac{\hbar^2}{2\mu}k^2 e^{ikx} + \rho_0 \int_{-\infty}^{x} K(|x-x'|)\,e^{ikx'}\,dx' + \rho_0 \int_{x}^{\infty} K(|x'-x|)\,e^{ikx'}\,dx' = 0,$$

or, replacing the variables of integration in both integrals

$$\left(C + \frac{\hbar^2}{2\mu}k^2\right)e^{ikx} + \rho_0 \int_0^{\infty} K(s)\,e^{ik(x-s)}\,ds + \rho_0 \int_0^{\infty} K(s)\,e^{ik(x+s)}\,ds,$$

from which, eliminating e^{ikx}, we have

$$C + \frac{\hbar^2}{2\mu}k^2 + 2\rho_0 \int_0^{\infty} K(s)\cos ks\,ds = 0. \tag{6.26}$$

This condition is analogous to our criterion of crystallization (see below), if constant C is identified with the temperature and the member containing the Planck constant is omitted. In this way, merely a special generalization of the Hartree-Fock method can lead to solutions analogous to those interesting us.

It should be stressed, however, that we have no other bases for this generalization of the Hartree-Fock equation: it has a merely formal character. We introduce it here in order for it to illustrate the inadequacy of the Hartree-Fock method. Equation (6.26) may easily be generalized to the case of processes periodic in time. It will easily be seen, however, that in that event it will have no relation to temperal processes in our theory.

7. The conclusion we must draw from what has been said is that classical statistics and quantum mechanics cannot at the present time explain either the process of crystal formation or its periodic structure. It can be said that a thoroughgoing complete description of the crystal is quite impossible within the framework of these theories.

The results indicated above show the limitations and inadequacy of the apparatus of quantum mechanics even over macro-intervals. This refutes the basic physical argument of Bohr's interpretation of quantum mechanics, the principles of which, according to Bohr, have a universal, and hence an epistemological character.

One remark must be made in conclusion. The connection, established above, between the ψ- and the f-functions in the particular cases indicated furnishes the possibility of an analogy in interpreting them. That is, we may retain the concept of localization of the particle and interpret the function f as the probability that a particle has definite values for its coordinates and velocities. The difference in principle from the interpretation of the ψ-function in quantum mechanics consists in the following. In the theory that has been set forth, the causes of the dispersion of the point particle in coordinate space and the presence of the distribution of velocities coincide exactly. These causes consist in the presence of the connection of the particle with the surrounding medium. The magnitude of the surface integrals that appear upon passing from the extended form to the point form (see Chapter III) quantitatively exhibits and makes concrete these causes. The fact that these causes of dispersion can be discovered and controlled in advance in the theory being set forth is an advantage of this theory as compared with quantum mechanics, where it is impossible to subject these causes to control or eliminate them by physical conditions.

It is of importance that the conception of probability here takes on an essentially different form, since the fact of a dispersion of particles occurs before including force interactions in the theory. This circumstance is the reason why the multiplicativity of the functions is not an approximation of the statistical or quantum problem of many particles, but expresses really existing and precise interrelationships in the theory being set forth.

The possibility of expressing the motion of each particle by its own function enables us to make the step forward to a new conception of the particle: to describe it by an extended function of distribution in real space with a specifically new feature, a diversity of velocities at each point of the "cloud" (see Point 6 of the following section). This step forward cannot be taken either in statistical or quantum mechanics, inasmuch as there are no individu functions expressing the behavior of each particle in these theories.

Section 7. The Physical Ideas of the Theory

In this section, we shall briefly dwell on the physical and methodologic problems characteristic of the theory being set forth.

1. Individual and collective properties. The question of the relation- ship between the individual properties of particles and the properties of th entire system of particles (the collective) has not yet been adequately elucidated in modern physics.

What properties of particles can we call specifically individual? Is it at all possible to point out properties of particles that their individua characteristics are not subject to the influence of the remainder of the collective?

We know that in most contemporary theories the charge is considered such an individual characteristic of the particle. Formally, the apparatus of these theories also quite permits the case, devoid of meaning, in which there is a single completely isolated particle which nevertheless has a char And yet, it is quite clear that we can only introduce the mere conception of the charged particle if we consider interactions.

It appears to us that there is another point of view that is more fruit and in any event more consistent, according to which the charge of the eleme ary particle is a characteristic that is essentially bound up with the prese of a system, a collective of particles.

2. Distant space-time connections. We also mention a second property of the connections between physical properties in a collective of particles. As we know, in the usual theories, these connections have a differential character. That is, the values for these magnitudes at infinitely near poin in space and time are connected.

A necessary step in generalization must be the introduction of distant interactions, in which each particle interacts simultaneously with all the other particles of the collective. Moreover, it is not only the state of th system at a given moment of time that must be taken into account but also th effect of all its evolution in time over a finite or infinitely great interv of time. There should correspond to this, in the mathematical apparatus of the theory, a special integral-differential method of expressing the connections.

It should be expressly stated here that introducing integral-differenti distant connections does not in any sense mean a return to the conception of action at a distance, but generalizes the customary conceptions concerning differential connections between physical magnitudes.

3. The problem of the relationship between "micro" and "macro", it see to us, may be set forth and to a certain extent explained if we express in o

theory the direct connection of each particle with the collective as a whole. This problem cannot be solved by means of present statistical methods. Attempts to solve it in statistics are bound up with the introduction, in addition to paired interactions, of triple, quadruple, etc. interactions, which is of little effect, since even the problem of free bodies has not been solved in classical mechanics. The introduction of integral connections with the entire collective as a whole for each particle appears to be an unavoidable step if we are to go forward in the field in question.

4. Collective interactions. Up to the present, there have been no physical theories known in which the incorporation of each new particle into a system of particles would evoke changes in the interaction and behavior of the particles initially present in the ensemble. And yet, the addition of a third particle to two others should change the law of the interactions of these two particles. Finding the most general quantitative expression of non-additive interactions of the collective type occurring for any particles is essential for expressing the role of the collective in the many-particle problem.

5. Motion in a collective of particles. The idea of motion as an objective process organically inherent in matter should be at the basis of every physical theory and be concretely expressed. We shall set forth where the inadequacies of the fundamental physical theories consist, from this point of view, and we shall likewise explain the way in which these inadequacies may be eliminated.

a) In the classical and relativistic mechanics of material points, there are no concepts corresponding to the temperature. Nonetheless, the thermal form of motion is widespread, so that there is no reason not to suppose that this form of motion is inherent in the very nature of particles.

b) The equations of classical electrodynamics do not describe the fluctuational changes of the field. As we know, these equations admit of solutions in which the field stresses vanish over all space (in the absence of charges and currents). The mere fact that such solutions are permissible is a defect in the theory.

c) Thermal phenomena are expressed in the kinetic theories of many particles, an example of which is the kinetic theory of gases. But it is taken for granted as obvious in these theories that thermal motions are derived from the equations of mechanics, that the equations of mechanics "are prior" to thermal motions - a point of view that arose during the historical course of scientific development, but not methodologically obvious. Is not the cause for this the fact that kinetic theories lead to the one-sided assertion that processes must have only an irreversible character?

A rational theory should correctly express not only effects connected with disorientation, but also processes of an ordering motion that does not disappear in the course of time, something which constitutes an insoluble problem within the framework of these theories. Hence, for example, it is impossible to obtain an explanation of the phenomenon of superfluidity from the pattern of kinetic theory.

d) In Gibbs' statistical theory, the "source" of thermal motion is transferred to a heat reservoir, whose nonmechanical nature, however, is not disclosed. This apparatus has been very useful and will continue to be useful, but it cannot pretend to give a rational unification of thermal and mechanical phenomena and in particular with electromagnetic phenomena (see Sec. 4 of this chapter). Furthermore, we come up here against difficulties of principle even in particular questions (see Sec. 5). Gibbs' theory, which includes a temperature parameter, is at the same time based on precise equations of mechanics, which, however, do not, from the point of view of the conception set forth, contain the thermal form of motion.

In the theories referred to above, motion is always conditioned by the presence of certain "sources" (for example, external forces, charges and currents, the heat reservoir). The absence of such sources leads to a state of complete absence of motion, something which has no physical meaning.

But, should there not exist in addition to the sources, more general factors expressing the inseparability of motion in physical objects? A theory based on considering such factors, which exist independently of the concrete form of interactions, should naturally synthesize the various physical theories that operate only with concrete forms of sources (for example, electrodynamics and mechanics).

In the theory being set forth, the factors referred to are made concrete in the following manner.

1) The idea that matter is unthinkable without motion is expressed in the very model of the particle. The laws of motion of this particle are of such a nature that they contain a parameter corresponding to the temperature as one of the particular states of motion.

2) There appears in a collective of particles an effect of spontaneous activity, which expresses the special effect of the collective on the behavior of each particle.

The connection of the theory being set forth with quantum mechanics should be specially considered, and still awaits its further refinement.

6. <u>Generalization of the particle concept</u>. The establishment of a conception of the particle is the initial point of any many-particle theory. Must point localization in coordinate and velocity space be an unalterable property of the particle, irrespective of the external physical conditions in which it is placed? Or, in other words, should the idea of the localization of particles (in the sense of classical mechanics) be primary in constructing the theory, that is, should it precede the introduction of force interactions? Undoubtedly, in making such an assumption we naturally proceed to introduce some sources of motion (by means of the acting forces). It seems to us that the contrary point of view is more correct, according to which the initial model of the particle (before the introduction of any forces) already comprises the possibility that the particle may have an infinite number of different velocities and spatial positions.

Classical mechanics, statistical physics, electrodynamics, and in a certain sense (see Sec. 7), quantum mechanics as well, comprise the vaguely formulated presupposition that localization of particles occurs independently of the existence of connections of the particle with the surrounding medium and with the remaining particles. The basic defects of such a localizing formulation of the particle come down to the following:

1) The point has no spatial dimensions.

2) The point does not contain within it any motions or the possibility of any internal development.

3) The localization of particles leads to difficulties in electrodynamics and statistical physics.

We eliminate the first inadequacy if we recognize that the particle has a volume in space, the magnitude of which, however, need not be given once for all, but must be defined by the interaction with other particles.

If we want to include within the particle inner motions as well, we must assign a definite velocity to each point of space where $f \neq 0$. However, this is an unacceptable conception, since it involves the assumption that elements adjacent to each other cannot go apart from each other to any distance whatever. Each infinitely small element moves as a whole. This limitation deprives the particle, as we have seen, of many of its properties. We introduce a deeper-reaching postulate, according to which, a space element of the particle contains infinitely many different elements, each of which moves with its own definite velocity.

Making what has been said concrete leads to setting up a function $f(\underline{r}, \underline{v}, t)$, which expresses the fact of the objective existence of the particle. This generalization not only reveals the particular cases in which the previous picture of localized particles in the sense of classical mechanics applies, but also makes it possible to establish new properties of many particles, which cannot be "caught" by previous methods. The new dynamics unites the ideas of continuity and corpuscularity, since the initial mode of describing the motion of the particle is connected with an extended function, and the localization of the particle appeared only in particular cases.

We do not have adequate bases for carrying over to the separate elements of the "cloud" characterizing each separate particle, the concepts of interaction derived from the classical scheme. In the theories based on the principles that have been set forth, classical interactions can occur only between different clouds, describing different particles.

This new model of the particle is essentially different from the particles considered in electron theory where, as is known, the forces of interaction between particles are assumed to be active as well between separate elementary volumes of a single particle. This view, in particular, leads to the conception of the charge as an essentially individual property of the particle, a conception we have already rejected.

The new conceptions concerning the particle liberate us from the well-known difficulties connected with the point localization of charged particles.

Under certain physical conditions (e.g., the boundary conditions for the f-function), the particle may occupy as small a region of space as desired, but inasmuch as the charge always appears only as the result of interaction of particles, a completely isolated particle has no electrostatic energy of its own.

We note once again that in addition to the interpretation set forth above, which generalizes the particle concept, the theory set forth admits a different, statistical interpretation of the f-function in certain cases (see Sec. 6). Although at the present state of the theory's development, the results are the same in both cases, the author, for a number of physical and methodological reasons, maintains the point of view developed immediately preceding.

7. "Test corpuscles." In explaining the limits within which a physical theory is applicable, the most difficult aspect is usually the discovery of those elements of the theory whose presence was required by the limitations linked up with the experimental establishment of the initial physical concepts on which the theory is based.

For example, classical electrodynamics is essentially based on the conception of test corpuscles. We can only measure the strength of fields by a suitable test body. We cannot at will introduce these test corpuscles into the theory or remove them from it. Electrodynamics can be formulated given at least two bodies, the state of one of which, however, the test body, is always fixed.

On the other hand, we are required in electrodynamics to ascribe special properties to these test corpuscles. Thus, they must have precise space and velocity localizations (and consequently cannot be themselves subject to fluctuation in their positions and velocities). In addition, the test bodies must have as small a charge as desired. The question remains open as to the extent to which these postulates can be carried over in a field with dimensions on the order of those of the elementary particles, where it is not possible to effectuate the test corpuscle procedure.

A further refinement of the theory should be a formulation of the laws of electrodynamics that would make it possible to eliminate and introduce these corpuscles according to the character of the objects under study. We have already pointed out that the apparatus of classical and relativistic dynamics, operating with strictly localized particles, contains the postulate that localization of the description of the particles' behavior is effected independently of the physical conditions in which the particle ensemble being studied is situated. The discovery that this postulate is valid only under certain definite physical conditions and therefore is not universal is important for an understanding of the limits within which these theories apply. In both cases (test corpuscles and the principle of localization) the cause for the existence of a limitied field within which the theory applies consists in the initial conditions, realization of which occurs only under special experimental conditions. It is impossible in both these cases to eliminate or modify these postulates without going outside the framework of the theory. It is of importance, however, that discovery of the fact referred to above makes it possible to work out a more general point of view, which explains precisely in what particular cases the previous condition hold true, and what the nature should be of the description of the behavior of many particles, when these conditions do not apply.

8. Closure of physical systems. In investigating any physical object, there is conflict between two competing factors: on the one hand, the fact that all the phenomena of nature are interconnected, and on the other, the need of separating out or isolating what is basic, principal, essential to the given object. This makes it a matter of urgent necessity to raise and to solve the question as to the criterion for the closure of a physical system.

In classical mechanics, the closure of a dynamic system is guaranteed (from the point of view of mechanics) by the requirements of conservation of the number of particles and the absence of external forces and fictional forces. In electrodynamics it is required that the field vanish identically at the boundaries of the region of space, that is, in a certain sense a second requirement is set up. In statistical physics, the position is less clear. On the one hand, Gibbs' ensembles are constructed for isolated (in the sense of classical mechanics) dynamical systems, the behavior of which is described by Liouville's dynamic equation, while, on the other hand, the interpretation of the particular solution - Gibbs' canonical distribution - requires a connection with a reservoir of heat.

In quantum mechanics realization of precise spatial limitation of systems requires the introduction of infinitely high potential barriers, which cannot be experimentally realized.

We observe in general here that it is impossible to give a universal physical definition of closure of a system, since it is impossible to foresee in advance all the possible infinitely diverse types of connection of the system under consideration with the surrounding medium, including modes of such connection that have not yet been discovered.

In this way we see that

a) An essential cause of the limitation of the field within which a theory is applicable is the treatment of the concept of "closure" accepted in the theory.

b) There can be no universal theories (either "macro" or "micro") if only because it is impossible to give a universal definition of closure.

c) Since the concept of closure is different in different theories, abandonment of this concept implies connection of the system with the medium, the type of which is also variable.

9. Cauchy's problem. States of a system that we consider initial states may in general be of two types. Cases are possible in which the initial state is the result of the development of this system itself or of external action, the result of which may be achieved by the natural evolution of the system. On the other hand, there are cases in which the initial state is the result of external actions on the system that are no longer reducible to the natural development of the system. Such actions may not be defined and described by

the initial principles on which the equations of motion of the system under consideration are based. In the second case, the apparatus of the theory must contain something expressing the uniqueness of the initial moment of time as compared with all the other moments.

Although setting up Cauchy's problem sets apart a certain initial moment of time, we in solving this problem cannot, of course, explain all the temporal processes possible in the given system, since those temporal processes taking place without this setting apart of the initial moment do not fall within our sphere of consideration. It is therefore of importance to know in what cases the initial moment can be isolated in solving Cauchy's problem and in what cases it cannot. In other words, it is necessary to give a formula for defining solutions no dependent on setting apart a privileged moment of time. These solutions should represent those processes whose natural evolution is not disturbed by physical actions on the system.

These two types of solutions are clearly represented in the theory being set forth. The solutions of the first type (with a unique initial moment of time) describe temporal processes in a many-particle system, characterized by the asymmetry of time that is inherent in irreversible processes. Solution of the second type are symmetrical with respect to time, which is a property of reversible processes (see Chapter V).

The establishment and solution of Cauchy's problem have an additional feature as well, which reduces the complex of time processes covered by it. In a number of cases in Cauchy's problem, assumptions are made as to fixed values of the function sought at the limiting spatial surfaces over the entire time interval t from $-\infty$ to $+\infty$. Hence, in addition to the initial conditions, conditions are also laid down as to the magnitude of certain surface integrals.

In the theory set forth, these conditions lead to a sharp limitation of the number of different types of temporal processes described by this theory. As will be seen later, the initial equations of the theory contain, on the one hand, solutions that include temperature effects and, on the other hand, the equations of classical mechanics, which has no parameter adequate to the temperature. We pass to the solutions of classical mechanics if we require that the surface integrals expressing the flow of particles out of a definite volume should vanish. This condition signifies elimination of the fluctuational emission of particles conditioned by temperature effects, which are included in the original equations. The transition to the classical pattern is connected with elimination of the temperature fluctuations. As a result, we obtain a pattern that does not contain any parameter adequate to the temperature.

The following general conclusions may be drawn from what has been said:

1. The solution of Cauchy's problem does not always cover all the temporal processes included in the initial equations of this or that theory. Establishing the problem sometimes involves postulating physical conditions that sharply limit the range of possible solutions.

2. It is therefore impossible to identify fulfillment of the general principle of causality with the solution of Cauchy's problem. The principle of causality as a philosophical assumption is not, of course, equivalent to particular methods of explaining temporal processes connected with the customary formulation of Cauchy's problem. The cause for some processes may consist not in the initial conditions for the function sought, but in other physical conditions that likewise determine it. Thus in the theory being set forth, for example, there arise, with continuous changes of the temperature and density of the medium, discontinuous and qualitatively new solutions (spatially periodic, nondamping solutions in the form of waves, etc.). From the mathematical point of view, these solutions are connected not with Cauchy's problem but with the problem as to the branching of the solutions of nonlinear integral equations upon continuous variation of a parameter entering into the equation. What is involved is thus the transfer of the cause (for some phenomena) from the initial conditions for the function sought to the initial conditions for a parameter, which, however, radically alters both the physical and mathematical side of the question.

Herewith a definite step forward is taken towards a profounder knowledge of the physical significance of the principle of causality.

CHAPTER II

INITIAL ASSUMPTIONS, FUNDAMENTAL EQUATIONS AND THEIR PROPERTIES

Section 8. Fundamental Equations

The theory is based on the two following fundamental physical assumptions.

1. Abandonment of the principle of space and velocity localization of particles (in the sense of classical mechanics), occurring independently of force interactions.

2. Introduction of force interactions by analogy with classical mechanics, but taking into account the new principle of nonlocalization of particles. This new means of introducing the interactions should generalize the previous method, but at the same time give a concrete expression to the direct connection of each particle with the entire collective as a whole, of which we spoke in the preceding chapter.

We satisfy the first equation if we describe the behavior of each particle by an extended function of distribution f(r, v, t), which satisfies the continuity equation in space of six dimensions.

A more general point of view is possible, according to which function f is set up in a space of velocities and accelerations of as high order as desired:

$$f = f(\mathbf{r}, \mathbf{v}, \dot{\mathbf{v}}, \ddot{\mathbf{v}}, \ldots, t),$$

$$-\frac{\partial f}{\partial t} = \operatorname{div}_r \mathbf{v} f + \operatorname{div}_v \dot{\mathbf{v}} f + \operatorname{div}_{\dot{v}} \ddot{\mathbf{v}} f + \ldots;$$

but we pass to this generalization only on a following level, when the fruitfulness of this simplest method of description has been made clear.

It is also possible to assume on the basis of the equation of continuity in space of 6N dimensions

$$f = f(\mathbf{r}_1, \mathbf{v}_1, \ldots, \mathbf{r}_n, \mathbf{v}_n, t),$$

$$-\frac{\partial f}{\partial t} = \sum_i^N \operatorname{div}_{r_i} \mathbf{v}_i f + \sum_i^N \operatorname{div}_{v_i} \mathbf{g}_i f$$

with an as yet unknown significance for the vectors g_i. We consider this possibility somewhat later (see also Chapter I, Sec. 4).

The force interactions are taken into account by means of the vector of acceleration in the equation of continuity in space of six dimensions. According to the second condition, we must perform this inclusion by analogy with classical mechanics, but taking into account the fact that the dispersion of the position of the particles takes place before introducing the forces and is defined by function f itself.

There is also the possibility of carrying out the inclusion of the forces according to a pattern customary in electrodynamics. In contrast to mechanics, electrodynamics admits of an effect of spontaneous action. From the mathematical point of view, both methods lead to similar equations. For a large number of particles, the equations coincide. The physical considerations discussed in point 6 of Sec. 7 convince us that we should stay on the first path for developing the theory, although making use of the second as well.

We first consider the simplest case, the case of two particles interacting by means of arbitrary central forces, as well as electrodynamically. We have:

$$-\frac{\partial f_1}{\partial t} = \operatorname{div}_{r_1} \boldsymbol{v}_1 f_1 + \operatorname{div}_{v_1} \boldsymbol{g}_1 f_1, \tag{8.1}$$

where

$$\boldsymbol{g}_1 = -\frac{1}{m_1} \nabla_{r_1} \int K(|\boldsymbol{r}_1 - \boldsymbol{r}_2|)\, \rho_2(\boldsymbol{r}_2, t)\, d\boldsymbol{r}_2 + \frac{e_1}{m_1}\left(\boldsymbol{e}_2 + \frac{1}{c}[\boldsymbol{v}_1 \boldsymbol{h}_2]\right),$$

$$-\frac{\partial f_2}{\partial t} = \operatorname{div}_{r_2} \boldsymbol{v}_2 f_2 + \operatorname{div}_{v_2} \boldsymbol{g}_2 f_2,$$

where

$$\boldsymbol{g}_2 = -\frac{1}{m_2} \nabla_{r_2} \int K(|\boldsymbol{r}_2 - \boldsymbol{r}_1|)\, \rho_1(\boldsymbol{r}_1, t)\, d\boldsymbol{r}_1 + \frac{e_2}{m_2}\left(\boldsymbol{e}_1 + \frac{1}{c}[\boldsymbol{v}_2 \boldsymbol{h}_1]\right),$$

and

$$\operatorname{div} \boldsymbol{e}_i = 4\pi e_i \int f_i\, d\boldsymbol{v}_i,$$

$$\operatorname{rot} \boldsymbol{h}_i - \frac{1}{c}\frac{\partial \boldsymbol{e}_i}{\partial t} = \frac{4\pi}{c} e_i \int \boldsymbol{v}_i f_i\, d\boldsymbol{v}_i,$$

$$\left(\rho_i = \int f_i\, d\boldsymbol{v}_i,\ \operatorname{div} \boldsymbol{h}_i = 0,\ \operatorname{rot} \boldsymbol{e}_i + \frac{1}{c}\frac{\partial \boldsymbol{h}_i}{\partial t} = 0\right),$$

$$i = 1, 2.$$

We take note of some special features of the proposed method for describing the particles:

1. The fact that function f is extended in the space of six dimensions generalizes the classical concept of the particle, which is described merely by a single point in this space.

2. The motion of each element ($d\underline{r}\ d\underline{v}$) of the f-"cloud" characterizing the particle is not influenced by interactions with the remaining elementary volumes of the same cloud. This point constitutes a difference in principle from electron theory, which carries the motion of electric interactions over to the individual elements of the elementary particle as well. In our theory, the charge of the single particle appears only as the result of interaction with another. In the absence of a second particle, the motion of the first reduces to simple translation, without any internal interactions or perturbations:

$$-\frac{\partial f}{\partial t} = \operatorname{div}_r vf,$$

the solution to which is:

$$f = \Omega\,(t_0,\ r - v\,(t - t_0),\ v),$$

where   is an arbitrary function.

3. In equation (8.1), the electromagnetic interactions have a different sense than in usual theories. The initial point here is the existence of a distribution function. Hence, the strengths of the fields have the function merely of a connecting factor among the particles. The ordinary conception of the field, however, should appear only after excluding dependence on the velocities in the f-function and the transition to a space-time description (see Chapter III).

4. The effect of the equality of "action and reaction" for any central forces may be included in the theory by assuming antisymmetry for the expression $\operatorname{grad}_r K(|\underline{r}_1 - \underline{r}_2|)$ for permutation of the differentiation variable. Then the general total force acting on both particles vanishes, so that

$$\int dr_1\rho_1 \int \operatorname{grad}_{r_1} K\,(|r_1 - r_2|)\,\rho_2\,(r_2, t)\,dr_2 =$$

$$= - \int dr_2\rho_2 \int \operatorname{grad}_{r_2} K\,(|r_2 - r_1|)\,\rho_1\,(r_1, t)\,dr_1.$$

5. The proposed method of description allows for the possibility of abandoning the idealized concept of the impenetrability of particles.

Strictly speaking, no such concept exists either in classical, statistical or quantum mechanical apparatus, since realization of it is connected with infinitely high potential barriers.

In point of fact, functions f_1 and f_2 may coincide partially or completely. In the particular case, if f_1 is identically equal to $f_2 = f$ for any moment of time, there is only one equation for f, since the equation for f_1 and f_2 are identical.

6. System of equations (8.1), in which each particle is described by its own distribution function in six-dimensional space may be expressed by means of only a single distribution function depending on the coordinates and velocities of both particles. That is, multiplying the equations of type (8.1) by f_2 and f_1, respectively, and combining, we easily obtain, ignoring the electrodynamic interactions:

$$\left.\begin{aligned} -\frac{\partial f}{\partial t} &= \operatorname{div}_{r_1} v_1 f + \operatorname{div}_{r_2} v_2 f + \operatorname{div}_{v_1}\left(-\frac{1}{m_1}\operatorname{grad} V_{12}\right) f + \\ &\quad + \operatorname{div}_{v_2}\left(-\frac{1}{m_2}\operatorname{grad} V_{21}\right) f, \\ V_{12} &= \int K(|r_1 - r_2|)\, dr_2 \int f\, dv_1\, dr_1\, dv_2, \\ V_{21} &= \int K(|r_2 - r_1|)\, dr_1 \int f\, dv_2\, dr_2\, dv_1, \end{aligned}\right\} \quad (8.2)$$

where the general function of distribution is multiplicative, i.e.,

$$f(1,\ 2) = f_1(1) f_2(2), \quad (8.3)$$

and the normalization condition is employed. $\quad f(1,\ 2, \ldots, N_1) = f_1(1) f_2(2) \ldots f_N(N).$

Hence the method employed for describing the ensemble of particles is equivalent to a description of a system of particles in space of 6N dimensions by means of a multiplicative distribution function of the form:

$$\int f_1\, dr_1\, dv_1 = \int f_2\, dr_2\, dv_2 = 1$$

It is of impartance that (8.3) is an exact solution of equation (8.2), leading to equations (8.1).

The property of multiplicativity of the functions does not involve any defect or approximativity in the theory. Multiplicativity can only lead to conclusions of this kind if it is regarded from the point of view of strictly localized particles, but since this position is untenable for a variety of reasons, it is necessary to generalize them (see Chapter I).

Equation (8.2) has a somewhat greater degree of generality than (8.1), since it satisfies solutions in which the variables are not separated, as well. However, physical considerations connected with generalization of the particle concept itself require us to give the preference to equation (8.1).

7. The structure of equations (8.1) is such that they do not contain the spontaneous action effect. However, this effect, too, can easily be included by taking as our basis here the electrodynamic method of calculating the interactions. That is, adding suitable terms for the spontaneous action and hence automatically including the interaction of the various elementary volumes of one and the same cloud, we write:

$$\left.\begin{aligned} V_{12} &= \int K_{11}\rho_1\,dr_1' + \int K_{12}\rho_2\,dr_2, \\ V_{21} &= \int K_{22}\rho_2\,dr_2 + \int K_{21}\rho_1\,dr_1. \end{aligned}\right\} \tag{8.4}$$

We likewise introduce the general field strength for the separate particles:

$$\operatorname{div} e = 4\pi\left(e_1 \int f_1\,dv_1 + e_2 \int f_2\,dv_2\right)$$

etc. In this way, we achieve the construction of a theory with spontaneous action.

It is of importance to note that even with spontaneous action for each separate particle, it is possible, in the case of uniform particles, to formulate a single equation with a single function f. For, setting

$$f = f_1 + f_2 \qquad (m_1 = m_2,\ K_{11} = K_{12} = K_{21} = K_{22}),$$

we have from (8.1), taking into consideration (8.4), and omitting the electrodynamic terms for simplicity's sake:

$$\left.\begin{aligned} -\frac{\partial f}{\partial t} &= \operatorname{div}_r vf + \operatorname{div}_v gf, \\ g &= -\frac{1}{m}\nabla_r \int K(|r - r'|) \int f\,dv' \cdot dr', \end{aligned}\right\} \tag{8.5}$$

where the components $\underline{v}$ and $\underline{r}$ are independent variables that may differ only in the domain of variation. We assume here that this domain is infinite.

We now set up the equations of motion for any number of particles. For example, for three different particles with different masses, and for any law governing the forces of interaction, the equation should have the form:

$$\left.\begin{aligned}
-\frac{\partial f_1}{\partial t} &= \operatorname{div}_{r_1} v_1 f_1 + \operatorname{div}_{r_1} \frac{1}{m}(-\operatorname{grad}_{r_1}(V_{12}+V_{13}+V_{123}))f_1,\\
V_{12}(r_1,t) &= \int K_{12}(|r_1-r_2|)\rho_2(r_2,t)\,dr_2,\\
V_{13}(r_1,t) &= \int K_{13}(|r_1-r_3|)\rho_3(r_3,t)\,dr_3,\\
V_{123}(r_1,t) &= \iint K_{123}(|r_1-r_2|,|r_1-r_3|,|r_2-r_3|)\times\\
&\qquad\times\rho_2(r_2,t)\rho_3(r_3,t)\,dr_2\,dr_3
\end{aligned}\right\} \tag{8.6}$$

for the first particle, and fully analogous equations may be written down for the second and third particles with distribution functions f_2 and f_3.

In the theory that has been set forth, the problem of three bodies (if we omit the member V_{123}) does not take on any specifically new properties; it does not here lead to any mathematical difficulties in principle (abstracting from the case of the presence of nonadditive interactions). Thus, for similar particles, setting ($f_1 \equiv f_2 \equiv f_3 = f$), we once more obtain a single equation for f, which does not differ in its mathematical structure from the corresponding equation for the case of two particles.

For a system of N similar particles, which are in the same state of motion and interact electrodynamically as well (setting $f_1 \equiv f_r \equiv \ldots \equiv f_N = f$), we have:

$$\left.\begin{aligned}
-\frac{\partial f}{\partial t} &= \operatorname{div}_r vf + \operatorname{div}_v \frac{1}{m}\left\{-\operatorname{grad} V + e\left(e+\frac{1}{c}[vh]\right)\right\}f,\\
(Vr,t) &= \sum_n^{N-1}\int\ldots\int K_n(r,r_1,r_2,\ldots,r_n)\rho(r_1,t)\times\\
&\qquad\times\rho(r_2,t)\ldots\rho(r_n,t)\,dr_1\ldots dr_n,\\
\operatorname{div} e &= 4\pi e(N-1)\int f\,dv,\\
\operatorname{rot} h - \frac{1}{c}\frac{\partial e}{\partial t} &= \frac{4\pi}{c}e(N-1)\int fv\,dv\\
\operatorname{div} h = 0;\ \operatorname{rot} e &+ \frac{1}{c}\frac{\partial h}{\partial t} = 0,\\
\rho(r,t) &= \int f(r,v,t)\,dv.
\end{aligned}\right\} \tag{8.7}$$

The number of groups of s particles that can be formed from a total of N particles equals:

$$\frac{(N-1)!}{s!\,(N-1-s)!}=\binom{N}{s}.$$

The collective interactions arising in groups of s particles should be repeated $\binom{N}{s}$ times. We introduce the numerical factor $\binom{N}{s}$ into nucleus K_s, so that we have

$$K_1=(N-1)\,K_1\,(|r_1-r_2|);$$
$$K_2=\frac{(N-1)\,(N-2)}{2}\,K\,(|r-r_1|,\ |r-r_2|,\ |r_1-r_2|)$$

etc. Nonlinear functionals express in the most general form the non-additiveness of the forces of interaction in a collective of particles. The second term of the functional sum, depending on three different indices, occurs only when there are at least three particles, the third member when there are four, etc. Summation according to n is performed from 1 to N - 1. The total number of particles in the system thus enter into the equation in a nontrivial manner.

We have no reason to feel that such a method for calculating interactions is necessary only for complex particles; on the contrary, its natural mathematical and physical nature (a functional sum), expressing the interaction of the collective of particles, indicate that it is applicable as well to what are called elementary particles, for example, protons or neutrons.

It must be held that the presence of functional nonadditive interactions constitutes the expression of new and varied properties of the collectively interacting particles.

For N > 1, replacing N - 1 by N and omitting interactions of merely additive type, we introduce into equation (8.7) the function $f^{*}=Nf$. Then for the function f^{*}, we obtain a primary, and most simple form (omitting the asterisks)

$$\left.\begin{aligned}-\frac{\partial f}{\partial t}&=\boldsymbol{v}\,\mathrm{grad}_{r}f+\frac{1}{m}\boldsymbol{F}\,\mathrm{grad}_{v}f,\\ \boldsymbol{F}&=-\,\mathrm{grad}_{r}\int K\,(|\boldsymbol{r}-\boldsymbol{r}'|)\,d\boldsymbol{r}'\int f\,(\boldsymbol{r}',\boldsymbol{v}',t)\,d\boldsymbol{v}',\end{aligned}\right\}\qquad(8.8)$$

where f now include N, and K - the interaction energy of the pair - no longer contains the total number of particles. Equation (8.8) belongs to the class of functional equations of the following type:

$$-\frac{\partial u}{\partial t}=y\left(\frac{\partial u}{\partial x}\right)+\left\{\lambda\left(\frac{\partial}{\partial x}\right)\int K(|x-x'|)\,u(t,x',y')\,dx'\,dy'\right\}\frac{\partial u}{\partial y}$$

$$(u=u(x,y,t)).$$

It coincides with equation (8.5), which includes the spontaneous action effect. It may therefore be asserted that for a sufficiently large number of particles, ignoring the nonadditive interactions, the difference between the two modes of description vanishes. In equation 8.8 (insofar as it exists for a collective of particles) the "spontaneous action" effect appears as a specific feature of the collective.

It is clear in advance that this effect should be a source of entirely new dynamic properties of a collective of particles. In particular, the inertial properties of the particles (mass) enter into the equation differently than in the customary theories, namely, in the form of a parameter in the integral term. Accordingly, some new manifestations of inertial properties are possible in special solutions. At the same time, it should be kept in mind that equation (8.8), which is nonlinear, does not change when f is multiplied by any number.

We also note that since no parameter of the number of particles entered into equation (8.8), the many-body problem is not complicated here upon the addition of each new particle. This result, however, applies only in the case N $>$ 1, and when the interactions of nonadditive type are ignored.

Section 9. Particular Forms of the Equations

We consider some particular forms of the initial equations.

1. <u>In the stationary case</u> $\left(\frac{\partial}{\partial t}=0\right)$ it is convenient to give the initial equations another form.

In this connection, we consider the problem of two bodies. Setting in equation (8.1)

$$f_i=W_i(\varepsilon_i)\,\rho(r_i),\qquad \text{where}\qquad \varepsilon_i=\frac{m_i}{2}(\xi_i^2+\eta_i^2+\zeta_i^2)$$

and taking into account the fact that the components of the radius vector (x, y, z) and the velocities (ξ,η,ζ) are independent variables which leads to $\operatorname{div}_r v=0$), we find:

$$W_i(\varepsilon_i) = W_i(0)\, e^{-\frac{\varepsilon_i}{\varepsilon_0}},$$

where ε_0 is the constant for the separation of variables, which may be identified with the temperature $(\varepsilon_0 = \theta = kT)$. The basis for this is first, the possibility of deducing Maxwell's distribution law by velocities (see below) and, secondly, the role that his parameter will play in the further discussion.

For the density of the probability of localization of the first particle, we find:

$$\left.\begin{aligned} \rho_1(r_1) &= a e^{-\frac{V(r_1)}{\theta}}, \\ V(r_1) &= \int_{-\infty}^{\infty} K(|r_1 - r_2|)\, \rho_2(r_2)\, dr_2 \end{aligned}\right\} \quad (9.1)$$

and similar equations for the probability density of the second particle. We see that in the stationary case the mass of the particles enters only in the distribution function according to velocity. In the expression for the particle is not individualized, so that we may set $\rho_1(r_1) \equiv \rho_2(r_2) = \rho(r)$, and describe the state of many particles by means of a single equation. For the potential V($\underline{r}$), the equation is as follows:

$$V(r) = \int_{-\infty}^{\infty} K(|r - r'|)\, f(V(r'))\, dr'; \qquad f(V(r')) = a e^{-\frac{V(r')}{\theta}} \quad (9.2)$$

It is sometimes convenient to write the initial equation as follows: for two points, the position of one of which is taken as the origin of coordinates, we have:

$$\rho(r) = a e^{-\frac{V(r)}{\theta}}; \qquad \rho(0) = a e^{-\frac{V(0)}{\theta}},$$

whence

$$\rho(r) = \rho(0)\, e^{-\frac{[V(r) - V(0)]}{\theta}}$$

and consequently, for the potential

$$\left.\begin{aligned} V(r) &= \int_{-\infty}^{\infty} K(|r-r'|) f(V(r') - V(0))\, dr', \\ \text{where} \quad f(V) &= \rho(0)\, e^{-\frac{V(r)-V(0)}{\theta}}. \end{aligned}\right\} \quad (9.3)$$

For N similar particles, we similarly obtain:

$$V(r) = (N-1) \int_{-\infty}^{\infty} K(|r-r'|) f(V(r') - V(0))\, dr' \qquad (9.4)$$

and taking into account the functional interactions, we obtain for the potential:

$$V(r) = \sum_n \int \cdots \int_{-\infty}^{\infty} K_n(r, r_1, \ldots, r_n) f(V(r_1) - V(0)) \ldots$$
$$\ldots f(V(r_n) - V(0))\, dr_1 \ldots dr_n. \qquad (9.5)$$

The problem of stationary states in a many-particle system thus reduces to the problem of solving nonlinear integral equations of a definite class. The basic parameter in equations (9.1) - (9.5) contains the temperature. It is an essential fact that such a problem may be formulated even for two particles. Here we obtain the temperature effect without constructions involving the introduction of a heat reservoir. This is the realization of the ideas voiced in points 5 and 6 of Sec. 7.

The initial equation in the stationary state may be written down as:

$$\left.\begin{aligned} v_i \frac{\partial f}{\partial x_i} - \frac{1}{m} \frac{\partial V}{\partial x_i} \frac{\partial f}{\partial v_i} &= 0, \\ V = \int_{-\infty}^{\infty} K(|r-r'|) \int_{-\infty}^{\infty} f(r', v')\, dv'\, dr', & \end{aligned}\right\} \quad (9.6)$$

where summation is made according to the indexes i.

In this differential equation V($\underline{r}$) may be considered as a known function, and then the characteristics equation corresponding to this equation is written as:

$$\frac{dx_i}{v_i} = \frac{dv_i}{-\frac{1}{m}\frac{\partial V}{\partial x_i}}.$$

The integral of this equation is the expression

$$\frac{mv^2}{2} + V(r).$$

Consequently, any function of this equation is a solution of the initial differential equation of the first degree in partial derivatives:

$$f = \Omega\left(\frac{mv^2}{2} - V(r)\right).$$

Employing this function, we obtain a more general nonlinear integral equation in the form

$$V(r) = \int_{-\infty}^{\infty} K(|r - r'|) \int_{-\infty}^{\infty} \Omega\left(\frac{mv'^2}{2} + V(r')\right) dr'\, dv'. \tag{9.7}$$

We now consider the equation for the stationary state as applied to a system of N electrons. Here, in the stationary state, and taking into account the electric and magnetic interactions, we have:

$$\operatorname{div}_r vf + \frac{e}{m}\operatorname{div}_v\left(-\operatorname{grad}_r \varphi + \frac{1}{c}[vh]\right) f = 0.$$

But since

$$\operatorname{div}_r vf \equiv v \operatorname{grad}_r f$$

($\underline{v}$ - the independent variable) and

$$\operatorname{div}_v [vh] f \equiv [vh] \operatorname{grad}_v f,$$

then the previous equation may be written in the following form:

$$\boldsymbol{v}\,\mathrm{grad}_r f - \frac{e}{m}\left(-\mathrm{grad}_r \varphi + \frac{1}{c}[\boldsymbol{vh}]\right)\mathrm{grad}_r f = 0.$$

where

$$\Delta\varphi = -4\pi(N-1)e\int f\,d\boldsymbol{v},$$

$$\Delta A = -4\pi(N-1)\frac{e}{c}\int \boldsymbol{v}f\,d\boldsymbol{v} \qquad (9.8)$$

$$(\boldsymbol{h} = \mathrm{rot}\,\boldsymbol{A}).$$

The solutions to this equation do not depend on magnetic interaction.

For, the equation for fo has the integral

$$f(\boldsymbol{r},\boldsymbol{v}) = \Omega\left(\frac{mv^2}{2} + e\varphi(\boldsymbol{r})\right) \qquad (\Omega - \text{ being any function})$$

$$\mathrm{grad}_r f = m\boldsymbol{v}\frac{\partial\Omega}{\partial u}, \qquad u = \frac{mv^2}{2} + e\varphi,$$

and consequently,

$$[\boldsymbol{vh}]\,\mathrm{grad}_r f \equiv 0.$$

Thus we see that in the case of a system of charged particles in thermal equilibrium, the magnetic field (both the external field and the field conditioned by the interaction with other particles) does not play any role. In a given case, therefore, we may simply omit the magnetic terms. In this point, we have a generalization of a well-known result of statistical theory, since here this result was obtained by previously taking into account the magnetic interaction among particles.

It is important to note that what has been said is true only for points of the region sufficiently distant from its boundaries. The problem of including the boundary effects seems to us to go beyond the framework of the stationary problem.

2. For describing processes in time, another form of dynamic equation is sometimes convenient (one not containing time and coordinate derivatives).

We write the equation in the form

$$\frac{\partial f}{\partial t} + \boldsymbol{v}\,\mathrm{grad}_r f = F(\boldsymbol{r},\boldsymbol{v},t) \qquad (9.9)$$

or

$$\frac{\partial f}{\partial t} + Af = F,$$

where A is the operator: $\underline{v}$ grad_r.

Since operator A does not explicity contain the time, formal integration gives

$$f(t, r, \ldots) = e^{-A(t-t_0)} f(t_0, r, \ldots) + \int_{t_0}^{t} e^{-A(t-\tau)} F(\tau, r, \ldots)\, d\tau.$$

The meaning of operator $e^{-A(t-t_j)}$ is easily found by means of solving a homogeneous equation. For F = 0, the characteristic equation is:

$$dt = \frac{dx_i}{v_i}.$$

The integral is the expression

$$r - v(t - t_0).$$

Consequently, any function

$$f(t_0, r - v(t - t_0), v)$$

is a solution of the homogeneous equation.

Accordingly, the action of the operator $[\exp(-A(t-t_0))]$ corresponds to the displacement of the distribution function from the point t, $\underline{r}$, $\underline{v}$ to the point t_0, $\underline{r} - \underline{v}(t - t_0)$, $\underline{v}$ (free translation of particles). Employing this result, we may give the following form to the initial equations describing two points, for example:

$$\left.\begin{aligned} f_1(t, r, v) &= f_1(t_0, r - v(t - t_0), v) + \\ &\quad + \int_{t_0}^{t} F(\tau, r - v(t - \tau), v)\, d\tau, \\ F(t, r, v) &= \text{div}_v \left\{ \frac{1}{m_1} \text{grad}_r \int_{-\infty}^{\infty} K(|r - r'|)\, \rho_2(r', t)\, dr' \right\} f_1, \\ \rho_2(t, r) &= \int_{-\infty}^{\infty} f_2(t, r, v)\, dv, \end{aligned}\right\} \qquad (9.1\ldots$$

with a completely analogous expression for the second particle. This form is convenient for solution of the Cauchy problem, when the distribution function is given for $t = t_0$.

We may choose another form of solution for the equation

$$\frac{\partial f}{\partial t} + Af = F(t, r, v), \qquad (9.\ldots$$

namely:

$$f(t, r, v) = c_1 \int_{t_0}^{t} F(\tau, r - v(t-\tau), v)\, d\tau + c_2 \int_{t_1}^{t} F(\tau, r - v(t-\tau) v)\, d\tau,$$

where $t_0 \leqslant t \leqslant t_1$, and c_1 and c_2 are constants, the sum of which must equal 1:

$$c_1 + c_2 = 1.$$

If we are interested in the temporal processes, symmetrical with respect to time over an interval

$$t_0 \leqslant t \leqslant t_1,$$

then the role of the section from t to t_0 and from t to t_1 should be the same. This can be achieved by setting $c_1 = c_2 = \frac{1}{2}$.

Section 10. Laws of Conservation

We first define the concept of a closed system, which we shall employ further.

We shall consider a system enclosed within a space surface as a closed system if:

a) There are no external forces;

b) The distribution function vanishes identically over the entire closed surface surrounding the system;

c) The distribution function vanishes sufficiently rapidly as $|\mathbf{v}| \to \infty$ (if this condition were not fulfilled, it would be difficult to require that the f-function vanish over the above-mentioned surface at sufficiently high velocities)+.

1. To derive the law of conservation of momentum in a closed system, we consider for simplicity's sake the one-dimensional case and the problem of two particles.

The expression for the general momentum will be:

$$P = \int_{-\infty}^{\infty} m_1 \xi_1 f_1(x_1, \xi_1, t)\, dx_1\, d\xi_1 + \int_{-\infty}^{\infty} m_2 \xi_2 f_2(x_2, \xi_2, t)\, dx_2\, d\xi_2,$$

$$-\frac{\partial P}{\partial t} = \int_{-\infty}^{\infty} m_1 \xi_1 \left(-\frac{\partial f_1}{\partial t}\right) dx_1\, d\xi_1 + \int_{-\infty}^{\infty} m_2 \xi_2 \left(-\frac{\partial f_2}{\partial t}\right) dx_2\, d\xi_2,$$

+ For a discussion of some questions connected with the concept of "closure" see Chapter I, Sec. 7, point 8, and likewise Chapter III.

but since

$$-\frac{\partial f_1}{\partial t} = \frac{\partial \xi_1 f_1}{\partial x_1} + \frac{\partial F_{12} f_1}{\partial \xi_1}, \qquad -\frac{\partial f_2}{\partial t} = \frac{\partial \xi_2 f_2}{\partial x_2} + \frac{\partial F_{21} f_2}{\partial \xi_2},$$

we receive:

$$-\frac{\partial P}{\partial t} = \int_{-\infty}^{+\infty} m_1 \xi_1 \left\{ \frac{\partial \xi_1 f_1}{\partial x_1} + \frac{\partial F_{12} f_1}{\partial \xi_1} \right\} dx_1\, d\xi_1 + \int_{-\infty}^{+\infty} m_2 \xi_2 \left\{ \frac{\partial \xi_2 f_2}{\partial x_2} + \frac{\partial F_{21} f_2}{\partial \xi_2} \right\} dx_2\, d\xi_2.$$

We now make use of the condition of closure, which we formulate as:

$$f \Big|_{x=-\infty}^{x=+\infty} = 0; \qquad f \Big|_{\xi=-\infty}^{\xi=+\infty} = 0$$

for both particles. We now obtain:

$$-\frac{\partial P}{\partial t} = \int_{-\infty}^{+\infty} F_{12} f_1\, dx_1\, d\xi_1 + \int_{-\infty}^{+\infty} F_{21} f_2\, dx_2\, d\xi_2 = \int_{-\infty}^{+\infty} F_{12} \rho_1 (x_1)\, dx_1 + \int_{-\infty}^{\infty} F_{21} \rho_2 (x_2)\, dx_2.$$

But since in the absence of external forces

$$F_{12} = -\int_{-\infty}^{+\infty} \frac{\partial K(|x_1 - x_2|)}{\partial x_1} f_2(x_2, \xi_2, t)\, dx_2\, d\xi_2 =$$

$$= -\int_{-\infty}^{+\infty} \frac{\partial K(|x_1 - x_2|)}{\partial x_1} \rho_2(x_2)\, dx_2,$$

$$F_{21} = -\int_{-\infty}^{+\infty} \frac{\partial K(|x_1 - x_2|)}{\partial x_2} f_1(x_1, \xi_1, t)\, dx_1\, d\xi_1 =$$

$$= -\int_{-\infty}^{+\infty} \frac{\partial K(|x_1 - x_2|)}{\partial x_2} \rho_1(x_1)\, dx_1,$$

where

$$\rho_1(x_1, t) = \int_{-\infty}^{\infty} f_1(x_1, \xi_1, t)\, d\xi_1 \text{ и } \rho_2(x_2, t) = \int_{-\infty}^{\infty} f_2(x_2, \xi_2, t)\, d\xi_2.$$

we obtain for the first member of the right side:

$$\int_{-\infty}^{+\infty} F_{12}\rho_1(x_1, t)\, dx_1 = -\int_{-\infty}^{+\infty} \rho_1(x_1, t)\, dx_1 \int_{-\infty}^{+\infty} \frac{\partial K(|x_1 - x_2|)}{\partial x_1} \rho_2(x_2, t)\, dx_2 =$$

$$= -\int_{-\infty}^{+\infty}\int \frac{\partial K(|x_1 - x_2|)}{\partial x_1} \rho_1(x_1)\, \rho_2(x_2)\, dx_1\, dx_2,$$

and similarly for the second. Here we have assumed that the functions ρ_1 and ρ_2 satisfy the conditions of permutability of the limits of integration, as a result we obtain:

$$-\frac{dP}{dt} = -\int_{-\infty}^{\infty} \left\{ \frac{\partial K(|x_1 - x_2|)}{\partial x_1} + \frac{\partial K(|x_1 - x_2|)}{\partial x_2} \right\} \rho_1(x_1)\, \rho_2(x_2)\, dx_1\, dx_2,$$

but, since

$$\frac{\partial K(|x_1 - x_2|)}{\partial x_1} = -\frac{\partial K(|x_1 - x_2|)}{\partial x_2},$$

we may say that for a system that is closed in the sense indicated above, the law of conservation of the complete momentum is valid.

2. We further consider the expression for the energy

$$E = \int_{-\infty}^{\infty} \frac{m_1 \xi_1^2}{2} f_1(x_1, \xi_1, t)\, d\xi_1\, dx_1 + \int_{-\infty}^{\infty} \frac{m_2 \xi_2^2}{2} f_2(x_2, \xi_2, t)\, d\xi_2\, dx_2 +$$

$$+ \iint_{-\infty}^{\infty} K(|x_1 - x_2|)\, \rho_1(x_1, t)\, \rho_2(x_2, t)\, dx_1\, dx_2.$$

Similarly to the foregoing, we have:

$$-\frac{dE}{dt} = \int_{-\infty}^{\infty} \frac{m_1 \xi_1^2}{2} \left\{ \frac{\partial \xi_1 f_1}{\partial x_1} + \frac{1}{m_1} \frac{\partial F_{12} f_1}{\partial \xi_1} \right\} dx_1\, d\xi_1 +$$

$$+ \int_{-\infty}^{\infty} \frac{m_2 \xi_2^2}{2} \left\{ \frac{\partial \xi_2 f_2}{\partial x_2} + \frac{1}{m_2} \frac{\partial F_{21} f_2}{\partial \xi_2} \right\} dx_2\, d\xi_2 +$$

$$+ \int_{-\infty}^{\infty} K(|x_1 - x_2|) \left[\left\{ \frac{\partial \xi_1 f_1}{\partial x_1} + \frac{1}{m_1} \frac{\partial F_{12} f_1}{\partial \xi_1} \right\} f_2 + \right.$$

$$\left. + \left\{ \frac{\partial \xi_2 f_2}{\partial x_2} + \frac{1}{m_2} \frac{\partial F_{21} f_2}{\partial \xi_2} \right\} f_1 \right] dx_1\, d\xi_1\, dx_2\, d\xi_2.$$

Making use of the condition of closure, we find that only the following terms on the right side differ from zero:

$$-\int_{-\infty}^{\infty} m_1 \xi_1 F_{12} f_1\, dx_1\, d\xi_1,$$

$$-\int_{-\infty}^{\infty} m_2 \xi_2 F_{21} f_2\, dx_2\, d\xi_2,$$

$$-\int_{-\infty}^{\infty} \xi_1 \frac{\partial K(|x_1 - x_2|)}{\partial x_1} f_1 f_2\, dx_1\, d\xi_1\, dx_2\, d\xi_2,$$

$$-\int_{-\infty}^{\infty} \xi_2 \frac{\partial K(|x_1 - x_2|)}{\partial x_2} f_1 f_2\, dx_1\, d\xi_1\, dx_2\, d\xi_2.$$

Setting in the first two terms instead of F_{12} the expression

$$F_{12} = \int_{-\infty}^{\infty} - \frac{\partial K(|x_1 - x_2|)}{\partial x_1} f_2 \, dx_2 \, d\xi_2$$

and similarly for F_{21}, we see that the first two terms differ from the last two only in sign, and consequently give zero in the sum. Hence, the law of the conservation of energy is satisfied for a closed system.

3. To derive the law of conservation of the moment of momentum, we consider the plane case for two particles.

Then

$$M = m_1 \int_{-\infty}^{\infty} (x_1\eta_1 - y_1\xi_1) f_1 \, ds_1 \, d\sigma_1 + m_2 \int_{-\infty}^{\infty} (x_2\eta_2 - y_2\xi_2) f_2 \, ds_2 \, d\sigma_2,$$

where $ds \, d\sigma$ is an element of four-dimensional phase space,

$$-\frac{dM}{dt} = m_1 \int_{-\infty}^{\infty} (x_1\eta_1 - y_1\xi_1) \left\{ \frac{\partial \xi_1 f_1}{\partial x_1} + \frac{\partial F_{12x} f_1}{\partial \xi_1} + \frac{\partial \eta_1 f_1}{\partial y_1} + \frac{\partial F_{12y} f_1}{\partial \eta_1} \right\} ds_1 \, d\sigma_1 +$$

$$+ m_2 \int_{-\infty}^{\infty} (x_2\eta_2 - y_2\xi_2) \left\{ \frac{\partial \xi_2 f_2}{\partial x_2} + \frac{\partial \eta_2 f_2}{\partial y_2} + \frac{\partial F_{12x} f_2}{\partial \xi_2} + \frac{\partial F_{12y} f_2}{\partial \eta_2} \right\} ds_2 \, d\sigma_2.$$

Thanks to the conditions of closure, the following members vanish:

$$\int_{-\infty}^{\infty} x_i\eta_i \frac{\partial \eta_i f_i}{\partial_i y} \, ds_i \, d\sigma_i = 0; \qquad \int_{-\infty}^{\infty} y_i\xi_i \frac{\partial \xi_i f_i}{\partial x_i} \, ds_i \, d\sigma_i = 0$$

$$(i = 1, \ 2),$$

$$\int_{-\infty}^{\infty} x_i\eta_i \frac{\partial F_{12} f_1}{\partial \xi_i} \, ds_i \, d\sigma_i = 0; \qquad \int_{-\infty}^{\infty} y_i\xi_i \frac{\partial F_{12} f_1}{\partial \eta_i} \, ds_i \, d\sigma_i = 0$$

$$(i = 1, \ 2).$$

In addition, there are terms differing only in their sign and giving zero as their sum:

$$\int_{-\infty}^{\infty} x_1\eta_1 \frac{\partial \xi_1 f_1}{\partial x_1} ds_1\, d\sigma_1 = -\int_{-\infty}^{\infty} \xi_1\eta_1 f_1\, ds_1\, d\sigma_1,$$

$$\int_{-\infty}^{\infty} y_1\xi_1 \frac{\partial \eta_1 f_1}{\partial y_1} ds_1\, d\sigma_1 = -\int_{-\infty}^{\infty} \xi_1\eta_1 f_1\, ds_1\, d\sigma_1,$$

and similarly for the second particle. As a result, there remain on the right side only the terms

$$\int_{-\infty}^{\infty} x_1\eta_1 \frac{\partial F_{12x} f_1}{\partial \eta_1} ds_1\, d\sigma_1 = -\int_{-\infty}^{\infty} x_1 F_{12x} f_1\, ds_1\, d\sigma_1,$$

$$-\int_{-\infty}^{\infty} y_1\xi_1 \frac{\partial F_{12y} f_1}{\sigma \xi_1} ds_1\, d\sigma_1 = \int_{-\infty}^{\infty} y_1 F_{12y} f_1\, ds_1\, d\sigma_1,$$

and analogously for the second particle. We therefore obtain;

$$-\frac{dM}{dt} = -\int_{-\infty}^{\infty} (x_1 F_{12x_1} - y_1 F_{12y_1}) f_1\, ds_1\, d\sigma_1 -$$

$$-\int_{-\infty}^{\infty} (x_2 F_{21x_2} - y_2 F_{21y_2}) f_2\, ds_1\, d\sigma_1 =$$

$$= -\int_{-\infty}^{\infty} (x_1 F_{12x_1} - y_1 F_{12y_1}) \rho_1\, ds_1 - \int_{-\infty}^{\infty} (x_2 F_{12x_2} - y_2 F_{21y_1}) \rho_2\, ds_2.$$

But if there are no external forces, we have:

$$F_{12x_1} = -\int_{-\infty}^{\infty} \frac{\partial K(|s_1 - s_2|)}{\partial x_1} \rho_2(s_2)\, ds_2$$

and similarly for

$$F_{12y_1} \text{ и } F_{21x_2},\ F_{21y_2}.$$

This substitution yields: $$-\frac{dM}{dt} = \int_{-\infty}^{\infty} \left(x_1 \frac{\partial K_{12}}{\partial x_1} - y_1 \frac{\partial K_{12}}{\partial y_2} \right) \rho_1\rho_2\, ds_1\, ds_2 +$$

$$+\int_{-\infty}^{\infty}\left(x_2\frac{\partial K_{12}}{\partial x_2}-y_2\frac{\partial K_{12}}{\partial y_2}\right)\rho_1\rho_2\,ds_1\,ds_2,$$

but, since the forces are central, the expressions in the parentheses vanish identically (for any K) and consequently the law of conservation of the complete moment of momentum holds true for a closed system.

We emphasize the following points:

a. The examination above is easily generalized to the case of any number of particles.

b. Calculation of the collective connections of nonadditive type does not disturb the laws of conservation under a condition of the central nature of a nonadditive interactions.

c. Passing to the case $N > 1$ and to the case of uniform particles, likewise does not destroy the conservation laws. Actually, the entire force acting on a system of uniform particles vanishes automatically.

$$\iint_{-\infty}^{\infty}\operatorname{grad}_{r_1}K(|r_1-r_2|)\,\rho(r_1)\,\rho(r_2)\,dr_1\,dr_2=0,$$

if we consider changing the $\underline{r_1}$ to $\underline{r_2}$, the grad operation changes its sign, and $\rho(1)\rho(2)\equiv\rho(2)\rho(1)$ because of the fact that the distribution functions are uniform.

Section 11. Some Cases of Exact Solutions

1. Uniform distribution in coordinate space. We consider the equation for the potential distribution in the problem of two or more particles interacting according to any central law. Equation (9.2) takes the form

$$\left.\begin{aligned}V(r)&=\int_{-\infty}^{\infty}K(|r-r'|)\,ae^{-\frac{V(r')}{\theta}}\,dr'\\ &\text{or}\\ \varphi(r)&=\lambda\int_{-\infty}^{\infty}K^*(|r-r'|)\,e^{\varphi(r')}\,dr',\end{aligned}\right\}\qquad(11.1)\qquad .1)$$

where

$$\varphi = -\frac{V(r)}{\theta}, \quad \lambda = -\frac{4\pi a}{\theta}\int_0^\infty K(\rho)\rho^2 d\rho; \quad K^* = \frac{K(|r-r'|)}{4\pi\int_0^\infty K(\rho)\rho^2 d\rho}.$$

Setting $\varphi(r) = \text{const.} = c$, we obtain for the definition of c the transcendent equation

$$c = \lambda e^c \int_{-\infty}^{\infty} K^*(|r-r'|)\, dr'$$

or, since the nucleus is normalized as unity, simply

$$c = \lambda e^c. \tag{11.}$$

For $\lambda > 0$, this equation has two real solutions for c, as will easily be seen, if $\lambda < \frac{1}{e}$, and no real solution if $\lambda > \frac{1}{e}$. As λ increases and approaches the critical value, one of the solutions increases from zero and the other solution decreases from $+\infty$. At the point $\lambda = \frac{1}{e}$, both solutions coincide (Fig. 1). The point of contact is defined by the position

$$1 = \lambda e^c,$$

which together with (11.2) gives $c = 1$ и $\lambda = \frac{1}{e}$.

There thus exists a solution with a potential that is constant over the entire coordinate space. The uniform potential distribution is matched by a uniform distribution of density, since

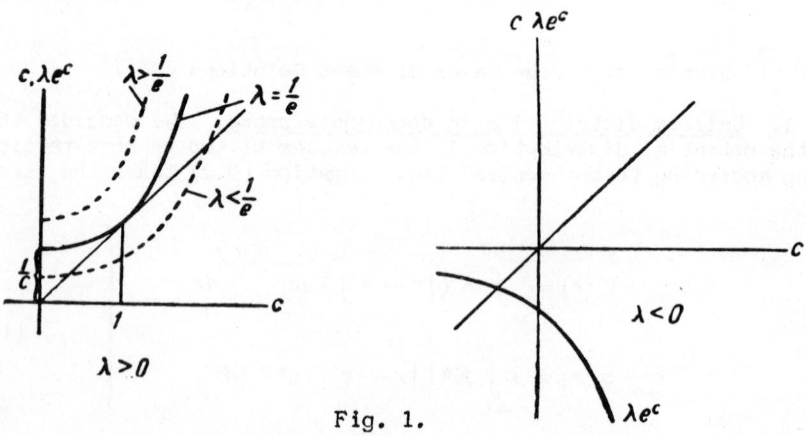

Fig. 1.

$$\rho(r) = \rho(0)e^{-\frac{V(r)-V(0)}{\theta}},$$

which is likewise constant for $V(r) \equiv V(0) = \text{const.}$ ρ

We define the boundary for the existence of solutions with constant density. We find this boundary as follows:

$$\lambda = \frac{1}{e}, \tag{11.3}$$

$$-\frac{4\pi a}{\theta}\int_0^\infty K(\rho)\rho^2\,d\rho = \frac{1}{e}$$

or, since

$$\rho = ae^{-\frac{V}{\theta}} = ae^{c}, \quad \text{for} \quad c = 1 \quad \rho = a\cdot e = \rho_0,$$

we have:

$$\int_0^\infty K(\rho)\rho^2\,d\rho < 0.$$

This condition was first obtained in 1944 as the condition for the occurrence of periodic solutions (see part II).

There is a critical point in the case where the forces of attraction predominate over the forces of repulsion:

$$-\frac{4\pi\rho_0}{\theta}\int_0^\infty K(\rho)\rho^2\,d\rho = 1. \tag{11.4}$$

In the inverse case $\lambda < 0$, we do not have any such critical point. A uniform distribution occurred over the entire interval of temperatures and densities, and in the stationary case constitutes the <u>only</u> solution, as is immediately seen from Fig. 1 ($\lambda < 0$).

We consider the equation

$$\varphi(r) = \lambda \int_{-\infty}^{\infty} K(|r - r'|)(e^{\varphi(r')} - 1)\, dr'.$$

multiplying on both sides by $(e^{\varphi(r)} - 1)$ and integrating for r, we write this as

$$\int_{-\infty}^{\infty} \varphi(r)(e^{\varphi(r)} - 1)\, dr = \lambda \iint_{-\infty}^{\infty} K^* (e^{\varphi(r)} - 1)(e^{\varphi(r')} - 1)\, dr\, dr'.$$

If kernel K is positive, which we assume here, then the double integral will always be greater than or equal to zero, but in view of the fact that $\lambda < 0$ the right portion always is negative or equal to zero. Consequently,

$$\int_{-\infty}^{\infty} \varphi(r)(e^{\varphi(r)} - 1)\, dr \leqslant 0,$$

while the expression under the integral sign

$$\varphi(r)(e^{\varphi(r)} - 1) \geqslant 0,$$

since in the region where φ is positive, the expression in the parentheses is likewise positive, and vice versa. We therefore must set:

$$\varphi(r)(e^{\varphi(r)} - 1) = 0.$$

It is easily seen that the equation obtained for φ has only a unique solution

$$\varphi(r) \equiv 0.$$

Thus, for forces that satisfy the condition

$$\int_0^{\infty} K(\rho)\rho^2\, d\rho < 0$$

with a positive K, the base of uniform density is actually the sole solution which we have just proved analytically as well.

The results obtained for the solution with uniform density are generalized as well to the equation

$$V(r)=\sum_{n} a^{n} \int_{-\infty}^{\infty} \cdots \int_{-\infty}^{+\infty} K_{n}(r, r_{1}, \ldots, r_{n}) e^{-\frac{V(r_{1})}{\theta}} \cdots e^{-\frac{V(r_{n})}{\theta}} dr_{1} \ldots dr_{n},$$

where the value of c is in a similar way defined by the equation

$$c=\sum_{n} \lambda_{n} e^{nc}, \tag{11.5}$$

the number of the solutions of which depends in essence on the signs and magnitudes of λ_n .

2. <u>The Maxwell velocities distribution as one exact solution of the initial equations.</u> For simplicity's sake, we consider the case of a single variable

$$\xi \frac{\partial f}{\partial x}-\frac{1}{m} \frac{\partial V(x)}{\partial x} \frac{\partial f}{\partial \xi}=0. \tag{11.6}$$

We shall seek the solution of this equation by the method of separation of variables:

$$f(x, \xi)=w(\xi) \rho(x).$$

Substitution in (11.6) gives:

$$\frac{\partial w}{\xi \partial \xi}=\frac{\partial \rho}{\partial x} \frac{1}{\rho \frac{\partial V}{\partial x}}=-\frac{1}{\theta}, \tag{11.7}$$

where θ is a constant, from which

$$w(\xi)=w(0) e^{-\frac{m \xi^{2}}{2 \theta}}. \tag{11.8}$$

The result obtained is easily generalized to the case of three dimensions and any number N of particles with interactions of any functional type, since these interactions will determine only the form of V, and will not influence the equation in w.

It is of importance to show that the Maxwell distribution is an exact solution not only for any central forces of interaction between particles, but also in the presence of electrodynamic forces. We need only analyze the equation describing the behavior of a single particle:

$$v \operatorname{grad}_r f + \frac{1}{m}\left\{-\operatorname{grad}_r V + e\left(e + \frac{1}{c}[vh]\right)\right\} \operatorname{grad}_v f = 0, \qquad (11.9)$$

where V, e, h are in turn defined by means of the distribution function of the remaining particles (and they also include the external field as well).

Setting

$$f = W(v^2)\,\rho(r)$$

and observing that in this case the magnetic term vanishes automatically, as was shown above, we obtain

$$\operatorname{grad}_r \rho = \frac{1}{m}\{-\operatorname{grad}_r V + ce\}\frac{\partial W}{\partial v^2}\rho. \qquad (11.1$$

If we now take into consideration the fact that in the static case vector e is a potential, we come to the previous result: a Maxwellian distribution for $W(v^2)$.

3. ___-type solutions. We consider the equation

$$V(r) = \rho(0)\int_{-\infty}^{\infty} K(|r - r'|)\, e^{-\frac{V(r) - V(0)}{\theta}}\, dr'. \qquad (11.1$$

In the neighborhood of the point $\theta = 0$, the expression under the integral sign is not expanded into a Taylor series in θ, so that the ordinary methods for solving nonlinear integral equations, based on the expansion of the functions into holomorphic series, are not applicable for the present case with $\theta = 0$.

We assume that at some arbitrarily chosen point in space, taken as the origin of coordinates, the following conditions for the first and second derivatives of the potential are realized:

$$\left.\begin{aligned} V_x(0) = V_y(0) = V_z(0) = 0, \\ V_{xx}(0) = V_{yy}(0) = V_{zz}(0) > 0. \end{aligned}\right\} \qquad (11.1$$

The possibility of this assumption will be tested after we have obtained a solution.

We solved the function V(r) at the point r = 0 into a Taylor series, employing the assumption made above. Considering point θ as close to $\theta = 0$, because of the δ - type of the function

$$\exp\left\{-\frac{x^2+y^2+z^2}{2\theta}V''(0)\right\},$$

we may write:

$$\int_{-\infty}^{\infty} K(|r-r'|)\, e^{-\frac{(x^2+y^2+z^2)V''(0)}{2\theta}}\, dr' \to \left(\frac{2\pi\theta}{V''(0)}\right)^{\frac{3}{2}} K(|r|) \tag{11.13}$$

under the condition $\theta \to 0$.

On the other hand, to define the density we may likewise employ the expansion referred to above

$$\rho(r) = \rho(0)\, e^{-\frac{[V(r)-V(0)]}{\theta}} \to \rho(0)\, e^{-\frac{(x^2+y^2+z^2)V''(0)}{2\theta}}.$$

We now impose on function ρ (r) the normalization condition:

$$\int_{-\infty}^{\infty} \rho(r')\, dr' = 1. \tag{11.14}$$

Then

$$\rho(0) = \left(\frac{V''(0)}{2\theta}\right)^{\frac{3}{2}}.$$

In this way, in the limit, for $\theta \to 0$, we obtain the solution

$$\rho(r) = \delta(|r|), \qquad V(r) = K(|r|). \tag{11.15}$$

The initial assumption, as we shall now see, is realized under the following condition, imposed on the nature of the forces of interaction:

$$K'(0) = 0; \qquad K''(0) > 0. \tag{11.16}$$

The physical sense of the solution obtained is obvious.

If the forces of interaction satisfy the condition for K (the existence of a parabolic potential sink at the point $\underline{r} = 0$), then for $\theta \to 0$, the particles are concentrated at the point $\underline{r} = 0$. It is an essential point that the result obtained applies for any number of interacting uniform particles, since the form of the initial equation is independent of the number of particles.

The equation obtained may be interpreted as the merging of many particles into a single one.

Section 12. Invariance and Relativity Form of the Fundamental Equations*

The initial continuity equation

$$-\frac{\partial f}{\partial t} = \operatorname{div}_r \boldsymbol{v} f + \operatorname{div}_v \boldsymbol{g} f$$

(where $\underline{g}$ is the acceleration vector, which may depend on x_k, v_k, t) expresses a conservation law. The operations entering into it retain their significance obviously in any systems of calculation. It should therefore be expected that this equation should be invariant both in the classical and the relativity cases. We write the initial equation in the form

$$\frac{\partial f}{\partial t} + v_k \frac{\partial f}{\partial x_k} + g_k \frac{\partial f}{\partial v_k} + f \frac{\partial g_k}{\partial v_k} = 0 \tag{12.1}$$

and consider first transformations of the form

$$x'_i = c_{ik} x_k + c_i, \quad \text{where} \quad c_{ik} c_{ij} = \delta_{kj},$$
$$v'_i = c_{ik} v_k, \quad g'_i = c_{ik} g_k \quad (i, k, j = 1, 2, 3), \tag{12.2}$$

and likewise

$$x'_i = x_i - v_i^0 t \quad \text{и} \quad v'_i = v_i - v_i^0, \quad g'_i = g_i. \tag{12.3}$$

Inasmuch as the element of six-dimensional phase space

$$(d\underline{r}\ d\underline{v}) = (d\underline{r}'\ d\underline{v}')$$

* The questions of this section were considered in collaboration with A. Luch

is invariant with respect to transformations (12.2) and (12.3), then from the law of the conservation of the number of particles

$$f(r, v, t)(dr\, dv) = f'(r', v', t)(dr'\, dv')$$

there follows the invariance of function f:

$$f(\underline{r}, \underline{v}, t) = f'(\underline{r}', \underline{v}', t)$$

under the transformations in question. Direct substitution of the transformation formulas in

$$\frac{\partial f}{\partial t} + v_k' \frac{\partial f}{\partial x_k'} + g_k' \frac{\partial f}{\partial v_k'} + f \frac{\partial g_k'}{\partial v_k'} = 0$$

shows that the equation is invariant to transformations (12.2), and covariant to transformations (12.3).

To prove the invariance of the equations when integral interactions are included, it is sufficient to show that the expression

$$\frac{\partial}{\partial x_k} \int\int_{-\infty}^{\infty} K(|r - r'|) f(r', v', t)\, dr'\, dv'$$

is invariant to transformations (12.3) and is a vector with reference to (12.2). The proof is obvious. We may therefore say that the initial equations with the functional inclusions are covariant under Galilean transformations.

We now show the invariance of the continuity equation under Lorentz transformation, defined by the relations

$$x_1' = \frac{x_1 - v_1^0 t}{\left(1 - \frac{(v_1^0)^2}{c^2}\right)^{\frac{1}{2}}}; \quad x_2' = x_2; \quad x_3' = x_3; \quad t' = \frac{t - x_1 \frac{v_1^0}{c^2}}{\left(1 - \frac{(v_1^0)^2}{c^2}\right)^{\frac{1}{2}}}. \qquad (12.4)$$

We take into account the fact that the velocities and accelerations are transformed into

$$
\left.
\begin{aligned}
&v_1' = \frac{v_1 - v_1^0}{1 - \frac{v_1 v_1^0}{c^2}}, \quad v_2' = \frac{v_2\left(1 - \frac{(v_1^0)^2}{c^2}\right)^{\frac{1}{2}}}{1 - \frac{v_1 v_1^0}{c^2}}, \quad v_3' = \frac{v_3\left(1 - \frac{(v_1^0)^2}{c^2}\right)^{\frac{1}{2}}}{1 - \frac{v_1 v_1^0}{c^2}}.\\
&g_1' = \frac{\left(1 - \frac{(v_1^0)^2}{c^2}\right)^{\frac{3}{2}}}{\left(1 - \frac{v_1^0 v_1}{c^2}\right)^3} g_1;\\
&g_2' = \frac{1 - \frac{(v_1^0)^2}{c^2}}{\left(1 - \frac{v_1^0 v_1}{c^2}\right)^2} g_2 + \frac{1 - \frac{(v_1^0)^2}{c^2}}{\left(1 - \frac{v_1^0 v_1}{c^2}\right)^3} \cdot \frac{v_2 v_1^0}{c^2} g_1;\\
&g_3' = \frac{1 - \frac{(v_1^0)^2}{c^2}}{\left(1 - \frac{v_1^0 v_1}{c^2}\right)^2} g_3 + \frac{1 - \frac{(v_1^0)^2}{c^2}}{\left(1 - \frac{v_1 v_1^0}{c^2}\right)^3} \cdot \frac{v_3 v_1^0}{c^2} g_1.
\end{aligned}
\right\} \tag{12.5}
$$

It will readily be seen that as we pass from the primed to the unprimed variables, the element of phase volume ($d\underline{r}\ dv$) is not invariant. For, we have, employing the transformation formulas for coordinates and velocities:

$$
(dr'\, dv') = \frac{\left(1 - \frac{(v_1^0)^2}{c^2}\right)^{\frac{5}{2}}}{\left(1 - \frac{v_1 v_1^0}{c^2}\right)^5} \cdot (dr\, dv).
$$

But the law of conservation requires:

$$
f'(\underline{r}', v', t) \cdot (d\underline{r}'\ dv') = f(\underline{r}, v, t) \cdot (d\underline{r}\ dv).
$$

Hence, the transformational properties of the function (12.6) are univocally determined.

$$
f'(r', v', t') = f(r, v, t) \frac{\left(1 - \frac{v_1 v_1^0}{c^2}\right)^5}{\left(1 - \frac{(v_1^0)^2}{c^2}\right)^{\frac{5}{2}}}. \qquad (12.
$$

We now convert the expression

$$\frac{\partial f'}{\partial t'}+v_k'\frac{\partial f'}{\partial x_k'}+g_k'\frac{\partial f'}{\partial v_k'}+f'\frac{\partial g_k'}{\partial v_k'}$$

to new variables, making use of the relationships

$$\frac{\partial}{\partial t'}=\frac{\partial}{\partial x_1}\frac{\partial x_1}{\partial t'}+\frac{\partial}{\partial t}\frac{\partial t}{\partial t'},\quad \frac{\partial}{\partial x_1'}=\frac{\partial}{\partial x_1}\frac{\partial x_1}{\partial x_1'}+\frac{\partial}{\partial t}\frac{\partial t}{\partial x_1'},\quad \frac{\partial}{\partial x_2'}=\frac{\partial}{\partial x_2},\quad \frac{\partial}{\partial x_3'}=\frac{\partial}{\partial x_3},$$

$$\frac{\partial}{\partial v_1'}=\frac{\partial}{\partial v_1}\frac{\partial v_1}{\partial v_1'}+\frac{\partial}{\partial v_2}\frac{\partial v_2}{\partial v_1'}+\frac{\partial}{\partial v_3}\frac{\partial v_3}{\partial v_1'},\quad \frac{\partial}{\partial v_2'}=\frac{\partial}{\partial v_2}\frac{\partial v_2}{\partial v_2'},\quad \frac{\partial}{\partial v_3'}=\frac{\partial}{\partial v_3}\frac{\partial v_3}{\partial v_3'};$$

and transformation formulas for coordinates, time, velocities and accelerations, as well as formula (12.6). We obtain:

$$\frac{\partial f'}{\partial t'}+v_k'\frac{\partial f'}{\partial x_k'}+g_k'\frac{\partial f'}{\partial v_k'}+f'\frac{\partial g_k'}{\partial v_k'}=\frac{\partial f}{\partial t}\cdot\left\{\frac{\left(1-\frac{v_1v_1^0}{c^2}\right)^5}{\left(1-\frac{(v_1^0)^2}{c^2}\right)^3}+\right.$$

$$+\frac{\left(1-\frac{v_1v_1^0}{c^2}\right)^4}{\left(1-\frac{(v_1^0)^2}{c^2}\right)^3}\cdot\frac{v_1v_1^0}{c^2}-\frac{\left(1-\frac{v_1v_1^0}{c^2}\right)^4}{\left(1-\left(\frac{v_1^0}{c}\right)^2\right)^2}-\left.\frac{\left(1-\frac{v_1v_1^0}{c^2}\right)^4}{\left(1-\frac{(v_1^0)^2}{c^2}\right)^3}\cdot\frac{v_1^{0^2}}{c^2}\right\}+$$

$$+v_k\frac{\partial f}{\partial x_k}\frac{\left(1-\frac{v_1v_1^0}{c^2}\right)^4}{\left(1-\frac{(v_1^0)^2}{c^2}\right)^2}+\frac{\partial f}{\partial x_1}\left\{\frac{\left(1-\frac{v_1v_1^0}{c^2}\right)^5}{\left(1-\frac{(v_1^0)^2}{c^2}\right)^3}v_1^0+\frac{\left(1-\frac{v_1v_1^0}{c^2}\right)^4}{\left(1-\frac{(v_1^0)^2}{c^2}\right)^3}v_1-\right.$$

$$\left.-\frac{\left(1-\frac{v_1v_1^0}{c^2}\right)^4}{\left(1-\frac{(v_1^0)^2}{c^2}\right)^3}v_1^0-\frac{\left(1-\frac{v_1v_1^0}{c^2}\right)^4}{\left(1-\frac{(v_1^0)^2}{c^2}\right)^2}v_1\right\}+$$

$$+\frac{\partial f}{\partial t}\frac{\left(1-\frac{v_1v_1^0}{c^2}\right)^4}{\left(1-\frac{(v_1^0)^2}{c^2}\right)^3}+g_k\frac{\partial f}{\partial v_k}\frac{\left(1-\frac{v_1v_1^0}{c^2}\right)^4}{\left(1-\frac{(v_1^0)^2}{c^2}\right)^2}+$$

$$+g_1\frac{\partial f}{\partial v_2}\left\{-\frac{\left(1-\frac{v_1v_1^0}{c^2}\right)^3}{\left(1-\frac{(v_1^0)^2}{c^2}\right)^2}\frac{v_2v_1^0}{c^2}+\frac{\left(1-\frac{v_1v_1^0}{c^2}\right)^3}{\left(1-\frac{(v_1^0)^2}{c^2}\right)^2}\frac{v_3v_1^0}{c^2}\right\}+$$

$$+g_1\frac{\partial f}{\partial v_3}\left\{-\frac{\left(1-\frac{v_1v_1^0}{c^2}\right)^3}{\left(1-\frac{(v_1^0)^2}{c^2}\right)^3}\frac{v_3v_1^0}{c^2}+\frac{\left(1-\frac{v_1v_1^0}{c^2}\right)^3}{\left(1-\frac{(v_1^0)^2}{c^2}\right)^2}\frac{v_3v_1^0}{c}\right\}+$$

$$+fg_1\left\{-5\frac{\left(1-\frac{v_1v_1^0}{c^2}\right)^3}{\left(1-\frac{(v_1^0)^2}{c^2}\right)^2}\frac{v_1^0}{c^2}+3\frac{\left(1-\frac{v_1v_1^0}{c^2}\right)^3}{\left(1-\frac{(v_1^0)^2}{c^2}\right)^2}\frac{v_1^0}{c^2}+\frac{\left(1-\frac{v_1v_1^0}{c^2}\right)^3}{\left(1-\frac{(v_1^0)^2}{c^2}\right)^2}\frac{v_1^0}{c^2}+\right.$$

$$\left.+\frac{\left(1-\frac{v_1v_1^0}{c^2}\right)^3}{\left(1-\frac{(v_1^0)^2}{c^2}\right)^2}\frac{v_1^0}{c^2}\right\}+f\frac{\partial g_k}{\partial v_k}\frac{\left(1-\frac{v_1v_1^0}{c^2}\right)^4}{\left(1-\frac{(v_1^0)^2}{c^2}\right)^2}+$$

$$+f\frac{\partial g_1}{\partial v_2}\left\{-\frac{\left(1-\frac{v_1v_1^0}{c^2}\right)^3}{\left(1-\frac{(v_1^0)^2}{c^2}\right)^2}\frac{v_2v_1^0}{c^2}+\frac{\left(1-\frac{v_1v_1^0}{c^2}\right)^3}{\left(1-\frac{(v_1^0)^2}{c^2}\right)^2}\frac{v_2v_1^0}{c^2}\right\}+$$

$$+f\frac{\partial g_1}{\partial v_3}\left\{-\frac{\left(1-\frac{v_1v_1^0}{c^2}\right)^3}{\left(1-\frac{(v_1^0)^2}{c^2}\right)^2}\frac{v_3v_1^0}{c^2}+\frac{\left(1-\frac{v_1v_1^0}{c^2}\right)^3}{\left(1-\frac{(v_1^0)^2}{c^2}\right)^2}\frac{v_3v_1^0}{c^2}\right\}=$$

$$=\left\{\frac{\partial f}{\partial t}+v_k\frac{\partial f}{\partial x_k}+g_k\frac{\partial f}{\partial v_k}+f\frac{\partial g_k}{\partial v_k}\right\}\frac{\left(1-\frac{v_1v_1^0}{c^2}\right)^4}{\left(1-\frac{(v_1^0)^2}{c^2}\right)^2}.$$

In this way, we can obtain results (12.1). The continuity equation is covariant to Lorentz transformations, if g_k satisfies the transformation formulas (12.5).

We must now carry out the inclusion of the functional expression of th force in a relativity manner. The relativity dynamics of a point gives the equations

$$m_0\frac{du_\mu}{d\tau}=f_\mu,\quad (\mu=1, 2, 3, 4) \tag{12.7}$$

where f_μ is a four-dimensional vector, whose first three components corres to the force and the fourth component to the energy.

$$u_k = \frac{v_k}{\sqrt{1-\beta^2}}, \quad u_4 = \frac{ic}{\sqrt{1-\beta^2}}, \quad d\tau = dt\sqrt{1-\beta^2} = inv$$

$$\left(\beta^2 = \frac{v_1^2 + v_2^2 + v_3^2}{c^2}\right).$$

Equation (12.7) may be written as:

$$\frac{du_\mu}{dt} = \frac{1}{m_0}\sqrt{1-\beta^2}\, f_\mu = G_\mu. \tag{12.8}$$

For electromagnetic interactions, we have

$$G_i = \frac{e}{m}\left(e + \frac{1}{c}[vh]\right)_i = \frac{1}{m_0 c} F_{i\nu} j_\nu, \tag{12.9}$$

where $F_{i\nu}$ is the tensor of the electromagnetic field, and $j_k = ev_k$, $j_4 = ice$ are the four-dimensional current. The formulas for G_μ cannot be used directly, since the continuity equation contains another vector g_k. We can therefore produce self-consistency only by expressing g_k in terms of G_k.

Relationship (12.5) gives, after direct differentiation:

$$\left.\begin{aligned}
G_1 &= \frac{1}{(1-\beta^2)^{\frac{3}{2}}}\left\{\left(1-\beta^2+\frac{v_1^2}{c^2}\right) g_1 + \frac{v_1 v_2}{c^2} g_2 + \frac{v_1 v_3}{c^2} g_3\right\},\\
G_2 &= \frac{1}{(1-\beta^2)^{\frac{3}{2}}}\left\{\frac{v_2 v_1}{c^2} g_1 + \left(1-\beta^2+\frac{v_2^2}{c^2}\right) g_2 + \frac{v_2 v_3}{c^2} g_3\right\},\\
G_3 &= \frac{1}{(1-\beta^2)^{\frac{3}{2}}}\left\{\frac{v_3 v_1}{c^2} g_1 + \frac{v_3 v_2}{c^2} g_2 + \left(1-\beta^2+\frac{v_3^2}{c^2}\right) g_3\right\}.
\end{aligned}\right\} \tag{12.10}$$

From this we obtain the formulas of the converse transformation:

$$\left.\begin{aligned}
g_1 &= (1-\beta^2)^{\frac{1}{2}}\left\{\left(1-\frac{v_1^2}{c^2}\right) G_1 - \frac{v_1 v_2}{c^2} G_2 - \frac{v_1 v_3}{c^2} G_3\right\},\\
g_2 &= (1-\beta^2)^{\frac{1}{2}}\left\{-\frac{v_2 v_1}{c^2} G_1 + \left(1-\frac{v_2^2}{c^2}\right) G_2 - \frac{v_2 v_3}{c^2} G_3\right\},\\
g_3 &= (1-\beta^2)^{\frac{1}{2}}\left\{-\frac{v_3 v_1}{c^2} G_1 - \frac{v_3 v_2}{c^2} G_2 + \left(1-\frac{v_3^2}{c^2}\right) G_3\right\}.
\end{aligned}\right\} \tag{12.11}$$

Hence, g_k is defined by means of (12.11), and replacing expressions G_k by the field strengths, we obtain the relativity covariant and self-consistent relations.

CHAPTER III

CONNECTION WITH CLASSICAL THEORIES+

Section 13. Transition to Classical Mechanics

For simplicity's sake, we consider the problem of two particles. The initial equation may be written as:

$$\left.\begin{aligned}&\frac{\partial f_1}{\partial t}+\frac{\partial v_i f_1}{\partial x_i}-\lambda_1\frac{\partial f_1}{\partial v_i}\times\\&\quad\times\left\{\frac{\partial}{\partial x_i}\int\limits_{(r',v'\|t)}\int K(|r-r'|)f_2(r',v',t)(dr'\,dv')\right\}=0,\\&\frac{\partial f_2}{\partial t}+\frac{\partial v_i f_2}{\partial x_i}-\lambda_2\frac{\partial f_2}{\partial v_i}\times\\&\quad\times\left\{\frac{\partial}{\partial x_i}\int\limits_{(r',v'\|t)}\int K(|r-r'|)f_1(r',v',t)(dr'\,dv')\right\}=0\end{aligned}\right| \qquad (13.1)$$

(summation over i = 1, 2, 3). Here we integrate over the entire region of the phase space, where $f_2 \neq 0$ and $f_1 \neq 0$, respectively. The integration limits may thus depend on t.

We explain the functional dependence between the quantities of the classical type $\int\int x_k f\,dr\,dv$ and $\int\int v_k f\,dr\,dv$, which defines the law of motion of the center of gravity of the body. To this end, we multiply the first equation by x_k and integrate it over a certain definite region. We shall assume in this that the necessary conditions are fulfilled for partial integration of multiple integrals and differentiation according to a parameter under the integral sign, the limits of which do not depend on this parameter.

The terms obtained in this way are equal to:

$$\left.\begin{aligned}&\int\limits_{(r,v)}\int x_k\frac{\partial f_i}{\partial t}\,dr\,dv=\frac{d}{dt}\int\limits_{(r,v)}\int x_k f_i\,dr\,dv=\frac{d}{dt}x_k^i(t),\\&\int\limits_{(r,v)}\int x_k\frac{\partial v_n f_i}{\partial x_n}\,dr\,dv=\int\limits_{(s,v)}\int x_k f_i v_n\,ds_n\,dv-\int\limits_{(r,v)}\int v_k f_i\,dr\,dv,\\&\int\limits_{(r,v)}\int x_k\frac{\partial f_i}{\partial v_n}\left\{\frac{\partial}{\partial x_n}\int\limits_{(r',v'\|t)}\int K(|r-r'|)f_{\overline{i+1}}(r',v',t)\,dr'\,dv'\right\}\times\\&\quad\times dr\,dv=-\int\limits_{(r,\sigma)}\int x_k f_i F_{n,\overline{i+1}}\,dr\,d\sigma_n,\end{aligned}\right| \qquad (13.1')$$

+ The questions in this chapter were considered in collaboration with A. Luchina. Also see the author's article in Zh. E.T.F. 18, 840 (1948).

where

$$F_{n,\overline{i+1}} = -\frac{\partial}{\partial x_n} \int\limits_{(r', v' \| t)}\int K(|r-r'|) f_{\overline{i+1}}(r', v', t)\, dr'\, dv'$$

($i = 1, 2$; k and $n = 1, 2, 3$; $\overline{i+1}$ is a symbol designating the index obtained by i circular substitution).

Here ds_n and $d\sigma_n$ are, respectively, projections of elements of surface in three-dimensional coordinate and velocity spaces.

Next, we multiply the equations of system (13.1) by v_k and likewise integrate for $\underline{r}$ and $\underline{v}$ in this region ($\underline{r}$, $\underline{v}$), and obtain:

$$\left.\begin{aligned}
&\int\limits_{(r,v)}\int v_k \frac{\partial f_i}{\partial t}\, dr\, dv = \frac{d}{dt}\int\limits_{(r,v)}\int v_k f_i\, dr\, dv = \frac{d}{dt}\overline{v}_k^i(t),\\
&\int\limits_{(r,v)}\int v_k \frac{\partial v_n f_i}{\partial x_n}\, dr\, dv = \int\limits_{(s,v)}\int v_k f_i v_n\, ds_n\, dv,\\
&\int\limits_{(r,v)}\int v_k \frac{\partial f_i}{\partial v_n}\left\{\frac{\partial}{\partial x_n}\int\limits_{(r',v'\|t)}\int K(|r-r'|) f_{\overline{i+1}}(r', v', t)\, dr'\, dv'\right\}\times\\
&\qquad\times dr\, dv = -\int\limits_{(r,\sigma)}\int v_k f_i F_{n,\overline{i+1}}\, dr\, d\sigma_n +\\
&\qquad+\int\limits_{(r,v)}\int F_{k,\overline{i+1}}(r, t) f_i(r, v, t)\, dr\, dv =\\
&= -\int\limits_{(r,\sigma)}\int v_k f_i F_{n,\overline{i+1}}(r, t)\, dr\, d\sigma_n + \overline{F_{k,\overline{i+1}}(r, t)^i},
\end{aligned}\right\} \quad (13.2)$$

where

$$\overline{F_{k,\overline{i+1}}(r, t)^i} = \int\limits_{(r,v)}\int F_{k,\overline{i+1}}(r, t) f_i(r, v, t)\, dr\, dv.$$

We then pass to the following system of equations:

$$\left.\begin{aligned}
&\frac{d}{dt}\overline{r}^i(t) = \overline{v}^i(t) - \int\limits_{(s,v)}\int r f_i v_n\, ds_n\, dv - \lambda_i \int\limits_{(r,\sigma)}\int r f_i F_{n,\overline{i+1}}(r, t)\, dr\, d\sigma_n,\\
&\frac{d}{dt}\overline{v}^i(t) = \lambda_i \overline{F}^i_{\overline{i+1}} - \int\limits_{(s,v)}\int v f_i v_n\, ds_n\, dv - \lambda_i \times\\
&\qquad\times \int\limits_{(r,\sigma)}\int v f_i F_{n,\overline{i+1}}(r, t)\, dr\, d\sigma_n \quad (i = 1, 2),\\
&F_{\overline{i+1}} = -\nabla_r \int\limits_{(r',v'\|t)}\int K(|r-r'|) f_{\overline{i+1}}(r', v', t)\, dr'\, dv'.
\end{aligned}\right\} \quad (13.$$

It should be noted that equations (13.2) are exact relations obtained from the system of equations (13.1). They show that the theory that has been set forth contains something more than the practical mechanics of discrete points, since additional curves appear in addition to the usual connections between mean values.

We now consider the conditions for the transitions to ordinary classical mechanics. This transition will take place if the surface integrals in (13.2) vanish for over the entire space interval $t\,(-\infty < t < +\infty)$ independently of the other terms. This is a radical requirement; it is introduced into the theory from outside, and indicates to us that the classical scheme may be derived from the theory that has been set forth only by choosing a narrow class of solutions of the initial equations. The solutions corresponding to classical mechanics belong to the class of functions satisfying the requirement that the surface integrals vanish.

It is an essential feature that here exact vanishing of these integrals is required, since the existence of even small magnitudes may greatly distort the classical theory. In fact, the surface integrals represent a kind of flow of matter, so that in the case of even small values for them within the volume, there may be great concentrations of matter at individual points, which will lead to a sharp deviation from the classical theory. The requirement of identical vanishing of the surface integrals means physically the elimination from the theory of a possibility of any fluctuational dispersion of the particles as a temperature effect. In this way, we make the transition to the classical theory, ignoring the fluctuational movements. Naturally, we obtain as a result, a scheme containing no parameter adequate to the temperature.

It should be kept in mind in this connection that the equations have other solutions as well, which are realized under different physical conditions, for which the thermal effects are automatically included in the theory. We pass to the analysis of a second condition:

$$\bar{F}^{i}_{k,\overline{i+1}} = -\int\limits_{(r)} \rho_i(r, t)\left\{\frac{\partial}{\partial x_k}\int\limits_{(r'\,\|\,t)} K(|r-r'|)\,\rho_{\overline{i+1}}(r', t)\,dr'\right\}dr = -\frac{\partial}{\partial \bar{x}^{i}_{k}} K(|\bar{r}^{i} - \bar{r}^{\overline{i+1}}|). \tag{13.3}$$

Obviously, in the case of sufficient concentration of densities, of distributions ρ_i and $\rho_{\overline{i+1}}$ in the neighborhood of points $\bar{r}^{i}$ and $\bar{r}^{\overline{i+1}}$, respectively, the equality indicated above is satisfied even if there is no overlapping of these clouds. An evaluation of the necessary degree of localization can be made.

We expand $-\frac{\partial}{\partial x_k} K(|r-r'|)=\Phi_k$ as a power series around the point $r-\overline{r^{i+1}}$ in the following manner:

$$\Phi_k(r-r')=\Phi_k(r-\overline{r^{i+1}}+\overline{r^{i+1}}-r')=\Phi_k(r-\overline{r^{i+1}})+ \\ +\frac{\partial\Phi_k}{\partial x_l}(\overline{x_l^{i+1}}-x_l')+\frac{1}{2}\frac{\partial^2\Phi_k}{\partial x_l\,\partial x_m}(\overline{x_l^{i+1}}-x_l')(\overline{x_m^{i+1}}-x_m')+\dots \tag{13.4}$$

We require that the series converge uniformly in **r** with respect to $(r^{i+1}-r')$ over a region, where $\rho_{\overline{i+1}}\neq 0$ and $\rho_i\neq 0$. This condition is equivalent to the requirements of adequate localization of cloud $\rho_{\overline{i+1}}$ and the inpenetrability of the clouds. Integrating for **r'** does not change the region of convergence of the series or the character of uniformity of convergence.

We choose the size of the $\overline{i+1}$ -th cloud in such a way that after integration we may confine ourselves to the first term. We have:

$$\overline{F}^i_{k,\overline{i+1}}(r,t)=\int \rho_i(r,t)\,\Phi_k(r-\overline{r^{i+1}})\,dr. \tag{13.5}$$

We now again expand $\Phi_k(r-\overline{r^{i+1}})$ to a series at point $\underline{r}=\underline{r}^{-t}$:

$$\Phi_k(r-\overline{r^{i+1}})=\Phi_k(\overline{r}^i-\overline{r^{i+1}}+r-\overline{r}^i)=\Phi_k(\overline{r}^i-\overline{r^{i+1}})+ \\ +\frac{\partial\Phi_k}{\partial\overline{x}_l^i}(x_l-\overline{x}_l^i)+\frac{1}{2!}\frac{\partial^2\Phi_k}{\partial\overline{x}_l^i\,\partial\overline{x}_m^i}(x_l-\overline{x}_l^i)(x_m-\overline{x}_m^i)+\dots \tag{13.6}$$

(summation for l and m).

If here, too, we require sufficient localization of the i-th cloud, so that we may limit ourselves to the first member of the expansion, we have+:

$$\overline{F}_k^i=-\frac{\partial}{\partial\overline{x}_k^i}K(|\overline{r}^i-\overline{r^{i+1}}|). \tag{13.7}$$

+ In passing from (13.4) to (13.7), the normalization condition for f_i is introduced.

In confining ourselves to the first terms, we have essentially required that the following condition for the dimensions of the cloud be satisfied:

$$\overline{\Delta x_l^i \Delta x_m^i} \ll 2 \left| \frac{\frac{\partial}{\partial \bar{x}_k^i} K(|\bar{r}^i - \bar{r}^{i+1}|)}{\frac{\partial^3}{\partial \bar{x}_k^i \, \partial \bar{x}_l^i \, \partial \bar{x}_m^i} K(|\bar{r}^i - \bar{r}^{i+1}|)} \right|. \tag{13.8}$$

This condition is an expression of the degree of space localization of the particles that is necessary for going over to classical mechanics.

The limitations on the dispersion of velocities derive from the requirement that: $\overline{\Delta v_k^{2i}} \ll (\bar{v}_k^i)^2$.

A completely analogous transition is possible for any number of particles as well. To this end, we consider the case of uniform particles with spontaneous action:

$$\left.\begin{aligned} &\frac{\partial f}{\partial t} + \operatorname{div}_r vf + \operatorname{div}_{\dot{v}} \frac{1}{m} Ff = 0, \\ &F = -\operatorname{grad}_r \int K(|r - r'|) f(r', v', t)\, dr'\, dv'. \end{aligned}\right\} \tag{13.9}$$

The transition to form (13.2) is made in the manner essentially analogous to the preceding. We seek an equation for f in the class of functions of the type of the δ-functions:

$$f(r, v, t) = \sum_{i=1}^{N} \delta(r - r_i(t))\, \delta(v - v_i(t)),$$

where $\underline{r}_i(t)$ and $\underline{v}_i(t)$ are subject to the definition of the time function. Then we find, from (13.2):

$$\frac{d}{dt} \sum_{i=1}^{N} r_i(t) = \sum_{i=1}^{N} v_i(t),$$

$$\frac{d}{dt} \sum_{i=1}^{N} v_i(t) = -\sum_{i=1}^{N} \nabla_{r_i} \sum_{j=1}^{N} K(|r_i(t) - r_j(t)|).$$

The expression for the force includes the continuous action effect, which is expressed by the presence of the term

$$\nabla_{r_i} K(|r_i - r_j|)|_{r_i = r_j}.$$

In classical mechanics, this term should not be present. In this way we see that in addition to the conditions to which reference has been made concerning the vanishing of all the surface integrals (this condition applies as well when δ -functions are present) and the requirement of an adequate degree of localization of particles, the transition to classical mechanics for particles with spontaneous action requires an additional condition for the nature of the forces of interaction:

$$\nabla_r K(r)|_{r=0} = 0.$$

We may regard this fact as an argument in favor of abandoning calculation of the spontaneous action effect for the individual particle in the theory set forth.

Section 14. Transition to Theory of Continuous Media

Here, too, we shall take as our starting point a system of equations for two particles. We multiply each of them Q(v) (a function of the velocity components that is thus arbitrary) and integrate for v over a definite region (v); we thus obtain:

$$\frac{\partial}{\partial t} \rho_i \overline{Q}^i + \frac{\partial}{\partial x_k} \rho_i \cdot \overline{v_k Q}^i - \lambda_i \int_{\sigma} Q f_i \frac{\partial}{\partial x_k} \int_{(r', t)} K(|r + r'|) \rho_{\overline{i \mp 1}}(r', t)\, dr'\, d\sigma_k +$$
$$+ \lambda_i \rho_i \frac{\overline{\partial Q}^i}{\partial v_k} \cdot \frac{\partial}{\partial x_k} \int K(|r - r'|) \rho_{\overline{i+1}}(r', t)\, dr'. \quad (14.1)$$

Here the summation is only over the index k (k = 1, 2, 3; i = 1, 2) and we introduce the designations:

$$\overline{Q}^i = \frac{\int_{(r)} Q(v) f_i(r, v, t)\, dv}{\int_{(r)} f_i(r, v, t)\, dv},$$

$$\overline{v_k Q}^i = \frac{\int_{(r)} v_k Q(v) f_i(r, v, t)\, dv}{\int_{(r)} f_i(r, v, t)\, dv},$$

$$\frac{\overline{\partial Q}^i}{\partial v_k} = \frac{\int_{(r)} \frac{\partial Q}{\partial v_k} f_i(r, v, t)\, dv}{\int_{(r)} f_i(r, v, t)\, dv}.$$

We may ignore the surface integral in velocity space appearing in (14.1), if we confine ourselves to a velocities distribution function that decreases with sufficient rapidity as the velocity modulus increases.

Setting Q = 1, we obtain a system of continuity equations:

$$\frac{\partial \rho_i}{\partial t} + \frac{\partial}{\partial x_k} (\rho_i \bar{v}_k^i) = 0 \tag{14.2}$$

(i = 1, 2; summation for k; k = 1, 2, 3).

From the first equation of system (14.1), we calculate the first equation of system (14.2), multiplied by $\bar{Q}^1$, and from the second equation of system (14.1) the second equation of system (14.2), multiplied by $\bar{Q}^2$; we then obtain:

$$\rho_i \frac{\partial \bar{Q}^i}{\partial t} = \bar{Q}^i \frac{\partial}{\partial x_k} (\rho_i \bar{v}_k^i) - \frac{\partial}{\partial x_k} (\rho_i \overline{v_k Q^i}) - \\ - \lambda_i \rho_i \frac{\partial \bar{Q}^i}{\partial v_k} \cdot \frac{\partial}{\partial x_k} \int\limits_{(r' \| t)} K(|r - r'|) \rho_{\overline{i+1}}(r', t)\, dr' \tag{14.3}$$

(i = 1, 2; summation for k, k = 1, 2, 3).

We introduce new independent variables instead of $\underline{v}_1$:

$$v_1 = u_0 + u, \quad v_2 = v_0 + v, \quad v_3 = w_0 + w,$$

where

$$u_0 = \bar{v}_1^i, \quad v_0 = \bar{v}_2^i, \quad w_0 = \bar{v}_3^i$$

for the i-th equation.

Then, making use of

$$\bar{u}^i = 0, \quad \bar{v}^i = 0, \quad \bar{w}^i = 0,$$

that is

$$\overline{v_1 Q^i} = u_0 \bar{Q}^i + \overline{u Q^i}, \quad \overline{v_2 Q^i} = v_0 \bar{Q}^i + \overline{v Q^i}, \quad \overline{v_3 Q^i} = w_0 \bar{Q}^i + \overline{w Q^i}$$

and (14.4)

$$\frac{\partial \bar{Q}^i}{\partial v_1} = \frac{\partial \bar{Q}^i}{\partial u_0}, \quad \frac{\partial \bar{Q}^i}{\partial v_2} = \frac{\partial \bar{Q}^i}{\partial v_0}, \quad \frac{\partial \bar{Q}^i}{\partial v_3} = \frac{\partial \bar{Q}^i}{\partial w_0},$$

we have:

$$\frac{\partial\bar{Q}^i}{\partial t}+u_0\frac{\partial\bar{Q}^i}{\partial x_1}+v_0\frac{\partial\bar{Q}^i}{\partial x_2}+w_0\frac{\partial\bar{Q}^i}{\partial x_3}=$$

$$=-\frac{1}{\rho_i}\left[\frac{\partial(\rho_i\overline{uQ^i})}{\partial x_1}+\frac{\partial(\rho_i\overline{vQ^i})}{\partial x_2}+\frac{\partial(\rho\overline{wQ^i})}{\partial x_3}\right]-$$

$$-\frac{\partial}{\partial x_1}\int\limits_{(r'\parallel t)}K(|r-r'|)\,\rho_{\overline{i+1}}(r',\,t)\,dr'\,\frac{\partial\bar{Q}^i}{\partial u_0}\lambda_i-$$

$$-\frac{\partial}{\partial x_2}\int\limits_{(r'\parallel t)}K(|r-r'|)\,\rho_{\overline{i+1}}(r',\,t)\,dr'\,\frac{\partial\bar{Q}^i}{\partial v_0}\lambda_i-$$

$$-\frac{\partial}{\partial x_3}\int\limits_{(r'\parallel t)}K(|r-r'|)\,\rho_{\overline{i+1}}(r',\,t)\,dr'\,\frac{\partial\bar{Q}^i}{\partial w_0}\lambda_i. \qquad (14.5)$$

Setting in (14.5) $Q=v_1$, $Q=v_2$, $Q=v_3$, consecutively, and calculating $\lambda_i=\frac{1}{m_i}$,, we obtain two systems of equations for two clouds:

$$\left.\begin{aligned}
&\frac{\partial u_0}{\partial t}+u_0\frac{\partial u_0}{\partial x}+v_0\frac{\partial u_0}{\partial y}+w_0\frac{\partial u_0}{\partial z}=\\
&\qquad=X^i+\frac{1}{\sigma_i}\left(\frac{\partial X^i_x}{\partial x}+\frac{\partial X^i_y}{\partial y}+\frac{\partial X^i_z}{\partial z}\right),\\
&\frac{\partial v_0}{\partial t}+u_0\frac{\partial v_0}{\partial x}+v_0\frac{\partial v_0}{\partial y}+w_0\frac{\partial v_0}{\partial z}=\\
&\qquad=Y^i+\frac{1}{\sigma_i}\left(\frac{\partial Y^i_x}{\partial x}+\frac{\partial Y^i_y}{\partial y}+\frac{\partial Y^i_z}{\partial z}\right),\\
&\frac{\partial w_0}{\partial t}+u_0\frac{\partial w_0}{\partial x}+v_0\frac{\partial w_0}{\partial y}+w_0\frac{\partial w_0}{\partial z}=\\
&\qquad=Z^i+\frac{1}{\sigma_i}\left(\frac{\partial Z^i_x}{\partial x}+\frac{\partial Z^i_y}{\partial y}+\frac{\partial Z^i_z}{\partial z}\right)\\
&\qquad\qquad(i=1,\,2).
\end{aligned}\right\} \qquad (14.6)$$

Here

$$\sigma_i = m_i \rho_i(r, t)$$

$$\left.\begin{aligned} X^i &= -\frac{1}{m_i}\frac{\partial}{\partial x}\int\limits_{(r' \| t)} K(|r-r'|)\,\rho_{\overline{i+1}}(r', t)\,dr', \\ Y^i &= -\frac{1}{m_i}\frac{\partial}{\partial y}\int\limits_{(r' \| t)} K(|r-r'|)\,\rho_{\overline{i+1}}(r', t)\,dr', \\ Z^i &= -\frac{1}{m_i}\frac{\partial}{\partial z}\int\limits_{(r' \| t)} K(|r-r'|)\,\rho_{\overline{i+1}}(r', t)\,dr', \end{aligned}\right\} \quad (14.7)$$

$$\left.\begin{aligned} X^i_x &= -\sigma_i\overline{u^2}^i, & Y^i_x &= -\sigma_i\overline{uv}^i, & Z^i_x &= -\sigma_i\overline{uw}^i, \\ X^i_y &= -\sigma_i\overline{vu}^i, & Y^i_y &= -\sigma_i\overline{v^2}^i, & Z^i_y &= -\sigma_i\overline{vw}^i, \\ X^i_z &= -\sigma_i\overline{wu}^i, & Y^i_z &= -\sigma_i\overline{wv}^i, & Z^i_z &= -\sigma_i\overline{w^2}^i. \end{aligned}\right\} \quad (14.8)$$

The main quantities $X^i_x, X^i_y, \ldots$ form a symmetrical tensor, as will easily be seen if we perform the calculation of these magnitudes in a new system of coordinates, which is defined by the relationship:

$$x'_i = a_{ik}x_k + a_i, \qquad a_{ik}a_{il} = \delta_{kl},$$
$$x'_i = x_i - a_i t,$$

employing $\left|\frac{\partial(u', v', w')}{\partial(u, v, w)}\right| = 1$ for the Galilean transformations and the property of the invariance of $f_i(\underline{r}, \underline{v}, t)$ to the same transformations.

It will easily be seen that the average velocity $\{u_0, v_0, w_0\}$ has the transformation properties of a velocity under Galilean transformations, since the transformations are linear, the distribution function is invariant, and $\left|\frac{\partial(v')}{\partial(v)}\right| = 1.$

The system of equations (14.6) corresponds to the equations for the classical theory of continuous media. In the particular case when the distribution function is symmetrical with respect to the velocity, the system of equations that has been written down goes over into the equations of the hydrodynamics of a nonviscous liquid.

Thus the classical theory of continuous media is included in the theory being considered, and is a particular case of it. It is of importance to stress the following points:

1. The hydrodynamic equations include the space force of spontaneous action, and the components of the tensor on the right side likewise depend on the distribution function.

2. The transition is subject to the condition that the surface integra vanish in the space of velocities. The result obtained is in essence a differential law of conservation for a magnitude.

3. Since in the theory set forth the means for describing two or many particles are alike, the transition to hydrodynamics does not in practice depend on the number of particles, nor consequently on the density. This fact sheds new light on the old problem of the connection between discreteness and the continuum. It is of practical importance that it justifies the application of hydrodynamics in the sense of a form of the fundamental equations for as rarefied a gas as desired.

4. One essential point is that the hydrodynamic form (14.6), (14.7), (14.8), strictly speaking, cannot serve as the apparatus of the theory, since the elements of tensor (14.8) are themselves defined only by means of an unknown initial function f.

Section 15. Transition to Electrodynamics

We consider an equation for N uniform particles, interacting electrodynamically:

$$-\frac{\partial f^{(i)}}{\partial t}=\frac{\partial v_k f^{(i)}}{\partial x_k}+(N-1)\frac{\partial}{\partial v_k}\left\{\frac{1}{mc}F^{(i)}_{k\nu}j_\nu f^{(i)}\right\},$$

$$\frac{\partial F^{(i)}_{\mu\nu}}{\partial x_\nu}=\frac{4\pi}{c}J^{(i)}_\mu,\quad \frac{\partial F^{(i)}_{\mu\nu}}{\partial x_\lambda}+\frac{\partial F^{(i)}_{\lambda\mu}}{\partial x_\nu}+\frac{\partial F^{(i)}_{\nu\lambda}}{\partial x_\mu}=0,\quad (i=1,2,\ldots,N),\qquad (1\ldots$$

$$J^{(i)}_k=e\int_{(v)} v_k f^{(i)}(r,v,t)\,dv,\quad J^{(i)}_4=eic\int_{(v)} f^{(i)}(r,v,t)\,dv,\qquad (1\ldots$$

where $F^{(i)}_{\mu\nu}$ — is a tensor of the electromagnetic field, relating to the i-th particle and j is defined by the components $j_k = ev_k (k = 1, 2, 3)$, $j_4 = ice$ (the Greek indexes go through the values 1, 2, 3, 4).

Let N be large (we replace N - 1 by N); then, designating Nf(i) by f $NF^{(i)}_{\mu\nu}\to F_{\mu\nu}$, $Nj_\nu\to j_\nu$ and $NJ^{(i)}_\nu\to J_\nu$, we write the initial equation in the form

$$-\frac{\partial f}{\partial t}=\frac{\partial v_k f}{\partial x_k}+\frac{\partial}{\partial v_k}\left\{\frac{1}{mc}F_{k\nu}j_\nu f\right\},\qquad (1\ldots$$

$$\frac{\partial F_{\mu\nu}}{\partial x_\nu}=\frac{4\pi}{c}J_\mu,\quad \frac{\partial F_{\mu\nu}}{\partial x_\lambda}+\frac{\partial F_{\nu\lambda}}{\partial x_\mu}+\frac{\partial F_{\lambda\mu}}{\partial x_\nu}=0,$$
$$J_\nu=\int_{(v)} j_\nu f(r,\ v,\ t)\,dv,\quad j_k=ev_k,\quad j_4=eic, \tag{15.4}$$

where $F_{\nu\mu}$ — is now a tensor of the electromagnetic field that is set up by all N particles.

Here, too, we can go over to the classical system of the electron theory of charged localized particles, if we describe the charged particle by the adequately localized function f in space of six dimensions. We shall assume that $f^{(i)}$ is normalized to 1:

$$\int_{(r,v)} f^{(i)}(r,\ v,\ t)\,dr\,dv=1.$$

Multiplying equation (15.3) in turn by x_k and v_k, integrating for r and v over a given region (r, v), we obtain:

$$\begin{aligned}\frac{d}{dt}\int_{(r,v)} x_k f\,dr\,dv &= \\ &=\bar{v}_k-\int_{(s,v)} x_k f v_n\,ds_n\,dv-\frac{1}{mc}\int_{(r,\sigma)} x_k f F_{n\nu} j_\nu\,dr\,d\sigma_n,\end{aligned} \tag{15.5}$$
$$\frac{d}{dt}\int_{(r,v)} v_k f\,dr\,dv=\frac{1}{mc}\overline{F_{k\nu}j_\nu}-\int_{(s,v)} v_k f v_n\,ds_n\,dv-\frac{1}{mc}\int_{(r,\sigma)} v_k f F_{n\nu} j_\nu\,dr\,d\sigma_n,$$

where the symbols are analogous to those used in Sec. 13.

If

1) We omit the surface integrals,

2) We require that

$$\overline{F_{k\nu}j_\nu}\equiv e\overline{F_{kn}j_n}+ice\bar{F}_{k4}=eF_{kn}(\bar{r},\ t)\,\bar{v}_n+iceF_{k4}(\bar{r},\ t), \tag{15.6}$$

then we have the ordinary electronic theory of localized charged particles with their essential characteristics. The electromagnetic interactions take place even for the individual elements of the particle, which leads to the existence of spontaneous action. However, from the point of view of the theory being set forth, equations of electron theory (with spontaneous action) are obtained only for complex particles (in the derivation the assumption was made that $N \gg 1$).

A single isolated particle, as has been emphasized repeatedly, has no charge, and therefore there can be no spontaneous action for it, nor any characteristic electrostatic energy. The effect of spontaneous action and of being charged is obtained only as a result of the macroscopic nature of the system. The transition to classical electrodynamics that was made clearly shows the superiority of the new point of view, which relieves us from the specific difficulties of the theory of localized charged particles. As will easily be seen, condition (15.6) can be satisfied only on condition of an adequate degree of localization of the distribution function. We can evaluate this degree by a method similar to that employed in Sec. 13. We write $\overline{F_{kn}v_n}$ as follows:

$$\overline{F_{kn}v_n} = \bar{v}_n \int\limits_{(r,\,v)} F_{kn}(r, t) f(r, v, t)\, dr\, dv + \int\limits_{(r,\,v)} (v_n - \bar{v}_n) F_{kn}(r, t) f(r, v, t)\, dr\, dv. \quad (15.$$

We assume that $F_{kn}(\underline{r}, t)$ is expanded in a uniformly converging series in power of $(\underline{r} - \bar{\underline{r}})$ over a region $(\underline{r})$, where $f(r, v, t) \neq 0$. We consider that over the remainder of the region, f is equal to zero. Then

$$\overline{F_{kn}v_n} = \bar{v}_n F_{kn}(\bar{r}, t) + \frac{1}{2!}\frac{\partial^2 F_{kn}}{\partial \bar{x}_i \partial \bar{x}_l}\overline{(x_i - \bar{x}_i)(x_l - \bar{x}_l)\,v_n} + \ldots + \frac{\partial F_{kn}(\bar{r}, t)}{\partial \bar{x}_i}\overline{(x_i - \bar{x}_i)(v_n - \bar{v}_n)} + \frac{1}{2}\frac{\partial^2 F_{kn}(\bar{r}, t)}{\partial \bar{x}_i^2}\overline{(x_i - \bar{x}_i)^2 (v_n - \bar{v}_n)} + \ldots \quad (15.$$

If we separate out of region $(\underline{r})$ a subregion $(\underline{r} - \bar{\underline{r}})$ (in the remaining $\underline{r}$, we assume $f = 0$), of such a nature that we can limit ourselves to the first term in the expansion of $\overline{F_{kn}v_n}$, so that at least the following condition should be satisfied

$$\overline{(x_i - \bar{x}_i)(x_l - \bar{x}_l)} \ll 2\frac{|F_{kn}(\bar{r}, t)|}{\left|\dfrac{\partial^2 F_{kn}(\bar{r}, t)}{\partial \bar{x}_i \partial \bar{x}_l}\right|}. \quad (15.$$

Confining ourselves to the first term in the expansion of $\overline{F_{kn}v_n}$, we find:

$$\overline{(x_i-\bar{x}_i)\,(v_n-\bar{v}_n)} \ll \frac{|F_{kn}(\bar{r},t)|}{\left|\frac{\partial F_{kn}(\bar{r},t)}{\partial \bar{x}_i}\right|}, \tag{15.10}$$

But from the condition $\overline{F}_{k4} = F_{k4}(\bar{r}, t)$, we have the following limitation:

$$\overline{(x_i-\bar{x}_i)\,(x_l-\bar{x}_l)} \ll 2\,\frac{|F_{k4}(\bar{r},t)|}{\left|\frac{\partial^2 F_{k4}(\bar{r},t)}{\partial \bar{x}_i \partial \bar{x}_l}\right|}, \tag{15.11}$$

which can be combined with condition (15.9), assuming there instead of the Latin index n, the Greek index $\nu = 1, 2, 3, 4$. The magnitude on the right side of the inequality plays the part of the "effective" radius of action of the field set up by the particle. The criterion we have obtained for the applicability of electrodynamics runs as follows: the effective radius of the particle must be sufficiently small as compared with the effective radius of the field it sets up.

Section 16. Transition to Relativity Dynamics of Point Particles

We consider the initial relativity relationships of Sec. 12, Chapter II. Here, too, the transition to the dynamics of localized particles does not create difficulties. We write down the equations for the means:

$$\int rf\,dv\,dr; \quad \int u_i f\,dr\,dv \qquad (i=1,\,2,\,3),$$

where $u_i = \frac{v_i}{\sqrt{1-\beta^2}}$. With this in mind, we multiply

$$\frac{\partial f}{\partial t} + v_k\frac{\partial f}{\partial x_k} + g_k\frac{\partial f}{\partial v_k} + f\frac{\partial g_k}{\partial v_k} = 0 \tag{16.1}$$

at first by $\underline{r}$, and then u_i, and finally by $\frac{m_0 c}{i}u_4$, and integrate for $\underline{r}$ and $\underline{v}$ over a definite region $(\underline{r}, \underline{v})$. We then have:

$$\left.\begin{aligned} \frac{d}{dt}\bar{r}(t) &= \bar{v}(t) - \int\limits_{(s,\,v)} rfv_n\,ds_n\,dv - \int\limits_{(r,\,\sigma)} rfg_n\,dr\,d\sigma_n, \\ \frac{d}{dt}u(t) &= \overline{\frac{\partial u}{\partial v_k}g_k} - \int\limits_{(s,\,v)} uf\,v_n\,ds_n\,dv - \int\limits_{(r,\,\sigma)} ufg_n\,dr\,d\sigma_n, \end{aligned}\right\} \tag{16.2}$$

$$\frac{d}{dt}\overline{\frac{m_0c^2}{\sqrt{1-\beta^2}}} = m_0\overline{v_n G_n} - \int\limits_{(s,\,v)} u_4\frac{m_0c}{i}fv_n ds_n\,dv - \int\limits_{(r,\,\sigma)} u_4\frac{m_0c}{i}g_nf\,dr\,d\sigma_n,$$

where $$G_\mu = \frac{du_\mu}{dt}.$$

But direct calculation gives the following simple relationships among u, g, and G:

$$\frac{\partial u}{\partial v_k} g_k = G. \tag{16.3}$$

The system of equations (16.2) include relativity dynamics of a point if the following conditions are satisfied:

1. All the surface integrals vanish.

2. $\overline{G_k(r, v, t)} = G_k(\bar{r}, \bar{v}, t), \ \overline{v_k G_k} = \bar{v}_k G_k(\bar{r}, \bar{v}, t).$ (16.4)

3. $\bar{U}_k = \dfrac{\bar{v}_k}{\sqrt{1 - \sum\limits_{i=1}^{3} \dfrac{(\bar{v}_i)^2}{c^2}}}, \quad \dfrac{m_0 c}{i} \bar{U}_4 = \dfrac{m_0 c^2}{\sqrt{1 - \sum\limits_{i=1}^{3} \dfrac{(\bar{v}_i)^2}{c^2}}}.$ (16.5)

4. $\bar{v}_k$ is transformed in the same way as v_k and $\bar{x}_k$ as x_k. The first equality of condition 16.4 is satisfied, as in the preceding sections, when the distribution function is sufficiently localized; the second equality requires that a new condition be satisfied, namely $\overline{\Delta v_k \Delta v_i} \ll \bar{v}_k \cdot \bar{v}_i$. Since

$\dfrac{z_i}{\sqrt{1 - \sum\limits_k \dfrac{z_k^2}{c^2}}}$ is an analytic function of three complex variables

z_k, in the region where $\operatorname{Re}(z_1^2 + z_2^2 + z_3^2) < c^2$, then it may be expanded into a series at any point for which the conditions are satisfied

$\operatorname{Re}(\bar{z}_1^2 + \bar{z}_2^2 + \bar{z}_3^2) < c^2$, in powers of $(z_1 - \bar{z}_1)^l (z_2 - \bar{z}_2)^m (z_3 - \bar{z}_3)^n$

for all z_k satisfying this condition. This series converges absolutely and uniformly, and it can therefore be integrated member by member. In this way, we have:

$$\overline{U}_i = \frac{\overline{v}_i}{\left(1-\sum\limits_{k=1}^{3}\frac{(\overline{v}_k)^2}{c^2}\right)^{1/2}} + \frac{3}{2}\frac{\overline{v}_i}{c^2}\,\frac{\left(1-\sum\limits_{k=1}^{3}\frac{(\overline{v}_k)^2}{c^2}+\frac{(\overline{v}_i)^2}{c^2}\right)}{\left(1-\sum\limits_{k=1}^{3}\frac{(\overline{v}_k)^2}{c^2}\right)^{5/2}}\,\overline{(v_i-\overline{v}_i)^2}\,+$$

$$+\sum_{k\neq i}\frac{\overline{v}_k}{c^2}\,\frac{\left(1-\sum\limits_{l=1}^{3}\frac{(\overline{v}_l)^2}{c^2}+3\frac{\overline{v}_k\overline{v}_i}{c^2}\right)}{\left(1-\sum\limits_{l=1}^{3}\frac{(\overline{v}_l)^2}{c^2}\right)^{5/2}}\,\overline{(v_k-\overline{v}_k)(v_i-\overline{v}_i)}\,+$$

$$+\frac{1}{2}\sum_{k\neq i}\frac{\overline{v}_i}{c^2}\,\frac{\left(1-\sum\limits_{l=1}^{3}\frac{(\overline{v}_l)^2}{c^2}+3\overline{v}_i\frac{(\overline{v}_k)^2}{c^2}\right)}{\left(1-\sum\limits_{l=1}^{3}\frac{(\overline{v}_l)^2}{c^2}\right)^{5/2}}\,\overline{(v_k-\overline{v}_k)^2}\,+$$

$$+\frac{3}{2c^2}\,\frac{\overline{v}_i\overline{v}_k\overline{v}_{l\neq k\neq i}}{\left(1-\sum\limits_{m=1}^{3}\frac{(\overline{v}_m)^2}{c^2}\right)^{5/2}}\,\overline{(v_k-\overline{v}_k)(v_l-\overline{v}_l)}+\dots \qquad (16.6)$$

In order to be able to confine ourselves to the first term of the expansion, it is necessary and sufficient for the following conditions to be satisfied:

$$\left.\begin{aligned}
&\overline{(v_i-\overline{v}_i)^2} \ll \frac{2c^2}{3}\,\frac{\left(1-\sum\limits_{k=1}^{3}\frac{(\overline{v}_k)^2}{c^2}\right)^2}{1-\sum\limits_{m\neq i}\frac{(\overline{v}_m)^2}{c^2}},\\
&\overline{(v_i-\overline{v}_i)(v_k-\overline{v}_{k\neq i})} \ll \frac{c^2\left(1-\sum\limits_{l=1}^{3}\frac{(\overline{v}_l)^2}{c^2}\right)^2}{\left(1-\sum\limits_{m=1}^{3}\frac{(\overline{v}_m)^2}{c^2}+3\frac{\overline{v}_i}{c^2}\right)}\,\frac{\overline{v}_i}{\overline{v}_k},\\
&\overline{(v_k-\overline{v}_k)(v_l-\overline{v}_{l\neq k\neq i})} \ll \frac{c^4}{3}\,\frac{\left(1-\sum\limits_{m=1}^{3}\frac{(\overline{v}_m)^2}{c^2}\right)^2}{\overline{v}_k\overline{v}_l}.
\end{aligned}\right\} \qquad (16.7)$$

These conditions impose limitations on the magnitude of the dispersion of the velocities. Condition 4 leads to analogous expressions.

As a result, we find that the relativity equations for the motion of localized particles are likewise obtained, as a particular case, from the theory being considered:

$$\frac{d}{dt}\left(\frac{\bar{v}_k}{\sqrt{1-\sum_{l=1}^{3}\frac{(\bar{v}_l)^2}{c^2}}}\right)=\frac{1}{m_0 c}F_{k\nu}(\bar{r}, t)\, j_\nu(\bar{v}),$$

$$\frac{d}{dt}\left(\frac{m_0 c^2}{\sqrt{1-\sum_{l=1}^{3}\frac{(\bar{v}_l)^2}{c^2}}}\right)=m_0 G_k(\bar{r}, \bar{v}, t)\, \bar{v}_k,$$

where

$$\bar{v}=\frac{d\bar{r}}{dt}; \quad G_k=\frac{1}{m_0 c}F_{k\nu}j_\nu.$$

CHAPTER IV

DEFINITION OF STATIONARY STATES OF SYSTEMS OF TWO AND OF MANY PARTICLES, AS THE PROBLEM OF THE EIGEN VALUES OF NONLINEAR INTEGRAL EQUATIONS

Section 17. Fundamental Equations

It was shown in Chapter II that the initial equations for the stationary state ($\frac{\partial}{\partial t} = 0$) lead to the following nonlinear integral equations for two particles:

$$V(r) = a \int_{-\infty}^{\infty} K(|r - r'|) e^{-\frac{V(r')}{\theta}} dr', \qquad (17.1)$$

and for N uniform particles:

$$V(r) = \sum_n a^n \underbrace{\int \dots \int}_{n} K_n(r, r_1, \dots, r_n) e^{-\frac{V(r_1)}{\theta}} \dots \times$$
$$\times e^{-\frac{V(r_n)}{\theta}} dr_1 \dots dr_n. \qquad (17.2)$$

We define their fundamental properties:

1. Equations of this kind do not occur in the ordinary theory, operating with point particles. The very fact that the new problem of finding the eigenvalues of nonlinear integral equations even for two particles, entails emerging into a new field as compared with the theories of localized particles.

2. From the mathematical point of view, these equations are close to the equations of the type of Hammerstein, familiar in the theory of nonlinear integral equations. However, equations (17.1) and (17.2) are different from them with respect to:

a) The infinite limits of integration, which indicates the singular character of the equations written down;

---

* See also the author's article in Zh. E. T. F. 18, 840 (1948).

b) The type of kernels, which do not satisfy the condition usually laid down in the theory of integral equations, namely:

$$\int\int K(r_1, r_2)\, dr_1 dr_2 < \infty,$$

since in our case

$$\int_{-\infty}^{\infty} K(|r_1 - r_2|)\, dr_2 = 4\pi \int_0^{\infty} K(\rho)\, \rho^2 d\rho = \text{const.}$$

and, consequently, repeated integration for r_2 gives

Unfortunately, the mathematical theory of nonlinear integral equations has not been worked out as thoroughly as the theory of linear equations. The following fundamental mathematical problems arise in applying the present theory to physical questions:

a) Finding characteristic numbers and characteristic functions. By characteristic numbers we mean here those values of some parameter (see below) for which equations (17.1) and (17.2) have solutions differing from trivial solutions. Among the trivial solutions are the solutions of equations (17.1) and (17.2) corresponding to the case of uniform density.

b) The problem of branching, which consists in the following: let there be known even a single exact solution of the initial equation, corresponding to some value of parameter . The problem consists in finding out how this solution will change as the parameter varies continuously and what new solutions will arise in the process. It is essential to note that the apparatus of the ordinary theory of branching of solutions of integral equations, based on the expansion of the desired function in series, applies even without the above limitations with respect to the type of kernel and domain of integration. Accordingly, this apparatus may be applied to equations (17.1) and (17.2). Study of the convergence of the series obtained must be made separately for each problem.

In solving the equations that have been written down, it is natural to use Fourier series, since this method has shown its fertility in the case of linear equations with infinite limits of integration and kernels, depending on the modulus of difference of the argument. It is this method that we employ to solve the problem of finding periodic solutions (see Part II, Chapter I).

We transform equations (17.1) and (17.2), introducing the nondimensional magnitudes

$$-\frac{V(r)}{\theta} = \varphi(r), \quad -\frac{a}{\theta}\sigma(0) = \lambda, \quad K^*(|r'|) = K\frac{(|r|)}{\sigma(0)},$$

$$\sigma(0) = \int_{-\infty}^{\infty} K(|r - r'|)\, dr' = 4\pi \int_0^{\infty} K(\rho)\, \rho^2 d\rho.$$

where now the kernel K^+ is normalized to unity:

$$\int_{-\infty}^{\infty} K^*(|r - r'|)\, dr' = 1$$

(hereafter we omit the $^+$ after K).

We obtain as our fundamental equations

$$\varphi(r) = \lambda \int_{-\infty}^{\infty} K(|\bar{r} - r'|) e^{\varphi(r')} dr', \qquad (17.3)$$

$$\varphi(r) = \sum_n \lambda_n \int_{-\infty}^{\infty} \cdots \int_{-\infty}^{\infty} K_n(r, r_1, \ldots, r_n)\, e^{\varphi(r_1)} \ldots e^{\varphi(r_n)} \times$$

$$\times\, dr_1, \ldots, dr_n. \quad (17.4)$$

These equations may also be written in a different form. In point of fact, as we saw in Sec. 11,

$$V(r) = \rho(0) \int_{-\infty}^{\infty} K(|r - r'|)\, e^{\frac{-[V(r') - V(0)]}{\theta}} dr' \qquad (17.5)$$

has a precise solution: $V(\underline{r}) = V(0) =$ const., which corresponds to the case of uniform spatial distribution of the particles.

But from equation (17.5), we have:

$$V(0) = \rho(0) \int_{-\infty}^{\infty} K(|r'|)\, dr' = \rho(0) \int_{-\infty}^{\infty} K(|r - r'|)\, dr' = \text{const.} \quad (17.5')$$

Therefore, if we now subtract (17.5') from equation (17.5), we have:

$$V(r) - V(0) = \rho(0) \int_{-\infty}^{\infty} K(|r - r'|) \{ e^{-\frac{V(r') - V(0)}{\theta}} - 1 \}\, dr'.$$

Once again introducing the nondimensional magnitudes:

$$-\frac{V(r) - V(0)}{\theta} = \varphi(r), \quad -\frac{\rho(0)}{\theta} \sigma(0) = \lambda,$$

we have:

$$\varphi(r)=\lambda\int_{-\infty}^{\infty} K(|r-r'|)\{e^{\varphi(r')}-1\}\,dr', \qquad (17.6)$$

where kernel K is normalized in the same way as in (17.3).

Similarly, in calculating the nonlinear functionals

$$\varphi(r)=\sum_n \lambda_n \int_{-\infty}^{\infty}\dots\int_{-\infty}^{\infty} K_n(r, r_1, \dots, r_n)\times \times\{e^{\varphi(r_1)}\dots e^{\varphi(r_n)}-1\}\,dr_1, \dots, dr_n, \qquad (17.7)$$

where n is summated from 1 to N - 1, and the λ_n may be either positive or negative depending on the nature of the forces of interaction, that is, the sign of $\sigma_n(0)$.

The two forms of equations (17.3), (17.4) and (17.6), (17.7), which are connected by a simple transformation, are mathematically equivalent. Nonetheless, they are, of course, distinguished by means of the λ.

For, since in equation (17.3)

$$-\frac{a}{\theta}\sigma(0)=\lambda, \quad a=\rho(0)\,e^{+\frac{V(0)}{\theta}},$$

while in equation (17.6)

$$-\frac{\rho(0)}{\theta}\sigma(0)=\lambda_0 \quad \text{и} \quad \sigma(0)\,\rho(0)=V(0),$$

the two magnitudes λ are related by the expression

$$\lambda=\lambda_0 e^{-\lambda_0}. \qquad (17.8)$$

In view of the fact that the domains in which λ and λ_0 apply coincide only in part, and also in view of the fact that equation (17.8) is not a single-valued (for a given λ, λ_0 has two values), we must regard equations (17.3), (17.4) and (17.6), (17.7) as different.

Equations of the type (17.3) are suitable for those cases in which conditions of the experiment enable λ to be varied over the entire interval $-\infty<\lambda<+\infty$; equations (17.6) are applied when this possibility applies to the other magnitude, λ_0.

For forces satisfying the condition $\lambda<0$, i. e. $\sigma(0)>0$ for $K>0$ (forces of repulsion predominant), the solution $\varphi\equiv 0$ for equation (17.6) is <u>unique</u>. This was proved above.

We now prove that in the general case of any kernels K in general, in the neighborhood of point $\lambda = 0$ over the finite interval $0\leqslant\lambda<\lambda_{max}$ equation (17.6) has a <u>unique</u> holomorphic solution, represented in the form of the sum of a convergent series within positive and integral power of We show likewise that this unique solution is trivial: $\varphi(r) \equiv 0$.

For the proof, we employ the method of successive approximations. We represent the function $\varphi(r)$ in the form of a series whose divergence is postulated

$$\varphi(r)=\lambda\varphi_1(r)+\lambda^2\varphi_2(r)+\lambda^3\varphi_3(r)+\dots;$$

substituting this series in (17.6) and developing the eigen function $e^{\varphi(a)}$, we arrive at the following system of equations:

$$\varphi_1\equiv 0,\ \varphi_2=\int K\varphi_1\,dr',\ \dots,$$

$$\varphi_h=\int K\left[\varphi_{h-1}+\dots+\frac{1}{(h-1)!}\varphi_1^{h-1}\right]dr',$$

from which it is obvious that all the coefficients

$$\varphi_1\equiv\varphi_2\equiv\varphi_3\equiv\dots\equiv 0.$$

Hence, although the given interval of variation of parameter $\lambda\,(0,\lambda_{max})$, and consequently, of the variation of the temperatures from ∞ to some θ_{min} - the state of uniform density is the only holomorphic solution. The value θ_{min} (corresponding to λ_{max}) is defined by the radius of conversion of the series in question.

Section 18. Method of Branching

The method of "branching" of the solution of nonlinear integral equations was first presented by Liapunov* in his classic work on the equilibrium figures of a rotating fluid, and was later developed by Schmidt, Hammerstein, et al.++

Recently Nazarov+++ systematically reconsidered the theory of branching. Hereinafter (see Part II, Chapter I) we shall apply the methods developed by him to the equations we have obtained. Here we consider only some basic propositions. Let $\varphi(r)=\varphi_0(r, \lambda_0)$ be a precise solution of the equation

$$\lambda\varphi = \int_{-\infty}^{\infty} K(|r-r'|)(e^{\varphi(r')}-1)\,dr' \tag{18.1}$$

or in a more complex form:

$$\lambda\varphi = \sum_n \lambda_n \int_{-\infty}^{\infty} \cdots \int_{-\infty}^{\infty} K_n(r, r_1, \ldots, r_n) \times \times \{e^{\varphi(r_1)} \ldots e^{\varphi(r_n)} - 1\}\,dr_1 \ldots dr_n. \tag{18.2}$$

Let this solution correspond to a characteristic term $\lambda = \lambda_0$. To elucidate the question as to the behavior of the solution in the neighborhood of $\lambda = \lambda_0$, we set $\varphi(r)=\varphi_0(r, \lambda_0)+\psi(r);\ \lambda=\lambda_0+\varepsilon$. Substituting this (18.1) and taking into consideration that by assumption φ_0 satisfies equation (18.1) for the value $\lambda = \lambda_0$, we obtain the following equation for determining $\psi(r)$:

$$K\left\{\psi+\frac{\psi^2}{2!}+\frac{\psi^3}{3!}+\cdots\right\}-(\lambda_0+\varepsilon)\psi=\varepsilon\varphi_0, \tag{18.3}$$

for

$$K\{\ \ldots\ \} = \int_{-\infty}^{\infty} K(|r-r'|)\,e^{\varphi_0(r')}\{\ \ldots\ \}\,dr' \tag{18.4}$$

* A. Liapunov, Izvestiya Akad. Nauk, St. Petersburg, 1906.

++ L. Lichtenstein, Vorlesungen über einige Klassen nichtlinearen integralgleichungen, Berlin, 1931.

+++ N. Nazarov, /Nelineinie integrälnie uravneniya tipa Gammersteina Uzgosizdat, 1941. /Nonlinear integral equations of Hammerstein's type/ Uzgosizdat, 1941.

is an integral operator with kernel Ke^{φ} . Equation (18.3) is equivalent to equation (18.1).

A similar transformation in the case of equation (18.2) gives

$$\sum_n \lambda_n K_n \left\{\sum_i^n \psi(r_i) + \sum_i^{n:n}\sum_j \alpha_{ij}\psi(r_i)\psi(r_j) + \dots\right\} - (\lambda_0 + \varepsilon)\psi = \varepsilon\varphi_0, \tag{18.5}$$

where

$$K_n\{\dots\} = \int_{-\infty}^{\infty} \dots \int_{-\infty}^{\infty} K_n(r, r_1, \dots, r_n) \times$$

$$\times (e^{\varphi(r_1)} \dots e^{\varphi(r_n)})\{\dots\}\, dr_1 \dots dr_n \tag{18.6}$$

is a multi-place integral operator and α_{ij} in (18.5) are constants, defined by the presence of factorial terms $\frac{1}{n!}$, arising out of the expansion of e^{φ} in a series according to φ.

We shall solve nonlinear equations (18.3) and (18.5) by the method of successive approximations, setting

$$\psi(r) = \varepsilon\psi_1(r) + \varepsilon^2\psi_2(r) + \varepsilon^3\psi_3(r) + \dots$$

Equating terms with the same powers of ε , we have a series of equations in the method of successive approximations. It may happen, however, that the system of equations obtained in this way is insoluble within the class of finite functions $|\psi_n(r)| < \infty$ (there are no holomorphic solutions). In this case, we must look for a solution in the form of the following series:

$$\psi(r) = \varepsilon^{\frac{1}{n}}\psi_1(r) + \varepsilon^{\frac{2}{n}}\psi_2(r) + \varepsilon^{\frac{3}{n}}\psi_3(r) + \dots,$$

where n is an integer. This constitutes the case of the branching of solutions, and n gives the possible number of branches.

For our purposes, it suffices to consider the case n = 2. In this case, we arrive at two systems of equations of successive approximation: the "holomorphic" sequence

$$\left.\begin{array}{l} K\{\psi_1\} - \lambda_0\psi_1 = \varphi_0,\ K\{\psi_2\} - \lambda_0\psi_2 = \psi_1 - K\left\{\frac{\psi_1^2}{2!}\right\}, \\ K\{\psi_3\} - \lambda_0\psi_3 = \psi_2 - K\left\{\frac{\psi_1^3}{3!} + \psi_1\psi_2\right\} \\ \dots\dots\dots\dots\dots\dots\dots\dots \end{array}\right\} \tag{18.7}$$

and the "nonholomorphic" sequence

$$\left.\begin{aligned} &K\{\psi_1\} - \lambda_0\psi_1 = 0,\ K\{\psi_2\} - \lambda_0\psi_2 = \varphi_0 - K\left\{\frac{\psi_1^3}{2!}\right\},\\ &K\{\psi_3\} - \lambda_0\psi_3 = \psi_1 - K\left\{\frac{\psi_1^3}{3!} + \psi_1\psi_2\right\},\\ &\dots\dots\dots\dots\dots\dots\dots \end{aligned}\right\} \qquad (18.8)$$

We obtain similar systems for equations (18.2) as well, with the exception that the integral operator must be expressed by means of a sum of the expressions (18.6).

The following important properties of the method of branching should be given particular emphasis:

1. The linear integral equation of the type

$$K\{\psi_1\} - \lambda_0\psi_1 = 0,$$

or written in another way:

$$\psi(r) - \lambda\int_{-\infty}^{\infty} K(|r-r'|)\,e^{\varphi_0(r',\lambda)}\,\psi(r')\,dr' = 0, \qquad (18.9)$$

is decisive for the solution of the system of successive approximations. In the particular case, when $\varphi_0(\underline{r}, \lambda) \equiv 0$, is taken as the initial solution from which the branching is sought, it has the form:

$$\psi(r) - \lambda\int_{-\infty}^{\infty} K(|r-r'|)\,\psi(r')\,dr' = 0. \qquad (18.10)$$

At first sight, it may seem that the linear equation is only a crude approximation to the original one, and cannot in any way reflect the properties of the nonlinear equation. This is not the case, however.

The solvability of the nonhomogeneous equations entering into sequences (18.8) and (18.8) depends on whether the linear homogeneous equation is solvable for the given value $\lambda = \lambda_0$. But this is the same for all the approximations. Hence, the criterion of the solvability of the linear homogeneous equation is applicable to all the approximations. In this sense, the linear equations are exact.

2. The values of λ that are the characteristic numbers of equation (18.9) or (18.10) are the branching parts of the initial equation. As a result, for the values of λ in question, the solution cannot be prolonged analytically, since nontrivial solutions of the homogeneous linear equations arise ("characteristic fluctuations" arise).

3. In principle, the apparatus that has been set forth gives the method for finding solutions for all values of λ, if the exact solution is known for any one value λ_0. It is necessary to find branching solutions, and then, following along the branches, to seek new solutions. In Chapter I, Part II, we employ this apparatus as applied to the theory of the crystalline state.

4. Equation (18.9) is more complex than (18.10), since parameter enters into it not only as a multiplier in the kernel, but in a nonlinear way in the expression under the integral sign as well. No general methods exist for solving such equations. However, in some particular cases definite results can be obtained (see Chapter I of Part II).

In conclusion, we should like to dwell briefly upon the work of Nazarov.

Nazarov's doctor's dissertation (1938) is devoted to a study of the solutions of Hammerstein type equations[+]

$$u(x) = \lambda \int_0^1 K(x,y) f[y, u(y)]\, dy. \qquad (\alpha)$$

We consider some theorems from this work. We first show that for a sufficiently small value of λ, there exists a solution for this equation.

Let there be the inequality $|K(x,y)f(y,z)| < A$ for all the x, y, z

satisfying the conditions $0 \leqslant x \leqslant 1$; $0 \leqslant y \leqslant 1$; $-M \leqslant z \leqslant M$,

where M is any number, $M > 0$. We shall say that x, y, z belong to region R.

We now define a constant positive number k such that $A < kM$. We likewise assume that for function $f(y,z)$ region R the conditions of Lipschitz are satisfied:

$$|f(y, z) - f(y, u)| < N\,|z - u|.$$

We introduce the designation $\int_0^1 |K(x, y)|\, dy = K$ for $0 \leqslant x \leqslant 1$

and we consider the auxiliary integral equation

$$\varphi(x) = \frac{\mu}{R} \int_0^1 K(x, y) f[y, \varphi(y)]\, dy,$$

where $|\mu| < 1$. To solve this auxiliary equation, we apply the method

[+] See reference above.

of successive approximations; in it we assume that $f(y,0) \equiv 0$. We now set up the sequence of functions:

$$\varphi_0(x) \equiv 0; \quad \varphi_1(x) = \frac{\mu}{R}\int_0^1 (x, y) f[y, \varphi(y)]\, dy;$$

$$\cdots\cdots\cdots\cdots\cdots\cdots\cdots$$

$$\varphi_n(x) = \frac{\mu}{R}\int_0^1 K(x, y) f[y, \varphi_{n-1}(y)]\, dy;$$

and since $|\varphi_0(x)| < M$, , we have:

$$|\varphi_1(x)| \leqslant \frac{|\mu|}{R}\int_0^1 |K(x, y)| \cdot |f(y, 0)|\, dy < \frac{1}{R} \cdot A < M \text{ и т. д.}$$

We now evaluate the inequalities:

$$|\varphi_1(x) - \varphi_0(x)| = |\varphi_1(x)| < |\mu| M,$$

$$|\varphi_2(x) - \varphi_1(x)| \leqslant \frac{|\mu|}{R}\int_0^1 |K(x,y)| \cdot |f(y, \varphi_1(y)) - f(y, \varphi_0(y)|\, dy <$$

$$< \frac{|\mu|}{R}\int_0^1 |K(x, y)| \cdot N |\varphi_1(y) - \varphi_0(y)|\, dy <$$

$$< \frac{|\mu|^2}{R} \cdot N \cdot M \int_0^1 |K(x, y)|\, dy \leqslant \frac{|\mu|^2}{R} MN \cdot K;$$

etc.

$$|\varphi_n(x) - \varphi_{n-1}(x)| < \frac{|\mu|^n}{R^{n-1}} MN^{n-1} K^{n-1}.$$

It is clear from this that if $|\mu| < \min\left(1, \frac{k}{NK}\right)$, then the sequence of functions $\varphi_0(x), \varphi_1(x), \ldots, \varphi_n(x), \ldots$ uniformly approaches a limiting function $\varphi(x)$, which obviously satisfies the equation

$$\varphi(x) = \frac{\mu}{R}\int_0^1 K(x, y) f[y, \varphi(y)]\, dy.$$

It is now evident immediately that initial equation (α) has a solution that may be obtained by the method of successive approximations, if

$$|\lambda| < \lambda^*, \quad \text{where} \quad \lambda^* = \min\left\{\frac{1}{R}, \frac{1}{NK}\right\}.$$

Thus, for the initial equation (α), we have $0 \neq (x)$0n и $0 \neq$ 0γ

such that

$$u_0(x) = \lambda_0 \int_0^1 K(x, y) f[y, u_0(y)]\, dy,$$

in which we assume that $\lambda_0 \neq 0$, the solution does not satisfy the initial equation.

We now show that if one of the solutions of the initial equation system,value λ_0 is known, then we can find a solution for any λ that satisfies the condition $|\lambda - \lambda_0| < \rho$, where ρ is a positive number. For this, it is requisite that certain conditions be satisfied.

Theorem 1. If λ_0 is not an eigenvalue of the kernel

$$K(x, y) \cdot \frac{\partial f[y, u_0(y)]}{\partial u_0(y)},$$

then in the neighborhood of λ_0, the initial equation ($\varkappa$) has the unique solution $u_0(x, \lambda)$, which is holomorphic (expansible by positive and integral powers of λ and λ_0) and approaches $u_0(x)$ or $\lambda - \lambda_0$.

2. We seek a solution of the initial equation in the form $u(x) = u_0(x) + v(x)$ and correspondingly $\lambda = \lambda_0 + \mu$. We assume further that there exists the expansion

$$f[y, u(y)] = f[y, u_0(y) + v(y)] = A_0(y) + A_1(y) v(y) + A_2(y) v^2(y) + \ldots$$

In this case the initial equation may be written as:

$$u_0(x) + v(x) =$$

$$= (\lambda_0 + \mu) \int_0^1 K(x, y) [A_0(y) + A_1(y) v(y) + A_2(y) v^2(y) + \ldots]\, dy,$$

but since by assumption

$$u_0(x) = \lambda_0 \int_0^1 K(x, y) f(y, u_0(y))\, dy = \lambda_0 \int_0^1 K(x, y) A_0(y)\, dy,$$

then, consequently, we have the following equation to determine v(x)

$$v(x) = \lambda_0 \int_0^1 K(x, y) \{A_1(y) v(y) + A_2(y) v^2(y) + \ldots\}\, dy +$$

$$+ \mu \int_0^1 K(x, y) \{A_0(y) v(y) + \ldots\}\, dy. \qquad (\beta)$$

This equation is obviously equivalent to the initial one. We seek a solution for this equation in the form of the following series:

$$v(x) = \mu v_1(x) + \mu^2 v_2(x) + \mu^3 v_3(x) + \ldots,$$

for $v_n(x)$ are functions as yet not known. If we now substitute this series in equation (β) and compare the expressions for similar powers of μ, we obtain:

$$v_1(x) = \lambda_0 \int_0^1 K(x, y) A_1(y) v_1(y) \, dy + \int_0^1 K(x, y) A_0(y) \, dy,$$

$$v_2(x) = \lambda_0 \int_0^1 K(x, y) A_1(y) v_2(y) \, dy +$$

$$+ \int_0^1 K(x, y) \{\lambda_0 A_2(y) v_1^2(y) + A_1(y) v_1(y)\} \, dy,$$

$$v_3(x) = \lambda_0 \int_0^1 K(x, y) A_1(y) v_3(y) \, dy + \int_0^1 K(x, y) \{\lambda_0 A_3(y) v_1^3(y) +$$

$$+ 2\lambda_0 A_2(y) v_1(y) v_2(y) + A_1(y) v_2(y) + A_2(y) v_1^2(y)\} \, dy,$$

. . . .

$$v_n(x) = \lambda_0 \int_0^1 K(x, y) A_1(y) v_n(y) \, dy + f_n(x),$$

where $f_n(x)$ depends on the functions $v_1(x)$, $v_2(x)$, ..., $v_{n-1}(x)$. Since by is not an eigenvalue of the kernel

$$K(x, y) A_1(y) \equiv K(x, y) \frac{\partial f(y, u_0(y))}{\partial u_0(y)},$$

then the solution of any of the equations of the given system of successive approximation should have the following form:

$$v_n(x) = f_n(x) + \lambda_0 \int_0^1 R(x, y; \lambda_0) f_n(y) \, dy,$$

where $R(x, y; \lambda_0)$ gives the resolvant of nucleus $K(x, y)A_1(y)$, correspond to the value λ_0, since

$$f_1(x) = \int_0^1 K(x, y) A_0(y) \, dy = \int_0^1 K(x, y) f(y, u_0(y)) \, dy = \frac{u_0(x)}{\lambda_0} \neq 0,$$

we can find, in only one way, a system of functions $v_1(x)$, $v_2(x)$, $v_3(x)$...., which are not simultaneously identically equal to zero. In this way, the solution in the form of a series

$$v(x) = \mu v_1(x) + \mu^2 v_2(x) + \mu^3 v_3(x) + \dots$$

has been constructed.

The convergence of this series will not be shown here, but we shall only give the results obtained by Nazarov.

If kernels $K(x, y)$ and $f(y, u_0(y))$ are such that the following conditions are fulfilled:

1. $\left| \int_0^1 K(x, y) A_p(y)\, dy \right| < A$ for all $p = 0, 1, 2, \dots$

2. $|R(x, y, \lambda_0)| < B$ for all $0 \leq x \leq 1$ and $0 \leq y \leq 1$, then for the series

$\mu v_1(x) + \mu^2 v_2(x) + \dots$ there exists a finite circle of convergence ρ.

Thus, we have proved the existence of a solution in the form of a theory

$$u(x, \lambda) = u_0(x) + (\lambda - \lambda_0) v_1(x) + (\lambda - \lambda_0)^2 v_2(x) + \dots$$

Theorem 2. If λ_0 is an eigen value of kernel

$$K(x, y) \frac{\partial f(y, u_0(y))}{\partial u_0(y)}$$

and, in addition $\int_0^1 A_0(y) \varphi_1(y)\, dy = 0,$ where $\varphi_1(x)$ —

is an eigenfunction of kernel $K(x, y)A_1(y)$, corresponding to eigenvalue λ_0, then in general there exist in the neighborhood of λ_0 two solutions (or none) $u_1(x, \lambda)$ and $u_2(x, \lambda)$, which are holomorphic with respect to $\lambda - \lambda_0$ and tend to $u_0(x)$ for $\lambda \to \lambda_0$.

Proof. Setting, as in the first theorem, $u(x) = u_0(x) + v(x)$, $\lambda = \lambda_0 + \mu$, we arrive at the same system system of equations, the general term of which we write somewhat differently than previously:

$$v_1(x) = \lambda_0 \int_0^1 K(x, y) A_1(y) v_1(y)\, dy + \int_0^1 K(x, y) A_0(y)\, dy,$$

$$v_2(x) = \lambda_0 \int_0^1 K(x, y) A_1(y) v_2(y)\, dy + \int_0^1 K(x, y) \{\lambda_0 A_2(y) v_1^2(y) + A_1(y) v_1(y)\}\, dy,$$

$$v_3(x) = \lambda_0 \int_0^1 K(x, y) A_1(y) v_3(y)\, dy + \int_0^1 K(x, y) \{\lambda_0 A_3(y) v_1^3(y) + 2\lambda_0 A_2(y) v_1(y) v_2(y) + A_1(y) v_2(y) + A_2(y) v_1^2(y)\}\, dy,$$

. . . .

$$v_n(x) = \lambda_0 \int_0^1 K(x, y) A_1(y) v_n(y)\, dy + \int_0^1 K(x, y) \{A_1(y) v_{n-1}(y) + M_n(y)\}\, dy + \int_0^1 K(x, y) 2\lambda_0 A_2(y) v_1(y) v_{n-1}(y)\, dy,$$

where $M_n(y)$ depends on $v_1(y)$, $v_2(y)$, ..., $v_{n-2}(y)$.

We now assume that $K(x, y)$ is a symmetrical kernel, and that function $A_1(y)$ keeps its sign at (0, 1), that is, in this case we have what is called a symmetrical kernel. Since λ_0 is an eigenvalue of kernel $K(x, y) A_1(y)$, then according to Fredholm's theory, generalized for the case of the symmetrized kernel, the necessary and sufficient condition for the solvability of the first equation of the system given above is that the right member of the first equation be orthogonal to function $\phi_1(x)A_1(x)$, where $\phi_1(x)$ is an eigenfunction of a homogeneous equation. Or,

$$\int_0^1 \int_0^1 K(x, y) A_0(y)\, dy \cdot \varphi_1(x) A_1(x)\, dx =$$

$$= \int_0^1 dy\, A_0(y) \int_0^1 K(x, y) A_1(x) \varphi(x)\, dx = \frac{1}{\lambda_0} \int_0^1 A_0(y) \varphi_1(y)\, dy = 0,$$

which was to be proved (under the assumptions of the theorem). The solutic of the first equation is written down in the form

$$v_1(x) = c_1\varphi_1(x) + P_1(x),$$

where $p_1(x)$ is a completely defined function

$$P_1(x) = \int_0^1 K(x, y) A_0(y)\, dy + \lambda_0 \int_0^1 \frac{D\begin{pmatrix} x & \xi_1 \\ s & \eta_1 \end{pmatrix}\lambda_0\Big)}{D\begin{pmatrix} \xi_1 \\ \eta_1 \end{pmatrix}\lambda_0\Big)} \left(\int_0^1 K(s, y) A_0(y)\, dy \right) ds,$$

a $D\left(\begin{matrix} \xi_1 \\ \eta_1 \end{matrix}\middle|\lambda_0\right)$ and $D\left(\begin{matrix} x & \xi_1 \\ s & \eta_1 \end{matrix}\middle|\lambda_0\right)$ are Fredholm minors of

the first and second order, ξ_1 and η_1 being so chosen that

$D\left(\begin{matrix} \xi_1 \\ \eta_1 \end{matrix}\middle|\lambda_0\right) \neq 0$, which can be done, since $\varphi_1(x)$ is an eigen-

function corresponding to an eigenvalue of λ_0 of the first rank.

To determine the unknown coefficient c_1 appearing in solution, we make use of the condition for the solvability of the second equation of the sequence

$$\int_0^1 dx A_1(x)\varphi_1(x) \int_0^1 K(x, y)\{A_1(y) v_1(y) + \lambda_0 A_2(y) v_1^2(y)\}\, dy = 0.$$

If we now take into account that $\lambda_0 \int_0^1 K(x, y) A_1(x)\varphi_1(x)\, dx = \varphi_1(y)$,

then we have:

$$\int_0^1 \{A_1(y) v_1(y) + \lambda_0 A_2(y) v_1^2(y)\}\varphi_1(y)\, dy = 0.$$

Substituting in this equation $v_1(y) = c_1\varphi_1(y) + P_1(y)$, we have a quadratic equation to determine c_1:

$$c_1^2 \int_0^1 \lambda_0 A_2(y)\varphi_1^3(y)\, dy + c_1 \int_0^1 \{A_1(y) + 2\lambda_0 A_2(y)\}\varphi_1^2(y)\, dy +$$

$$+ \int_0^1 \{A_1(y) P_1(y) + \lambda_0 A_2(y) P_1^2(y)\}\varphi_1(y)\, dy = 0.$$

Solving this equation, we find in general two different values c_{11} and c_{12}. Hence, the solution of the first equation has the form

$$v_{1,\alpha}(x) = c_{1,\alpha}\varphi(x) + P_1(x),$$

where $\alpha = 1, 2$.

The solution of the second equation of the sequence may be written in the form

$$v_{2,\alpha}(x) = c_{2,\alpha}\varphi_1(x) + P_{2,\alpha}(x),$$

where $P_{2,\alpha}(x)$ is a completely defined function (for every value of α), the form of which is similar to that of function $P_1(x)$. To determine $c_{2,\alpha}$ we employ the solvability condition for the third equation of the sequence

$$\int_0^1\int_0^1 K(x, y)\{A_1(y)v_2(y) + 2t_0A_2(y)v_1(y)v_2(y) + M_2\}\varphi_1(x)A_1(x)\,dx\,dy = 0,$$

where M_2 is independent of v_2, or after rearranging

$$\int_0^1 \{A_1(y)v_2(y) + 2t_0A_2(y)v_1(y)\cdot v_2(y) + M_2(y)\}\varphi_1(y)\,dy = 0.$$

Since $v_2(y)$ enters linearly into this equation, we have a linear equation for $c_{2,\alpha}$ which give a unique solution (naturally for each α):

$c_{2,\alpha} = c_{2,1}, c_{2,2}$. Hence,

$$v_{2,\alpha} = c_{2,\alpha}\varphi_1(x) + P_{2,\alpha}(x),$$
$$v_{3,\alpha} = c_{3,\alpha}\varphi_1(x) + P_{3,\alpha}(x),$$
$$\cdots\cdots\cdots\cdots$$
$$v_{n,\alpha} = c_{n,\alpha}\varphi_1(x) + P_{n,\alpha}(x),$$

where $c_{n,\alpha}$ is determined from the linear equation following from the solvability condition of the $(n + 1)$-th equation of the initial sequence.

Note. In the linear equations for $c_{n,\alpha}$, with $c_{n,\alpha}$ there is the expression

$$\int_0^1 \{A_1(y) + 2t_0A_2(y)v_{1,\alpha}(y)\}\varphi_1^2(y)\,dy,$$

which is assumed to be unequal to zero.

If we omit again the proof of the convergence of the series obtained, we have two functions:

$$v_\alpha = \mu v_{1,\alpha} + \mu^2 v_{2,\alpha} + \mu^3 v_{3,\alpha} + \ldots,$$

which are in general different, holomorphic with respect to $\lambda - \lambda_0$, and approach zero as $\lambda - \lambda_0$.

If we consider only real functions as constituting solutions of the initial nonlinear integral equation, then in the case of real $c_{1,\alpha}$, we have two solutions of the initial nonlinear equation, corresponding to one and the same value of λ and approaching $u_0(x)$ as $\lambda \to \lambda_0$, while if $c_{1,\alpha}$ are complex numbers, then in the neighborhood of λ_0 no solutions exist that are holomorphic with respect to $\lambda - \lambda_0$ and approach $u_0(x)$ as $\lambda \to \lambda_0$. Hence, the theorem is true.

Since $\int_0^1 A_0(y)\varphi_1(y)\,dy = 0$ is not always satisfied, it is of important to consider the case when $\int_0^1 A_0(y)\varphi_1(y)\,dy \neq 0$. The third theorem is devoted to this.

Theorem 3. If λ_0 is an eigenvalue of kernel $K(x, y)\,A_1(y)$, and $\int_0^1 A_0(y)\varphi_1(y)\,dy \neq 0$, where $\varphi_1(x)$ is an eigenfunction of the same kernel $K(x, y)\,A_1(y)$, corresponding to eigenvalue λ_0, then in the neighborhood of λ_0, there exist in general two solutions of the initial nonlinear equation, corresponding to one and the same λ.

2. Since

$$\int_0^1 A_0(y)\varphi_1(y)\,dy \neq 0,$$

we cannot satisfy $v(x)$ by a series of the form

$$\mu v_1 + \mu^2 v_2 + \mu^3 v_3 + \ldots,$$

since Fredholm's solvability conditions are not satisfied. We try to satisfy the equation

$$v(x)=\lambda_0\int_0^1 K(x,y)\{A_1(y)v(y)+A_2(y)v^2(y)+\dots\}dy+$$

$$+\mu\int_0^1 K(x,y)\{A_0(y)+A_1(y)v(y)+\dots\}dy$$

by a series of the form

$$v(x)=\mu^{\frac{1}{2}}v_1(x)+\mu v_2(x)+\mu^{\frac{3}{2}}v_3(x)+\dots.$$

Inserting this series in equation(β) and comparing the expressions for similar powers of μ, we have:

$$v_1(x)=\lambda_0\int_0^1 K(x,y)A_1(y)v_1(y)\,dy,$$

$$v_2(x)=\lambda_0\int_0^1 K(x,y)A_1(y)v_2(y)\,dy+\int_0^1 K(x,y)\{\lambda_0A_2(y)v_1^2(y)+A_0(y)\}\,dy,$$

$$v_3(x)=\lambda_0\int_0^1 K(x,y)A_1(y)v_3(y)\,dy+$$

$$+\int_0^1 K(x,y)\{2\lambda_0A_2(y)v_1(y)v_2(y)+M_3\}\,dy$$

$$v_n(x)=\lambda_0\int_0^1 K(x,y)A_1(y)v_n(y)\,dy+$$

$$+\int_0^1 K(x,y)\{2\lambda_0A_2(y)v_1(y)v_{n-1}(y)+M_n\}\,dy,$$

which M_n depends on $v_1, v_2, \dots, v_{n-2}$, but does not depend on v_{n-1}. The solution of the first equation of this sequence will be $v_1(x)=c_1\varphi_1(x)$. The coefficient c_1 is defined from the solvability condition of the second equation of the sequence:

$$\int_0^1\int_0^1 K(x,y)\{\lambda_0A_2(y)v_1^2(y)+A_0(y)\}A_1(x)\varphi_1(x)\,dx\,dy=0$$

or

$$\int_0^1 \{\lambda_0 A_2(y)\, v_1^2(y) + A_0(y)\}\, \varphi_1(y)\, dy = 0,$$

whence

$$c_1^2 = -\frac{\int_0^1 A_0(y)\, \varphi_1(y)\, dy}{\lambda_0 \int_0^1 A_2(y)\, \varphi_1^3(y)\, dy};$$

we assume that $c_1 = \pm c_0$, where c_0 is a real number. This is possible, if only $\int_0^1 A_2(y)\, \varphi_1^3(y)\, dy \neq 0,$ which was the condition assumed.

Thus, $v_1(x) = \pm c_0 \varphi_1(x)$ has two values where $c_0 \neq 0$, since by assumption, we have $\int_0^1 A_0(y)\, \varphi_1(y)\, dy \neq 0.$ The solution of the second equation of the sequence given above has the form

$$v_2(x) = c_2 \varphi_1(x) + P_2(x),$$

where $P_2(x)$ is a completely determined function.

Coefficient c_2 is defined from the solvability condition of a third equation:

$$\int_0^1 \int_0^1 K(x, y)\, \{2\lambda_0 A_2(y)\, v_1(y)\, v_2(y) + M_3\}\, A_1(x)\, \varphi_1(x)\, dx\, dy = 0,$$

or

$$\int_0^1 \{2\lambda_0 A_2(y)\, v_1(y)\, v_2(y) + M_3\}\, \varphi_1(y)\, dy = 0,$$

$$c_2 \int_0^1 2\lambda_0 \cdot c_1 A_2(y)\, \varphi_1^3(y)\, dy + \int_0^1 \{2\lambda_0 A_2(y)\, v_1(y)\, P_2(y) + M_3(y)\}\, \varphi_1(y)\, dy = 0,$$

from which c_2 is univocally defined for each c_1, since

$$2\lambda_0 c_1 \int_0^1 A_2(y)\, \varphi_1^3(y)\, dy \neq 0.$$

Now in turn $v_n(x) = c_n \varphi_1(x) + P_n(x),$, where c_n is determined from the solvability of the (n + 1)-th equation of the sequence, which reduces to the equation:

$$c_n \int_0^1 2\lambda_0 c_1 A_2(y)\, \varphi_1^3(y)\, dy + \int_0^1 \{2\lambda_0 A_2(y)\, v_1(y)\, P_n(y) + M_3\}\, \varphi_1(y)\, dy = 0.$$

Thus, a system of two functions $v_{1\alpha}(x), v_{2\alpha}(x), \ldots$ has been constructed. If c_1 is real, then the series $\sqrt{\lambda - \lambda_0}\, v_{1\alpha} + (\lambda - \lambda_0)\, v_{2\alpha} + \ldots$ will be real for $\lambda \geqslant \lambda_0$. It is easily shown that if c_1 is an imaginary number, the series will be real for

$$\lambda \leqslant \lambda_0.$$

We have thus shown that if $\int_0^1 A_2(y)\, \varphi_1^3(y)\, dy \neq 0$, the initial nonlinear integral equation, has a) two different solutions corresponding to a single $\lambda \geqslant \lambda_0$, and no solution corresponding to $\lambda < \lambda_0$ on the condition that $\int_0^1 A_0(y)\, \varphi_1(y)\, dy$ и $\lambda_0 \int_0^1 A_2(y)\, \varphi_1^3(y)\, dy$ are different in sign; b) two solutions for $\lambda < \lambda_0$ and none for $\lambda > \lambda_0$, if $\int_0^1 A_0(y)\, \varphi_1(y)\, dy$ и $\lambda_0 \int_0^1 A_2(y)\, \varphi_1^3(y)\, dy$ have the same signs. The third theorem has been proved (omitting the proof of the convergence of the series obtained).

To conclude, we may note the following basic differences of the solutio of nonlinear integral equations of the Hammerstein type from solutions of linear integral equations.

1) A linear integral equation has a discrete spectrum of eigenvalues for λ (neglecting special kernels and infinite limits of integration), whereas for nonlinear integral equations, the spectrum is continuous in both cases.

2) The solutions of nonlinear integral equations of Hammerstein's type have points of branching solutions. Employing the apparatus for finding these branches, the character of the new branchings can be explained.

The above is a brief summary of Nazarov's work.

CHAPTER V

SOLUTIONS DEPENDING ON TIME

In this chapter, we are to consider solutions of nonstationary problems, when f clearly is dependent on the time. We consider first the case of linearized equations. It is obvious from the outset that in solving some problems, we can consider only distribution functions differing little from those corresponding to a spatial-homogeneous distribution. In this case, linearization of the initial equation is possible. Without as yet specifying the character of the forces and the form of the initial distribution function, we pose two problems. First, Cauchy's problem: with a given initial state, to define the dependence of function f on the time. As we have already stated, the solution of this problem brings out only a limited class of functions with specified initial moment of time. The second problem involves the finding of solutions for which the initial moment of time is not specified. In particular, for instance, the purely periodic solutions of the type $\exp\{i\omega t - i\boldsymbol{k}\boldsymbol{r})$, and likewise the stationary solutions $\left(\frac{\partial}{\partial t} = 0\right)$ belong to the second class of solutions. The concrete application of both types of solutions to physical problems is given in the second part of the book.

Section 19. Linearization of the Equations

We consider the initial equation for particles of one kind:

$$-\frac{\partial f}{\partial t} = \operatorname{div}_r \boldsymbol{v} f - \frac{1}{m}\operatorname{div}_v(\operatorname{grad}_r V) f, \qquad (19.1)$$

where

$$V(\boldsymbol{r},\ t) = \int_{-\infty}^{\infty} K(|\boldsymbol{r} - \boldsymbol{r}'|)\,\rho(\boldsymbol{r}',\ t)\,d\boldsymbol{r}'.$$

This equation, as has been noted repeatedly, has an exact solution corresponding to a uniform distribution of the particles of space. We obtain the solution setting $f = f(v^2)$, and consequently,

$$\rho = \int_{-\infty}^{\infty} f(\boldsymbol{v}^2)\,d\boldsymbol{v} = \text{const.} = a;$$

we then have:

$$V(\boldsymbol{r}) = \int_{-\infty}^{\infty} K(|\boldsymbol{r} - \boldsymbol{r}'|)\,\text{const.}\,d\boldsymbol{r}' = 4\pi a\int_0^{\infty} K(\rho)\,\rho^2 d\rho = \text{const.},$$

whence

$$\boldsymbol{F} = -\operatorname{grad}_r V \equiv 0.$$

We have assumed in this that

$$\int_{-\infty}^{\infty} K(|r-r'|)\,dr' < \infty. \tag{19.2}$$

We now seek solutions little different from the uniform distribution; in this we may assume

$$f(r, v, t) = f_0(v^2) + \varphi(r, v, t),$$

where

$$\varphi \ll f_0.$$

We see that φ is a perturbation of function f_0. Substituting this expression in (19.1) and retaining only those terms that are linear in φ, we obtain:

$$\frac{\partial\varphi}{\partial t} + v\,\mathrm{grad}_r\varphi =$$

$$= \frac{1}{m}\,\mathrm{grad}_v f_0\,\mathrm{grad}_r \int_{-\infty}^{\infty} K(|r-r'|) \int_{-\infty}^{\infty} \varphi(r', v', t)\,dv'\,dr'. \tag{19.3}$$

Solution (19.2) is not satisfied in the case of Coulomb forces, since in that case

$$\int_{-\infty}^{\infty} \frac{1}{|r-r'|}\,dr' = \infty.$$

Consequently, an equation describing particles interacting electrodynamicall[y] must be studied separately. For this purpose, we consider the case of charged particles of two kinds, where for each kind $N > 1$.

$$\left.\begin{aligned}
-\frac{\partial f_1}{\partial t} &= \mathrm{div}_r\, v f_1 + \mathrm{div}_v \frac{e_1}{m_1}\left(e + \frac{1}{c}[vh]\right) f_1, \\
-\frac{\partial f_2}{\partial t} &= \mathrm{div}_r\, v f_2 + \mathrm{div}_v \frac{e_2}{m_2}\left(e + \frac{1}{c}[vh]\right) f_2, \\
\mathrm{div}\, e &= 4\pi\left\{e_1\int f_1\,dv + e_2\int f_2\,dv\right\}, \\
\mathrm{rot}\, h - \frac{1}{c}\frac{\partial e}{\partial t} &= \frac{4\pi}{c}\left\{e_1\int v f_1\,dv + e_2\int f_2\,dv\right\}
\end{aligned}\right| \quad (1\ldots$$

$$\left(\mathrm{div}\, h = 0,\ \mathrm{rot}\, e + \frac{1}{c}\frac{\partial h}{\partial t} = 0\right).$$

In the limiting case $m_2 \to \infty$, the equation may be greatly simplified. In this case, the influence of the particles of the first kind on a particle of the second kind disappears:

$$-\frac{\partial f_2}{\partial t} = \operatorname{div}_r \boldsymbol{v} f_2.$$

The solution of this equation is

$$f_2(\boldsymbol{r}, \boldsymbol{v}, t) = \Omega(t_0, \boldsymbol{r} - \boldsymbol{v}(t - t_0), \boldsymbol{v}),$$

where Ω is an arbitrary function. In a particular case, this function may depend only on the modulus of the velocity

$$f_2 = f_2(v^2),$$

constituting a uniform distribution of particles throughout the space with an isotropic and velocity distribution. Then

$$j_2 = e_2 \int \boldsymbol{v} f_2 \, d\boldsymbol{v} = 0, \quad \rho_2 = e_2 \int f_2 \, d\boldsymbol{v} = -\rho_2^0$$

and for $f_1 = f$, we have:

$$\left.\begin{aligned} -\frac{\partial f}{\partial t} &= \operatorname{div}_r \boldsymbol{v} f + \operatorname{div}_v \frac{e}{m}\left(\boldsymbol{e} + \frac{1}{c}[\boldsymbol{v}\boldsymbol{h}]\right) f, \\ \operatorname{div} \boldsymbol{e} &= 4\pi \left\{ e \int f \, d\boldsymbol{v} - \rho_2^0 \right\}, \\ \operatorname{rot} \boldsymbol{h} - \frac{1}{c}\frac{\partial \boldsymbol{e}}{\partial t} &= \frac{4\pi}{c} e \int \boldsymbol{v} f \, d\boldsymbol{v}. \end{aligned}\right\} \qquad (19.5)$$

Equation (19.5) describes the behavior of a system of particles of one kind moving in the field of uniform distribution of particles of the second kind. Equation (19.5) likewise has an exact solution with a uniform distribution, but on condition of complete compensation of the space charges (that is, in the absence of the field):

$$f = f_0(v^2), \quad e \int f_0 \, d\boldsymbol{v} = \rho_2^0.$$

Once again, assuming that the deviations from a uniform distribution are small $f = f_0(v^2) + \varphi(\boldsymbol{r}, \boldsymbol{v}, t)$ and neglecting all terms of higher order in φ, we obtain a linearized equation of the form

$$\left.\begin{aligned} -\frac{\partial \varphi}{\partial t} &= \operatorname{div}_r \boldsymbol{v}\varphi + \operatorname{div}_v \frac{e}{m}\left(\boldsymbol{e} + \frac{1}{c}[\boldsymbol{v}\boldsymbol{h}]\right) f_0, \\ \operatorname{div} \boldsymbol{e} &= 4\pi e \int \varphi \, d\boldsymbol{v}, \quad \operatorname{rot} \boldsymbol{h} - \frac{1}{c}\frac{\partial \boldsymbol{e}}{\partial t} = \frac{4\pi e}{c} \int \boldsymbol{v}\varphi \, d\boldsymbol{v}, \\ &\left(\operatorname{div} \boldsymbol{h} = 0, \quad \operatorname{rot} \boldsymbol{e} + \frac{1}{c}\frac{\partial \boldsymbol{h}}{\partial t} = 0\right). \end{aligned}\right\} \qquad (19.6)$$

A characteristic of this linearized system is that it is possible to divide fields obtained in the solution into a vortex-free field and a vortical field (longitudinal and transfer waves). Hence, the general solution is made up additively of the two solutions taken separately. In order to show the correctness of this in our case, and find the equations describing both the portions, we represent the vectors entering into the system (19.6) in the form of a sum of two vectors, one vortical and one non-vortical, which is always possible:

$$e = e^{(l)} + e^{(t)},$$

where

$$\operatorname{rot} e^{(l)} = 0 \quad \text{и} \quad \operatorname{div} e^{(t)} = 0,$$

$$h = h^{(t)},$$

since div $\underline{h} = 0$. Consequently, we divide the vector of mean velocity as well

$$\bar{v} = \bar{v}^{(l)} + \bar{v}^{(t)},$$

but inasmuch as $\bar{v} = \int v\varphi\, dv \Big/ \int f_0\, dv$, this conception of a mean velocity will be possible only if the distribution function φ itself is represented as in the form of a sum of two parts:

$$\varphi = \varphi^{(l)} + \varphi^{(t)},$$

for which

$$n\bar{v}^{(l)} = \int v\varphi^{(l)}\, dv, \quad n\bar{v}^{(t)} = \int v\varphi^{(t)}\, dv,$$

where

$$n = \int f_0\, dv.$$

In order to obtain equations each of which describes one of the two parts, we consider two particular solutions of the system, when there are present:

a) Only a vortex-free field, that is, $e^{(l)} \neq 0$, $e^{(t)} = h^{(t)} = 0$, and in addition, $\bar{v}^{(t)} = 0$, $\bar{v}^{(l)} \neq 0$, which is possible if $\varphi^{(t)} = 0$, $\varphi^{(l)} \neq 0$;

b) Only a vortical field $e^{(l)}, \varphi^{(l)}, \bar{v}^{(l)}$ being equal to 0, and $e^{(t)}$, $\varphi^{(t)}, \bar{v}^{(t)} \neq$

Here, in case a), we have

$$\left.\begin{aligned}
-\frac{\partial\varphi^{(l)}}{\partial t} &= \operatorname{div}_r \boldsymbol{v}\varphi^{(l)} + \operatorname{div}_v \frac{e}{m}\boldsymbol{e}^{(l)} f_0,\\
\operatorname{div}\boldsymbol{e}^{(l)} &= 4\pi e \int \varphi^{(l)}\, d\boldsymbol{v},\\
-\frac{\partial \boldsymbol{e}^{(l)}}{\partial t} &= 4\pi e \int \boldsymbol{v}\varphi^{(l)}\, d\boldsymbol{v},
\end{aligned}\right\} \tag{19.7}$$

where likewise by postulate $\operatorname{rot}\boldsymbol{e}^{(l)} = 0$. For the vortical portion, we obtain:

$$\left.\begin{aligned}
-\frac{\partial\varphi^{(t)}}{\partial t} &= \operatorname{div}_r \boldsymbol{v}\varphi^{(t)} + \frac{e}{m}\operatorname{div}_v \frac{e}{m}\left(\boldsymbol{e}^{(t)} + \frac{1}{c}[\boldsymbol{v}\boldsymbol{h}^{(t)}]\right) f_0,\\
\operatorname{rot}\boldsymbol{h}^{(t)} - \frac{1}{c}\frac{\partial \boldsymbol{e}^{(t)}}{\partial t} &= \frac{4\pi e}{c}\int \boldsymbol{v}\varphi^{(t)}\, d\boldsymbol{v},\\
\operatorname{rot}\boldsymbol{e}^{(t)} + \frac{1}{c}\frac{\partial \boldsymbol{h}^{(t)}}{\partial t} &= 0,\\
\operatorname{div}\boldsymbol{e}^{(t)} = 0,\ \operatorname{div}\boldsymbol{h}^{(t)} &= 0.
\end{aligned}\right\} \tag{19.8}$$

The sum of the solution of system (19.7) and (19.8) will, because of the linearity of the equations, constitute a solution of the initial system (19.6). In addition, the sum of solutions of these equations may be chosen in such a way as to satisfy any initial conditions set up for the general system, since these conditions, too, may be represented by a corresponding sum of vortical and vortex-free members.

Section 20. Cauchy's Problem

As we have already pointed out in Sec. 7, the formulation of the problem with arbitrary initial conditions, from the physical point of view, implies that the dynamic system is subjected to actions at the initial moment of time that may not previously have been comprised in the equations describing the dynamic system. There, too, the question was raised of finding solutions in which the initial moment of time is essentially specified. This occurs, for example, if Cauchy's problem is solved by means of the Laplace-Mellin transformation. For, if we introduce instead of the unknown function a new function:

$$\varphi(p) = \int_0^\infty e^{-pt} f(t)\, dt,$$

by the conversion formula, we find that the function

$$f(t) = \begin{cases} 0 & \text{for} \quad t < 0,\\ \dfrac{1}{2\pi i}\displaystyle\int_{\sigma - i\infty}^{\sigma + i\infty} e^{pt}\varphi(p)\, dp & \text{for} \quad t > 0. \end{cases}$$

if f has the meaning of a perturbation, then such a solution presupposes that for all $t < 0$ this perturbation is precisely equal to 0, and exists only for $t > 0$. Obviously, if we are interested in solutions of the type $\exp(i\omega t)$ over the entire t interval of $-\infty$ to $+\infty$, then the method of Laplace-Mellin cannot be used.

We consider Cauchy's problem as applied to linearized equation (19.3) - (19.6). For the time being, we shall not specify either the character of the forces of interaction or the form of the initial distribution function. We had:

$$\frac{\partial\varphi}{\partial t} + \boldsymbol{v}\,\mathrm{grad}_r\varphi =$$

$$= \frac{1}{m}\,\mathrm{grad}_v f_0 \cdot \mathrm{grad}_r \int_{-\infty}^{\infty} K(|\boldsymbol{r}-\boldsymbol{r}'|) \int_{-\infty}^{\infty} \varphi(\boldsymbol{r}', \boldsymbol{v}', t)\, d\boldsymbol{r}'\, d\boldsymbol{v}'. \quad (20.1)$$

As we did above (Sec. 9), we write down the equation in functional form, not containing any derivatives for the coordinates:

$$\varphi(\boldsymbol{r}, \boldsymbol{v}, t) = \varphi(t_0, \boldsymbol{r} - \boldsymbol{v}(t-t_0), \boldsymbol{v}) +$$

$$+ \int_{t_0}^{t} d\tau \frac{1}{m} \nabla_v f_0 \nabla_r \int_{-\infty}^{\infty} K(|\boldsymbol{r} - \boldsymbol{v}(t-\tau) - \boldsymbol{r}'|)\, \rho(\boldsymbol{r}', \tau)\, d\boldsymbol{r}'. \quad (20.2)$$

We pass to the "space density" $\rho(\boldsymbol{r}, t) = \int \varphi\, d\boldsymbol{v}$, etc.
that is, we integrate the equation with respect to the velocities. We obtain:

$$\left.\begin{aligned} &\rho(\boldsymbol{r}, t) = \Psi(t_0, t, \boldsymbol{r}) + \\ &+ \int_{-\infty}^{\infty} d\boldsymbol{v} \frac{1}{m} \nabla_v f_0 \int_{t_0}^{t} \nabla_r \int_{-\infty}^{\infty} K(|\boldsymbol{r} - \boldsymbol{v}(t-\tau) - \boldsymbol{r}'|)\, \rho(\boldsymbol{r}', \tau)\, d\boldsymbol{r}'\, d\tau, \\ &\Psi(t_0, t, \boldsymbol{r}) = \int_{-\infty}^{\infty} \varphi(t_0, \boldsymbol{r} - \boldsymbol{v}(t-t_0), \boldsymbol{v})\, d\boldsymbol{v}. \end{aligned}\right\} \quad (20.3)$$

For $t = t_0$, let there be given a distribution function $\varphi(\boldsymbol{r}, \boldsymbol{v}, 0)$. The solution for $\varphi(\boldsymbol{r}, \boldsymbol{v}, t)$ is then sought in the form of the superposition of $\varphi_k(\boldsymbol{v}, t)\, e^{ikr}$; here $\rho(\boldsymbol{r}, t)$ will be a superposition of functions of the type $\rho_k(t)\, e^{ikr}$. Upon substitution of (20.3), the following integrals will emerge in the expression:

$$\int_{-\infty}^{\infty} K(|r - v(t-\tau) - r'|)\, e^{ikr'}\, dr' = e^{ik(r - v(t-\tau))} \sigma(k). \qquad (20.4)$$

where

$$\sigma(k) = \int_{-\infty}^{\infty} K(|r - v(t-\tau) - r'|)\, e^{-ik(r - v(t-\tau) - r')} dr' =$$

$$= 4\pi \int_0^{\infty} K(\rho) \frac{\sin k\rho}{k\rho} \rho^2 d\rho, \qquad (20.5)$$

$$\int_{-\infty}^{\infty} \varphi_k(v_0)\, e^{-ikv(t-t_0)}\, dv = F_k(t), \qquad (20.6)$$

$$\frac{\sigma(k)}{m} \int_{-\infty}^{\infty} e^{-ikv(t-\tau)} i\,(k\nabla_v) f_0\, dv = G_k(t-\tau). \qquad (20.7)$$

Here we have postulated that $\rho_k(t)$ и $f_0(v^2)$ satisfy the conditions of permutability of the limits of integration with respect to and v.

To define $\rho_k(t) = q_k(t)$, we obtain a Volterra integral equation of the second class:

$$q_k(t) = F_k(t) + \int_{t_0}^{t} G_k(t-\tau)\, q_k(\tau)\, d\tau, \qquad (20.8)$$

where $F_k(t)$ is expressed by the initial perturbation $\varphi_k(v, 0)$. and $G_k(t)$ by an unperturbed function f_0.

The solution of Volterra's equation is well known from the theory of integral equations+. Its solution is given by the following integral:

$$q_k(t) = \frac{1}{2\pi i} \int_{\sigma_0 - i\infty}^{\sigma_0 + i\infty} e^{pt} \frac{\overline{F}_k(p)}{1 - \overline{G}_k(p)}\, dp, \qquad (20.9)$$

+ See, for instance, G. Müntz, Integral'nia uravneniya Volterra /Volterra integral equations/ GTTI, 1934.

where

$$\bar{F}_k(p) = \int_0^\infty e^{-pt} F_k(t)\, dt \quad \text{и} \quad \bar{G}_k(p) = \int_0^\infty e^{-pt} G_k(t)\, dt. \tag{20.10}$$

The integration is over the right semi-plane of the variable $p = -\sigma - i\infty$ along a straight line parallel to the axis of imaginaries.

We now consider a particular case, taking as the initial perturbation

$$\bar{\varphi}_k(v,\ 0) = a_k f_0(\xi),$$

where

$$f_0(\xi) = \frac{N}{\pi c}\frac{1}{1+\frac{\xi^2}{c^2}}, \quad c^2 = \frac{2kT}{m}. \tag{20.11}$$

This function may be considered as an approximate expression of Maxwell's distribution law, which applies for small velocities, since we may write:

$$e^{-x^2} = \frac{1}{1 - \frac{x^2}{1!} + \frac{x^4}{2!} + \ldots}.$$

The advantage of formula (20.11) appears at once in calculations, since all the integrals appearing in it are exact and we obtain a solution without supplementary approximations. For,

$$G_k(t-\tau) = \frac{\varphi(k)}{m}\int_{-\infty}^{\infty} e^{-ik\xi(t-\tau)}\, ik\, \frac{\partial}{\partial \xi}\left(\frac{N}{\pi c}\frac{1}{1+\frac{\xi^2}{c^2}}\right) d\xi =$$

$$= -\frac{\varphi(k)N}{m} k^2 (t-\tau)\, e^{-kc(t-\tau)}$$

$$F_k(t) = \int_{-\infty}^{\infty} a_k e^{-ik\xi t}\frac{N}{\pi c}\frac{d\xi}{1+\frac{\xi^2}{c^2}} = a_k N e^{-kct}.$$

and, consequently:

$$\bar{F}_k(p) = \int_0^\infty e^{-pt} F_k(t)\, dt = a_k N \frac{1}{p - kc}.$$

$$\bar{G}_k(p) = \int_0^\infty e^{-pt} G_k(t)\, dt = -\frac{\varphi(k)N}{m} k^2 \frac{1}{(p - kc)^2}$$

We consider the case $\sigma(k) > 0$ (this takes place for example for the Coulomb law of forces). Then, designating

$$\frac{\sigma(k)\, k^2 N}{m} = \omega_0^2,$$

we have:

$$\frac{1}{2\pi i} \int_{\sigma_0 - i\infty}^{\sigma_0 + i\infty} e^{pt} \frac{p + kc}{(p + kc)^2 + \omega_0^2}\, dp = e^{-kct} \cos \omega_0 t \tag{20.12}$$

and hence, finally:

$$q_k(t) = a_k N e^{-kct} \cos \omega_0 t. \tag{20.13}$$

Thus, the solution of Cauchy's problem in this case leads to a vibrational-diffusional-nature for the propagation of perturbations in density, in which the characteristic frequencies and the decrement of damping for any forces, in general, depend essentially on k.

In the case $\sigma(k) < 0$, writing $\frac{\sigma(k)\, Nk^2}{m} = -\omega_1^2,$

we have

$$q_k(t) = a_k N \frac{1}{2\pi i} \int_{\sigma_0 - i\infty}^{\sigma_0 + i\infty} e^{pt} \frac{p + kc}{(p + kc)^2 - \omega_1^2}\, dp = a_k N e^{-kct} \operatorname{ch} \omega_1 t. \tag{20.14}$$

Thus, in the second case the solution contains the hyperbolic cosine, and the perturbation increases as time goes on, indicating the instability of the initial state with uniform density.

We see from equations (20.12) and (20.14) that the properties of the resulting solution will be defined by the position of the poles of the function under the integral in the plane of the complex variable. The equations

$$(p + kc)^2 + \omega_0^2 = 0$$

and

$$(p + kc)^2 - \omega_1^2 = 0$$

define some dependence of p on k. Hereinafter, we shall call these equations the <u>dispersion equations</u>, since they define the nature of a dispersion of perturbations of the type

$$\exp(pt + ikr).$$

In the general case, the poles of the expression under the integral sign are defined by the equation (from 20.9)

$$\overline{G}_k(p) - 1 = 0$$

or in a developed form

$$\int_0^{\infty} e^{-pt}\, dt \frac{\sigma(k)}{m} \int_{\infty}^{\infty} e^{-ikvt}\, ik\nabla_v f_0\, dv = 1,$$

and since

$$\int_0^{\infty} e^{-pt-ikvt} dt = -\frac{1}{p+ikv} \qquad (\operatorname{Re} p > 0),$$

we have:

$$-\frac{i\sigma(k)}{m} \int_{-\infty}^{\infty} \frac{k\nabla_v f_0}{p+ikv}\, dv = 1. \tag{20.15}$$

Here we have assumed that the unperturbed function f_0 satisfies the conditions of permutability of the limits of integration with respect to v and t.

Equation (20.15) applies for any forces and any initial distribution function. If the initial function is precisely the Maxwell function:

$$f_0(v) = N\left(\frac{m}{2\pi\theta}\right)^{\frac{3}{2}} \exp\left(-\frac{\varepsilon}{\theta}\right),$$

then the dependence of p on k cannot be expressed by means of dispersion equation (20.15) in elementary functions. However, for sufficiently long waves (k) we may obtain the result in the following way: expanding the expression under the integral as a power series in k and confining ourselves to the first nonvanishing term, we have:

$$\int_{-\infty}^{\infty} \frac{k\nabla_v f_0}{p+ikv}\, dv \sim \int_{-\infty}^{\infty} (p - ikv) \frac{k\nabla_v f_0}{p^2}\, dv = -\frac{1}{p^2} \int_{-\infty}^{\infty} ikv\, k\nabla_v f_0\, dv \tag{20.16}$$

or, without any violation of generality, choosing a system of coordinates with an x axis along k:

$$-\frac{\sigma(k)}{m}\frac{k^2}{p^2}\int_{-\infty}^{\infty}\xi\frac{\partial f_0}{\partial\xi}\,d\xi\,d\eta\,d\zeta=1.$$

Finally, substituting the Maxwell distribution for f_0, we have:

$$-\frac{\sigma(k)k^2}{mp^2}\left(\frac{m}{2\pi\theta^3}\right)^{\frac{1}{2}}\int_{-\infty}^{\infty}\xi^2 e^{-\frac{m\xi^2}{2\theta}}\,d\xi=1$$

or

$$-\frac{N\sigma(k)}{m}\frac{k^2}{p^2}\left(\frac{m}{2\pi\theta^3}\right)^{\frac{1}{2}}\left(\frac{\theta}{m}\right)^{\frac{3}{2}}\int_{-\infty}^{\infty}x^2 e^{-\frac{x^2}{2}}\,dx=1$$

and, finally,

$$\left.\begin{aligned} p^2 &= -\frac{N\sigma(k)}{m}k^2, \\ \sigma(k) &= 4\pi\int_0^{\infty}K(\rho)\,\frac{\sin k\rho}{k\rho}\,d\rho. \end{aligned}\right\} \qquad (20.17)$$

In this way, there will correspond to $\sigma(k) > 0$ periodic solutions (p being purely imaginary), and to $\sigma(k) < 0$, exponentially decreasing or increasing solutions.

Section 21. Solutions of the Type $e^{i\omega t}\varphi(x, y, z)$

We now proceed to consider to consider solutions in which the initial moment of time is not specified.

In order to obtain solutions of the type exp $(i\omega t)$, we must refrain from posing Cauchy's problem, and seek the solutions over an infinte time interval. Three cases should be distinguished:

1. $-\infty < t$,
2. $t < +\infty$,
3. $-\infty < t < +\infty$.

Correspondingly, we must consider three types of solutions. Two ways of proceeding appear open to us. We may directly solve the initial equations in the class of functions of the type exp ($i\omega t$), where ω is a real number. It is also possible from the outside to represent the initial equations in a special way after the type of lagging, leading and, finally, the half sum of lagging and leading potentials with infinite limits of integration with respect to the time.

At the outset, we consider the case of particles interacting according to Coulomb's law, and then generalize the result to any central forces. In order to cover asymmetrical distribution functions of the velocities as well, we shall take

$$f = f_0(\xi + \xi_0),$$

where ξ_0 is the mean velocity of translation of the particle and $f_0(\xi)$ a symmetrical function. Confining ourselves to the plane problem, the initial linearized equation may be written in the form:

$$\frac{\partial\varphi}{\partial t} + \xi\frac{\partial\varphi}{\partial x} - \frac{e}{m}\frac{\partial V(x, t)}{\partial x}\frac{\partial f_0}{\partial\xi} = 0, \tag{21.1}$$

$$\Delta V = -4\pi e\int_{-\infty}^{\infty}\varphi(x, \xi, t)\,d\xi.$$

We shall seek its solution in the form

$$\varphi = g_p(\xi)\,e^{i\omega t - px}, \tag{21.2}$$
$$V = ae^{i\omega t - px}.$$

Substitution gives:

$$(i\omega - p\xi)\,g_p = \frac{e}{m}(-p)\,a\frac{\partial f_0}{\partial\xi}, \tag{21.3}$$

$$p^2 = -4\pi e\int_{-\infty}^{\infty} g_p(\xi)\,d\xi, \tag{21.4}$$

whence

$$p^2 = -\frac{4\pi e^2}{m}(-p)\int_{-\infty}^{\infty}\frac{\frac{\partial f_0}{\partial\xi}\,d\xi}{i\omega - p\xi}$$

or

$$p^2 = -\frac{4\pi e^2}{m}\int_{-\infty}^{\infty}\frac{\frac{\partial f_0}{\partial\xi}\,d\xi}{\xi - \frac{i\omega}{p}}. \tag{21.5}$$

This condition for the solvability of the initial equation in the class of functions $\exp(i\omega t - px)$ was earlier designated as a dispersion equation, inasmuch as it connects ω and p. This equation (for a given real ω) has different meaning depending on where the number $\frac{i\omega}{p}$ is situated on the complex plane: on the upper or lower part of the plane or on the real axis.

We write the integral in the dispersion equation in the form

$$\int_{-\infty}^{\infty} \frac{\frac{\partial f_0(\xi+\xi_0)}{\partial \xi} \cdot d\xi}{\xi - \frac{i\omega}{p}} = \int_{-\infty}^{\infty} \frac{\frac{\partial f_0(x)}{\partial x} dx}{x - \left(\frac{i\omega}{p} + \xi_0\right)} = \int_{-\infty}^{\infty} \frac{\Phi(x)\, dx}{x - z}, \tag{21.6}$$

where

$$\xi + \xi_0 = x, \quad z = \frac{i\omega}{p} + \xi_0, \quad \Phi(x) = \frac{\partial f_0}{\partial x}.$$

This integral, as is known from the theory of functions of a complex variable, is of the Cauchy type integral. First, we obtain its limiting value, when z approaches the real axis, remaining on the upper (lower) portion of the plane of the complex variable. Physically, this case is realized by the passing in the limit to the static case ($\omega \longrightarrow 0$, $\omega > 0$). To this end, we break the integral up into three parts:

$$\int_{-\infty}^{\infty} \frac{\Phi(x)\, dx}{x - z} = \int_{-\infty}^{-L} \frac{\Phi(x)\, dx}{x - z} + \int_{-L}^{+L} \frac{\Phi(x)\, dx}{x - z} + \int_{+L}^{\infty} \frac{\Phi(x)\, dx}{x - z}, \tag{21.7}$$

and requires z to approach ξ_0, assuming that ξ_0 lies in the interval (- L, + L). At $L \rightarrow \infty$, the extreme integrals disappear, since they do not contain any singularity in the denominator; the mean integral, according to Privalov's theorem[+], approaches the limit

$$\int_{-L}^{+L} \frac{\Phi(x)\, dx}{x - \xi_0} \pm i\pi\Phi(\xi_0). \tag{21.8}$$

The sign (+) occurs in the case when z approaches the real axis while remaining in the upper portion of the plane, the sign (-), in the lower.

[+] I. Privalov, Integral Koshi, /Koshi's Integral/, Izvestiya Saratovskovo universiteta (1918). N. Muskhelishvili, Singularnie integral'nie uravneniya /Singular integral equations/ (1946).

Thus, finally,

$$\lim_{z \to \xi_0} \int_{-\infty}^{\infty} \frac{\Phi(x)\,dx}{x - z} = \int_{-\infty}^{\infty} \frac{\Phi(x)\,dx}{x - \xi_0} \pm i\pi\Phi(\xi_0). \tag{21.9}$$

Here it is assumed that the curve described by the point z does not touch the real axis; the contrary case, the limiting value of the integral may be different.

In the formula obtained, the integral is understood in the sense of the principal significance:

$$\int_{-\infty}^{\infty} \frac{\Phi(x)\,dx}{x - \xi_0} = \lim_{\varepsilon \to 0} \left\{ \int_{-\infty}^{\xi_0 - \varepsilon} \frac{\Phi(x)\,dx}{x - \xi_0} + \int_{\xi_0 + \varepsilon}^{\infty} \frac{\Phi(x)\,dx}{x - \xi_0} \right\}. \tag{21.10}$$

This defines the limitations on the class of functions Φ (x), for which the limit that has been indicated exists and has a finite value. If $\frac{i\omega}{p}$ lies on the real axis (which occurs if, for example, p = ik is an imaginary number), the formula with the limiting value (21.9) makes it possible to calculate the integral:

$$\int_{-\infty}^{\infty} \frac{\Phi(x)\,dx}{x - \xi_0} = \frac{1}{2} \left(\lim_{\substack{z \to \xi_0 \\ \operatorname{Im} z > 0}} \int_{-\infty}^{\infty} \frac{\Phi(x)\,dx}{x - z} + \lim_{\substack{z \to \xi_0 \\ \operatorname{Im} z < 0}} \int_{-\infty}^{\infty} \frac{\Phi(x)\,dx}{x - z} \right). \tag{21.11}$$

Thus, in considering the dispersion equation, we should distinguish three cases:

$$p^2 = -\frac{4\pi e^2}{m} \left\{ \int_{-\infty}^{\infty} \frac{\frac{\partial f_0}{\partial x}\,dx}{x - \xi_0} \pm i\pi \left(\frac{\partial f_0}{\partial x} \right)_{x = \xi_0} \right\} \tag{21.12}$$

and

$$p^2 = -\frac{4\pi e^2}{m} \int_{-\infty}^{\infty} \frac{\frac{\partial f_0}{\partial x}\,dx}{x - \xi_0}. \tag{21.13}$$

In the first two cases, we are concerned with monochromatic oscillations of exp $(i\omega t-px)$, $\omega\to 0$ (with $\omega\to 0$), but a complex value p, which corresponds to wave forms oscillatorily increasing (or decreasing) in space. In the third case, from the form of f_0: $p^2 > 0$ or $p^2 < 0$. Since p^2 is a real number here, we have either purely exponential or purely periodic solutions. All these cases are discussed in the appendices.

We now pass to the nonstationary problem, that is, we assume $\omega \neq 0$. To this end, we transform the dispersion equation into a form more convenient for applications.

We assume at first Im $z > 0$; then

$$\int_{-\infty}^{\infty}\frac{\Phi(x)\,dx}{x-z}=\int_{-\infty}^{\infty}\Phi(x)\,dx\,(-i)\int_{\infty}^{0}e^{-i(x-z)\tau}\,d\tau=$$

$$=(-i)\int_{\infty}^{0}e^{iz\tau}\,d\tau\int_{-\infty}^{\infty}\Phi(x)\,e^{-ix\tau}\,dx. \qquad (21.14)$$

This is possible, since by assumption Im $z > 0$ and consequently

$e^{iz\tau}|_{\tau\to+\infty}\to 0$. Since $f_0(x)$ is a symmetrical function

by definition, then $\Phi(x)=\frac{\partial f_0}{\partial x}$ — is an odd function, and

therefore

$$\int_{-\infty}^{\infty}\Phi(x)\,e^{-i\tau x}\,d\tau=-i\int_{-\infty}^{\infty}\Phi(x)\sin\tau x\,dx \qquad (21.15)$$

and

$$\int_{-\infty}^{\infty}\frac{\Phi(x)\,dx}{x-z}=-\int_{0}^{\infty}e^{-iz\tau}\,d\tau\int_{-\infty}^{\infty}\Phi(x)\sin\tau x\,dx \qquad (21.16)$$

(for Im $z > 0$).

Thus the dispersion equation takes on the form

$$p^2=\frac{4\pi e^2}{m}\int_{0}^{\infty}e^{-iz\tau}\,d\tau\int_{-\infty}^{\infty}\frac{\partial f_0}{\partial x}\sin\tau x\,dx, \qquad (21.17)$$

$$z=\frac{i\omega}{p}-\xi_0$$

$$(\mathrm{Im}\,z>0).$$

If z is situated in the lower portion of half plane, then

$$p^2 = \frac{4\pi e^2}{m} \int_0^\infty e^{iz\tau}\, d\tau \int_{-\infty}^{\infty} \frac{\partial f_0}{\partial x} \sin \tau x\, dx.$$

$$z = \frac{i\omega}{p} - \xi_0 \tag{21.18}$$

$$(\operatorname{Im} z < 0).$$

In the event that Imz = 0, p = ik (nondamping waves), we have, setting $z = z_r + iz_i$ and going over to $z_i \longrightarrow 0$:

$$\int_{-\infty}^{\infty} \frac{\Phi(x)\, dx}{x - \xi_0} = \frac{1}{2}\left\{ \lim_{z_i \to 0} \int_0^\infty e^{-z_i\tau} e^{iz_r\tau}\, d\tau \int_{-\infty}^{\infty} \frac{\partial f_0}{\partial x} \sin \tau x\, dx + \right.$$

$$\left. + \lim_{z_i \to 0} \int_0^\infty e^{z_i\tau} e^{-iz_r\tau}\, d\tau \int_{-\infty}^{\infty} \frac{\partial f_0}{\partial x} \sin \tau x\, dx \right\}, \tag{21.19}$$

and hence,

$$k^2 = \frac{4\pi e^2}{m} \int_0^\infty \cos z_r\tau\, d\tau \int_{-\infty}^{\infty} \frac{\partial f_0}{\partial x} \sin \tau x\, dx,$$

$$z_r = \frac{\omega}{k} - \xi_0 \quad (z_i = 0). \tag{21.20}$$

We now show that the three types of solutions considered (Im z > 0, Im z < 0, Im z = 0) correspond to lagging, leading, and the half sum of lagging and leading solutions. To this end, we consider three partial representations of the initial linearized equations:

$$\varphi = \frac{e}{m} \frac{\partial f_0}{\partial \xi} \int_{-\infty}^{t} \frac{\partial V}{\partial x} (x - \xi(t - \tau), \tau)\, d\tau, \quad (-\infty < t), \tag{21.21}$$

$$\varphi = -\frac{e}{m} \frac{\partial f_0}{\partial \xi} \int_{+\infty}^{t} \frac{\partial V}{\partial x} (x - \xi(t - \tau), \tau)\, d\tau, \quad (t < +\infty), \tag{21.22}$$

$$\varphi = \frac{e}{m} \frac{\partial f_0}{\partial \xi} \left\{ \frac{1}{2} \int^{t} \frac{\partial V}{\partial x} (x - \xi(t - \tau), \tau)\, d\tau - \right.$$

$$\left. + \frac{1}{2} \int_{+\infty}^{t} \frac{\partial V}{\partial x} (x - \xi(t - \tau), \tau)\, d\tau \right\} \tag{21.23}$$

$$(-\infty < t < +\infty);$$

and we shall have, correspondingly:

$$\Delta V(x,t)=\frac{4\pi e^2}{m}\int_{-\infty}^{\infty}\frac{\partial f_0}{\partial \xi}d\xi\int_{-\infty}^{t}\frac{\partial V}{\partial x}(x-\xi(t-\tau),\tau)\,d\tau, \qquad (21.24)$$

$$\Delta V(x,t)=\frac{4\pi e^2}{m}\int_{-\infty}^{\infty}\frac{\partial f_0}{\partial \xi}d\xi\int_{+\infty}^{t}\frac{\partial V}{\partial x}(x-\xi(t-\tau),\tau)\,d\tau, \qquad (21.25)$$

$$\Delta V(x,t)=\frac{4\pi e^2}{m}\left\{\int_{-\infty}^{\infty}\frac{\partial f_0}{\partial \xi}d\xi\left(\frac{1}{2}\int_{-\infty}^{t}\frac{\partial V}{\partial x}(x-\xi(t-\tau),\tau)\,d\tau+\right.\right.$$
$$\left.\left.+\frac{1}{2}\int_{+\infty}^{t}\frac{\partial V}{\partial x}(x-\xi(t-\tau),\tau)\,d\tau\right)\right\}. \qquad (21.26)$$

We seek a solution in the form

$$V(x,t)=V_0e^{i\omega t-px},\quad p=\varkappa+ik,\quad \omega>0. \qquad (21.27)$$

Upon substitution in the first two cases, we have the integrals

$$(-p)\int_{-\infty}^{t}e^{i\omega(\tau-t)+p\xi(t-\tau)}\,d\tau \quad \text{и} \quad (-p)\int_{+\infty}^{t}e^{i\omega(\tau-t)+p\xi(t-\tau)}\,d\tau,$$

and substituting in the first integral $t-\tau=\eta>0$ and in the second $\tau-t=\eta'>0$, we obtain:

$$p\int_{\infty}^{0}e^{(-i\omega+p\xi)\eta}d\eta=$$
$$=p\left\{\frac{1}{-i\omega+p\xi}-\frac{e^{(-i\omega+p\xi)\eta}}{-i\omega+p\xi}\bigg|_{\eta=+\infty}\right\}, \qquad (21.28)$$

$$-p\int_{\infty}^{0}e^{(i\omega-p\xi)\eta'}d\eta'=-p\left\{\frac{1}{i\omega-p\xi}-\frac{e^{(i\omega-p\xi)\eta'}}{i\omega-p\xi}\bigg|_{\eta'=\infty}\right\}. \qquad (21.29)$$

In equation (21.28), the last member disappears when $\varkappa < 0$, and in formula (21.29) when $\varkappa > 0$. Thus, the lagging solutions increase as x increases, and the leading solutions decrease.

The first case corresponds to that considered above when the point lay in the upper half plane, since $z = \frac{i\omega}{p} - \xi_0 = \frac{\omega}{k^2+\varkappa^2}(i\varkappa - k) - \xi_0$, and the second, when it is in the lower, which was to be proved. As a result, we have the previous dispersion equation both for lagging and leading potentials:

$$p^2 = -\frac{4\pi e^2}{m}\int_{-\infty}^{\infty}\frac{\frac{\partial f_0}{\partial \xi}\,d\xi}{\xi - \frac{i\omega}{p}}, \tag{21.30}$$

where Im z > 0 for the first case, and Im z < 0 for the second. It is clear that both cases have a physical meaning, since in practice both increasing and decreasing solutions must be taken into account. In the case of the "half sum" the integrals given above should be taken as extreme, taking their half sum from the side of the upper and lower half planes. As a result we come back again to the dispersion equation, where the point z will lie on the real axis (Im z = 0). We now pass to consider the initial equation with any law for the forces of interaction between the particles.

We consider equation (20.1). It has a particular solution of this type:

$$\begin{aligned}\varphi(r, v, t) &= c_1\int_{t_0}^{t}\frac{1}{m}\nabla_v f_0 \nabla_r \int_{-\infty}^{\infty} K(|r - v(t-\tau) - r'|)\,\rho(r', \tau)\,dr'\,d\tau + \\ &+ c_2\int_{t_1}^{t}\frac{1}{m}\nabla_v f_0 \nabla_r \int_{-\infty}^{\infty} K(|r - v(t-\tau) - r'|)\,\rho(r', \tau)\,dr'\,d\tau \\ &\qquad (t_0 \leqslant t \leqslant t_1),\end{aligned} \tag{21.31}$$

where $c_1 + c_2 = 1$. We can see this by direct substitution. In order to obtain solutions with unspecified initial moment of time over the entire interval $t_0 < t < t$, we assume that

$c_1 = \frac{1}{2}$, $c_2 = \frac{1}{2}$.

Integrating (21.31) with respect to the velocities, we have the following expression to be studied:

$$\rho(r,\ t) = \int\limits_{-\infty}^{\infty} \frac{1}{m} \nabla_v f_0 \, dv \left\{ \frac{1}{2} \int\limits_{t_0}^{t} d\tau \, \nabla_r \int\limits_{-\infty}^{\infty} K(|r - v(t-\tau) - r'|) \rho(r',\ \tau) \, dr' + \right.$$

$$\left. + \frac{1}{2} \int\limits_{t_1}^{t} d\tau \, \nabla_r \int\limits_{-\infty}^{\infty} K(|r - v(t-\tau) - r'|) \rho(r',\ \tau) \, dr' \right\}. \qquad (21.32)$$

We seek a solution for this equation in the form:

$$r,\ t) = \sum_k q_k(t) \, e^{ikr} \qquad (21.33)$$

Substitution gives the following equation for determining q_k (t):

$$q_k(t) = \frac{1}{2} \int\limits_{t_0}^{t} G_k(t-\tau) q_k(\tau) \, d\tau + \frac{1}{2} \int\limits_{t_1}^{t} G_k(t-\tau) q_k(\tau) \, d\tau \qquad (21.34)$$

$$(t_0 < t < t_1).$$

Here we have obviously sought a solution in the class of functions for which the condition for the permutability of the limits of integration with respect to velocity components and time is fulfilled.

Kernel $G_k(t - \tau)$ coincides with the kernel of the Volterra equation, which we obtained in the preceeding section:

$$G_k(t-\tau) = \frac{\sigma(k)}{m} \int\limits_{-\infty}^{\infty} e^{-ikv(t-\tau)} \, i \, (k\nabla_v) f_0 \, dv. \qquad (21.35)$$

We note that G_k is an odd function of its argument, since f_0 is assumed even, and consequently,

$$G_k(-t) = -G_k(t). \qquad (21.36)$$

Making use of this property of the kernel G_k, we may write equation (21.34) in the form of a Fredholm integral equation:

$$q_k(t) = \frac{1}{2} \int\limits_{t}^{t_1} G_k(|t-\tau|) q_k(\tau) \, d\tau. \qquad (21.37).$$

We emphasize the basic difference in principle between this result and that obtained in the previous section.

Here we pass to solutions describing temporal processes that are not comprised in consideration of the Cauchy problem. As we shall see, the temporal processes described by equation (21.37) change their character in a saltatory manner under continuous variation of the parameters entering into the kernel. Among these parameters may be the temperature, the density, and likewise parameters characterizing the forces of interaction.

We shall consider later the concrete forms of solutions of this type.

It is important that we can obtain an approximate solution of equation (21.37) in a perspicuous form if we take into account the fact that the kernel decreases as the argument of the function increases, so that as $t \to \infty$ $G(t) \to 0$. This is connected with the fact that as $(t - \tau)$ increases, the expression under the integral sign, as will be seen from (21.35), oscillates rapidly, and consequently the integral vanishes. We may therefore extend the limits of integration with respect to t in Fredholm's equation to infinity, and obtain an equation of a singular type:

$$q_k(t) = \frac{1}{2}\int_{-\infty}^{\infty} G_k(|t-\tau|)\, q_k(\tau)\, d\tau. \qquad (21.38)$$

We shall seek its solution in the form of a superposition of the function $q_k(\omega)\, e^{i\omega t}$. Substitution gives:

$$q_k(\omega) = \frac{1}{2}\int_{-\infty}^{t} G_k(t-\tau)\, q_k(\omega)\, e^{-i\omega(t-\tau)}\, d\tau + \frac{1}{2}\int_{t}^{\infty} G_k(\tau-t)\, q_k(\omega)\, e^{-i\omega(t-\tau)}\, d\tau. \qquad (21.39)$$

Changing variables in both integrals, we obtain the condition for the solvability of integral equation (21.38) in the following form:

$$\int_{0}^{\infty} G_k(t)\cos\omega t\, dt = 1. \qquad (21.40)$$

This equation establishes a connection between ω and k, since the quantity k enters into G (t) by means of (21.35). It is thus likewise a dispersion equation defining the solutions in this case. This dispersion equation can also be written as follows:

$$-\frac{\sigma(k)}{m}\lim_{T\to\infty}\int_{-\infty}^{\infty} ik\xi f_0(\xi^2)\,d\xi\int_0^T e^{-ik\xi t}\cos\omega t\,dt = 1. \qquad (21.41)$$

Integrating with respect to t, we find that for T sufficiently large, the expression under the limit sign differs to as small an extent as desired from the principal value of integral $\int_{-\infty}^{\infty}\frac{k\xi f_0(\xi^2)\,d\xi}{k\xi-\omega}$, , so that dispersion equation (21.40) assumes the following form:

$$\frac{\sigma(k)}{m}\int_{-\infty}^{\infty}\frac{\frac{\partial f_0}{\partial\xi}}{\xi-\frac{\omega}{k}}\,d\xi = 1; \quad f_0 = \int_{-\infty}^{\infty}\int f_0\,d\eta\,d\rho, \qquad (21.42)$$

taking the principal value of the integral. Comparing this integral with (20.15), we see that two different types of temporal processes have dispersion equations of the same form. According as the pole of the denominator is in the plane of the complex variable or on the real axis, we have two different types of solutions. In the first case (20.15), there are periodically damping or increasing solutions, and in the second (21.42), purely periodic solutions. For the Maxwell distribution function

$$f_0 = N\left(\frac{m}{2\pi kT}\right)^{\frac{3}{2}} e^{-\frac{m(\xi^2+\eta^2+\zeta^2)}{2kT}} \qquad (21.43)$$

we have

$$G_k(t) = \frac{\sigma(k)}{m}\int_{-\infty}^{\infty} e^{-ikvt}\, i(k\nabla_v) f_0\,dv =$$

$$= -\frac{N\sigma(k)}{m}k^2 t e^{-\frac{k^2c^2t^2}{4}}\left(c^2 = \frac{2kT}{m}\right). \qquad (21.44)$$

This result is easily obtained if setting the axis of k in the x direction we put:

$$\int_{-\infty}^{\infty} \xi e^{-\frac{\xi^2}{c^2}} \cos k\xi t\, d\xi = 0,$$

$$\int_{-\infty}^{\infty} \xi e^{-\frac{\xi^2}{c^2}} \sin k\xi t\, d\xi = \frac{kc^3}{4}\sqrt{\pi}\, e^{-\frac{k^2c^2t^2}{4}}.$$

The dispersion equation (21.40) then takes on the form:

$$-\frac{N\sigma(k)}{m}k^2\int_0^{\infty} te^{-\frac{k^2c^2t^2}{4}} \cos \omega t\, dt = 1$$

or, calculating the value of $\sigma(k)$ and introducing new variables of integration:

$$\frac{1}{\frac{4\pi N}{kT}\int_0^{\infty} K(\rho)\frac{\sin k\rho}{k\rho}\rho^2 d\rho} = -4\int_0^{\infty} xe^{-x^2}\cos\frac{2\omega}{kc}x\, dx. \qquad (21.45)$$

This equation adumbrates the dependence of ω on k, and is used in a number of applications (see Part II). The special function

$$I(v) = -\int_0^{\infty} xe^{-x^2}\cos vx\, dx \qquad (21.46)$$

is found by numerical integration.

Solutions of the linear integral equation in the form exp $(i\omega t - ikr)$ where ω and k may be either real or complex numbers, may be obtained by another, direct method. In equation (20.1):

$$\frac{\partial\varphi}{\partial t} + v\,\mathrm{grad}_r\varphi = \frac{1}{m}\mathrm{grad}_v f_0 \cdot \mathrm{grad}_r \int\int_{-\infty}^{+\infty} K(|r - r'|)\,\varphi(r', v', t)\, dv\, dr'$$

we seek again a solution ϕ in functions of the "wave" type:

$$\varphi \sim e^{i\omega t - ikr} \cdot g_k(\xi, \eta, \zeta).$$

For $g_k(\xi, \eta, \zeta)$ we have:

$$g_k(\xi, \eta, \zeta) = \frac{\sigma(k)}{m} \frac{(k\nabla) f_0}{[kv - \omega]} \int_{-\infty}^{+\infty} g_k(\xi, \eta, \zeta)\, dv, \qquad (\alpha)$$

where

$$\sigma(k) = 4\pi \int_0^{\infty} K(\rho) \cdot \rho \cdot \frac{\sin k\rho}{k}\, d\rho.$$

The condition for the existence of a nontrivial solution of this class is obtained from (α), integrating with respect to the velocities on both right and left:

$$\frac{\sigma(k)}{m} \int_{-\infty}^{+\infty} \frac{(k\nabla) f_0}{[kv - \omega]}\, dv = 1. \qquad (\beta)$$

This equation constitutes a dispersion equation linking ω and k. Since the integral in formula (α) does not depend on the velocities, the solution for g_k has a form

$$g_k(\xi, \eta, \zeta) = c(k) \frac{\sigma(k)}{m} \frac{(k\nabla) f_0}{[\omega - kv]} \qquad (\gamma)$$

or

$$\varphi = \sum_i c_i e^{i\omega_i t} \sum_j c(k_{ij}) \frac{\sigma(k_{ij})}{m} \frac{(k_{ij}\nabla) f_0}{[k_{ij} v - \omega_i]} e^{-ik_{ij} r},$$

where the summation for j with a given i in the general case should be conducted for all the roots of the dispersion equation. We consider equation (β)(assuming k directed along the x axis):

$$\frac{\sigma(k)}{m} \int_{-\infty}^{\infty} \frac{\frac{\partial f_0}{\partial \xi}}{\xi - \frac{\omega}{k}}\, d\xi\, d\eta\, d\zeta = 1$$

or

$$\frac{\sigma(k)}{m} \int_{-\infty}^{\infty} \frac{\frac{\partial \Phi_0}{\partial \xi}}{\xi - \frac{\omega}{k}}\, d\xi = 1, \qquad \Phi_0 = \int_{-\infty}^{\infty}\!\!\int f_0\, d\eta\, d\zeta;$$

it coincides with equation (21.12) obtained above. Objections were raised to the possibility that the denominator in the expression under the integral sign in the dispersion equation might vanish. It should be kept in mind, however, that first, for a large class of distribution functions f_0, for which we have

$$\frac{\partial f_0}{\partial \xi} = 0,$$

at the point $\xi = \frac{\omega}{k}$, there is no singularity and there is no problem as to the principal value of the integral. This class of functions alone is so broad that it would be possible to confine ourselves to it alone. It should be kept in mind that experimentally initial distributions may exist having functions of the most diverse kind, employing, for example, the method of electron beams, and especially so in considering nonstationary problems, strictly speaking, there are no theoretical or experimental bases for precisely fixing f_0, inasmuch as the limitations on the form of f_0 should be given by the <u>nonlinear</u> equation itself.

If either ω or k is a complex number with no matter how small an imaginary portion, the question arises as to the principal value of the integral+. In those cases where the integral must be taken in the sense indicated above, rejecting the point $\xi = \frac{\omega}{k}$ in integration:

$$\int_{-\infty}^{\infty} \frac{\Phi(x)\,dx}{x-c} = \lim_{\epsilon \to 0}\left\{ \int_{-\infty}^{c-\epsilon} \frac{\Phi(x)\,dx}{x-c} + \int_{c+\epsilon}^{\infty} \frac{\Phi(x)\,dx}{x-c} \right\},$$

this process specifies a particular solution of the initial equation, belonging to the group of equations symmetrical with respect to time of the wave type exp i (ω t - kx), where ω and k are real numbers, as was shown above in the text in a different connection. In particular, the solutions with the half sum of the leading and lagging potentials actively correspond to the integral concept in the sense of the principal value. As we have already pointed out, and shall have occasion to express later, this solution does not belong to the type of solutions of the Cauchy problem.

In order to bring out the characteristic properties of the law of dispersion, we consider for simplicity's sake a particular form of the initial distribution function Φ_0, and then generalize the result obtained.

We choose the "dispersion distribution" as the distribution function Φ_0:

$$\Phi_0(\xi) = \frac{Nc}{\pi} \frac{1}{c^2 + \xi^2}; \quad c^2 = \frac{2kT}{m},$$

+ See formula (21.9), which was deduced strictly (work of I. Privalov et al.).

which again may be considered as an approximation to the Maxwell distribution, which holds good for not very large velocities. Then

$$\frac{1}{m}\int_{-\infty}^{+\infty}\left|\frac{\frac{\partial\Phi_0}{\partial\xi}}{\xi-\frac{\omega}{k}}\right|d\xi=-\frac{2N}{\pi mc^2}\int_{-\infty}^{+\infty}\frac{x\,dx}{x-v}\left(\frac{1}{1+x^2}\right)^2;\quad\left(v=\frac{\omega}{k}\cdot\frac{1}{c}\right).$$

The integral that appears is expressed by elementary functions of the parameter v:

$$\int_{-\infty}^{+\infty}\frac{x\,dx}{x-v}\left(\frac{1}{1+x^2}\right)^2=\frac{\pi}{2}\,\frac{1-v^2}{(1+v^2)^2}.$$

The equation being studied may be represented in the following parametric form:

$$\frac{1}{\frac{4\pi N}{mc^2}\int_0^{\infty}K(\rho)\,\rho\,\frac{\sin k\rho}{k}\,d\rho}=-\frac{1-v^2}{(1+v^2)^2}=\frac{I_1(v)}{\frac{N}{mc^2}};\quad\omega^2=k^2v^2$$

for $v=\frac{\omega}{k}$. In the general case, when Φ_0 is not specified:

$$\frac{1}{\frac{4\pi}{m}\int_0^{\infty}K(\rho)\,\frac{\sin k\rho}{k\rho}\,\rho^2\,d\rho}=\int_{-\infty}^{\infty}\frac{\frac{\partial\Phi_0}{\partial\xi}}{\xi-v}\,d\xi=I_1(v),$$

$$\omega^2=v^2k^2$$

$$\left(\sigma(k)=4\pi\int_0^{\infty}K(\rho)\,\frac{\sin k\rho}{k\rho}\,\rho^2\,d\rho\right).$$

The characteristic properties of the law of dispersion obtained by means of the distribution indicated above apply to an extremely broad class of distribution functions. It is only necessary that the functions be symmetrical with respect to the velocities and decrease monotonously sufficiently rapidly as $\xi\to\pm\infty$. We prove this.

The fundamental properties of the right member are:

1. $I_1(v)$ be bounded over the entire interval v from 0 to ∞.

2. Change of sign when going from small to large values on the function $I_1(v)$.

3. Existence of a maximum for function $I_1(v)$ for $I_1(v) > 0$.

4. Approaching 0 like $\frac{1}{v^2}$ as $v \to \infty$.

It is easy to show this, if we consider the integral:

$$-I_1(v) = \int \frac{\frac{\partial \Phi_0}{\partial \xi}}{v-\xi} d\xi = \int_{-\infty}^{+\infty} \frac{xF(x)\,dx}{v-x} = \int_{-\infty}^{+\infty} \frac{x^2 F(x)\,dx}{v^2-x^2}$$
$$(0 \leqslant v \leqslant +\infty),$$

where $xF(x) = \frac{\partial \Phi_0}{\partial \xi}$, and F(x) is an even function. We confine ourselves to proving the first and last of the properties, since the proof of the others is analogous.

We have:

$$\int_{-\infty}^{+\infty} \frac{\psi(x)\,dx}{x-v} = \lim_{\varepsilon \to 0} \left\{ \int_{-\infty}^{v-\varepsilon} \frac{\psi(x)\,dx}{x-v} + \int_{v+\varepsilon}^{+\infty} \frac{\psi(x)\,dx}{x-v} \right\} =$$
$$= \lim_{\varepsilon \to 0} \int_{\varepsilon}^{+\infty} \frac{\psi(v+\xi) - \psi(v-\xi)}{\xi} d\xi;$$

and consequently, in order for the integral to be bounded, we need merely assume $\psi(x) \sim 0\left(\frac{1}{x}\right)$ as $x \to \infty$ and require the existence of a bounded differential at the point $x \to 0$:

$$\psi(x) = xF(x) = \frac{\partial \Phi_0}{\partial \xi} \sim O\left(\frac{1}{\xi}\right)$$

as $\xi \to \pm\infty$;

$$\psi'(x) = (xF(x))' = \frac{\partial^2 \Phi_0}{\partial \xi^2}$$

exists as $\xi \to 0$.

To prove the fourth proposition, we write

$$\frac{1}{v^2 - x^2} = \frac{1}{v^2} + \frac{x^2}{(v^2 - x^2)\, v^2};$$

then

$$- I_1(v) = \frac{1}{v^2} \int_{-\infty}^{+\infty} x^2 F(x)\, dx + \frac{1}{v^2} \int_{-\infty}^{+\infty} \frac{x^4 F(x)\, dx}{v^2 - x^2} =$$

$$= \frac{1}{v^2} \int_{-\infty}^{+\infty} x^2 F(x)\, dx + \frac{\varepsilon(v^2)}{v^2}$$

and it is necessary to show that as $v^2 \to \infty$ $\varepsilon(v^2)$ approaches zero.

$$|\varepsilon(v^2)| = \left| \int_{-\infty}^{+\infty} \frac{x^4 F(x)\, dx}{v^2 - x^2} \right| = \left| \int_{-\infty}^{+\infty} \frac{x^3 F(x)\, dx}{v - x} \right| = \left| \int_{-\infty}^{+\infty} \frac{\Phi(x)\, dx}{v - x} \right| \leqslant$$

$$\leqslant \left| \int_{v-1}^{v+1} \frac{\Phi(x)\, dx}{v - x} \right| + \left| \int_{v+1}^{+\infty} \frac{\Phi(x)\, dx}{v - x} \right| + \left| \int_{-\infty}^{v-1} \frac{\Phi(x)\, dx}{v - x} \right|$$

[$x^3 F(x) = \Phi(x)$ being an odd function of x].

For the first integral, we have:

$$\int_{v-1}^{v+1} \frac{\Phi(x)\, dx}{v - x} = \int_0^1 \frac{\Phi(v + \xi) - \Phi(v - \xi)}{\xi} d\xi \leqslant 2 \max \Phi'(v + \theta); \quad (0 < \theta < 1)$$

and consequently, for it to vanish as $v \to \infty$, it is only necessary that $\Phi'(x)$ approach zero as $x \to \infty$.

The second integral approaches zero as $v \to \infty$, since

$$\left| \int_{v+1}^{\infty} \frac{\Phi(x)\, dx}{v - x} \right| \leqslant \frac{1}{v} \int_{v+1}^{\infty} \Phi(x)\, dx$$

and consequently approaches zero as $v \to \infty$ for any function $\Phi(x)$, for which the integral exists within infinite limits.

The third integral

$$\left|\int_{-\infty}^{v-1}\frac{\Phi(x)\,dx}{v-x}\right|\leqslant\left|\int_{\frac{1}{2}v}^{v-1}\frac{\Phi(x)\,dx}{v-x}\right|+\left|\int_{-\infty}^{+\frac{1}{2}v}\frac{\Phi(x)\,dx}{v-x}\right|\leqslant$$

$$\leqslant\int_{\frac{1}{2}v}^{v-1}\Phi(x)\,dx+\frac{2}{v}\int_{-\infty}^{+\infty}\Phi(x)\,dx$$

approaches zero as $v\to\infty$, since $\Phi(x)$ is an odd function. Consequently, $\varepsilon(v^2)\to 0$ as $v\to\infty$.

Thus, for large v:

$$I_1(v)\to 0 \text{ as } v\to\infty$$

as

$$\frac{1}{v^2}\int_{-\infty}^{+\infty}x^2F(x)\,dx.$$

It is only requisite that

$$(x^3F(x))'\to 0 \quad \text{as} \quad x\to\infty$$

or

$$\left(\xi^2\frac{\partial\Phi_0}{\partial\xi}\right)'=2\xi\frac{\partial\Phi_0}{\partial\xi}+\xi\frac{\partial^2\Phi_0}{\partial\xi^2}\to 0 \qquad \text{as} \qquad \xi\to\infty.$$

It is thus shown that the nature of the relationship $I_1(v)$, found in the case of the simplest distribution function remains correct for a very real class of distribution functions $f_0(v)$.

In the second part of the book, we shall apply the methods developed here to both the electrodynamic and the short-acting forces of interaction between particles.

Section 22. Solutions Representing "Self-Accelerating" Processes

We consider the equation

$$\frac{\partial f}{\partial t}+\xi\frac{\partial f}{\partial x}-\frac{\partial f}{\partial\xi}\frac{1}{m}\frac{\partial}{\partial x}\int_{-\infty}^{+\infty}K(|x-x'|)f(t,x',\xi')\,d\xi'\,dx'=0$$

and we endeavor to convert it into an integral equation, which considerably facilitates the investigation. To this end, we shall solve it formally by the method of successive approximations, so that the equation for n-th approximation has the form:

$$\frac{\partial f_n}{\partial t}+\xi\frac{\partial f_n}{\partial x}-\frac{\partial f_n}{\partial \xi}\frac{1}{m}\frac{\partial}{\partial x}\int_{-\infty}^{+\infty}K(|x-x'|)f_{n-1}(t,x',\xi')\,d\xi'\,dx'=0.$$

Assuming that $f_{n-1}(t, x, \xi)$ is defined by the preceding approximations, we have a linear homogeneous equation in partial derivatives, equivalent to the system

$$dt=\frac{dx}{\xi}=-\frac{d\xi}{\frac{1}{m}\frac{\partial}{\partial x}\int K(x,x')f_{n-1}(t,x',\xi)\,d\xi'\,dx'}.$$

The first integrals of this system have the form:

$$\alpha_n(\xi t-x)=c_{1,n},$$

$$\frac{\xi^2}{2\theta_n}m+\mu_n+\frac{1}{\theta_n}\int K(x,x')f_{n-1}(t,x',\xi')\,d\xi'\,dx'=c_{2,n},$$

where α_n, θ_n, μ_n — are arbitrary constants. The general solution of the system, as is known, is given by the expression

$$f_n(t,x,\xi)=\Phi_n\left\{\alpha_n(\xi t-x),\ \frac{\xi^2 m}{2\theta_n}+\mu_n+\frac{1}{\theta_n}\int K(x,x')f_{n-1}\,d\xi'\,dx'\right\},$$

where Φ_n is an arbitrary function, whose form is defined by giving the supplementary conditions. We further assume that for each problem there exists a limiting function of the sequence Φ_n that even in the limit we shall have:

$$\left.\begin{aligned} f(t,x,\xi)=\Phi\Big\{&\alpha(\xi t-x),\\ &\frac{\xi^2}{2\theta}m+\mu+\frac{1}{\theta}\int K(x,x')f(t,x',\xi')\,d\xi'\,dx'\Big\}\end{aligned}\right\} \qquad (22.1)$$

or, otherwise written:

$$f(t,x,\xi)=\Phi(c_1,c_2),$$

where c_1, c_2 have an obvious meaning. We show that such a function Φ exists and that equation (22.1) is equivalent to the initial equation. Writing

$$\eta(t, x) = \int_{-\infty}^{\infty} K(|x - x'|) f(t, x', \xi') \, d\xi' \, dx',$$ we insert (22.1)

in the initial equation and require that it vanish identically because of the substitution. It is not hard to see that we arrive at the equation

$$\Phi'_{c_1} \frac{1}{\theta} \frac{\partial \eta}{\partial t} - \Phi'_{c_2} \frac{\alpha}{m} t \frac{\partial \eta}{\partial x} = 0. \tag{22.2}$$

Since the right side of the last relationship is independent of ξ, identical equality is possible only on the conditions

$$\gamma \Phi'_{c_1} - \Phi'_{c_2} = 0, \quad \gamma \frac{\alpha}{m} t \frac{\partial \eta}{\partial x} - \frac{1}{\theta} \frac{\partial \eta}{\partial t} = 0,$$

where γ is an arbitrary constant.

These equations are solved independently, and we obtain the expressions:

$$\left. \begin{aligned} \Phi(c_1 c_2) &= \Phi(c_1 + \gamma c_2), \\ \eta(t, x) &= \eta\left(x + \gamma \frac{\alpha\theta}{2m} t^2\right). \end{aligned} \right\} \tag{22.3}$$

In this way, the initial equation reduces to the form

$$f(t, x, \xi) = \Phi \left\{ \frac{\xi^2}{2\theta} m + \mu + \right. \\ \left. + \frac{1}{\theta} \int K(x, x') f(t, x', \xi') \, d\xi' \, dx' + \gamma\alpha(\xi t - x) \right\} \tag{22.4}$$

or to an equation relatively to function $\eta(t, x)$:

$$\eta(t, x) = \int_{-\infty} K(|x - x'|) \Phi \left\{ \frac{\xi'^2}{2\theta} m + \mu + \right. \\ \left. + \frac{1}{\theta} \eta(t, x') + \gamma\alpha(\xi' t - x') \right\} d\xi' \, dx'. \tag{22.4'}$$

Equations (22.4), (22.4') represent Hammerstein's equation in the most general form. Setting $\xi\sqrt{\frac{m}{2\theta}}+\gamma\alpha\sqrt{\frac{\theta}{2m}}=\zeta$ and integrating with respect to ζ from $-\infty$ to ∞, we convert (22.4') into the form

$$\eta(z)=\int_{-\infty}^{\infty}K(|z-z'|)\,\Psi\left\{-\gamma\alpha z'+\frac{1}{\theta}\eta(z')\right\}dz' \qquad (22.5)$$

$$\left(z=x+\gamma\alpha\frac{\theta}{2m}t^2\right).$$

Further investigation of the equation obtained requires specification of function Φ. Observing that for $\gamma = 0$ (22.5) becomes equivalent to a stationary equation, we set $\Phi(c_1, c_2) = e^{-c_1+\gamma c_2}$ so that

$$V(z)=\frac{1}{\theta}\int_{-\infty}^{\infty}K(|z-z'|)\,e^{\gamma z'-V(z')}\,dz'. \qquad (22.6)$$

Here

$$V(z)=\frac{1}{\theta}\eta(z), \quad \gamma\alpha=\gamma.$$

These results are easily generalized to the case of the general initial equation for $f(t, r, v)$. In this form of the functions $\eta(t, r)$ is determined from the equation $\frac{\partial\eta}{\partial t}-\gamma\frac{\theta}{m}t\nabla_r\eta=0$, from which $\eta(t, r)=\eta\left(r+\gamma\frac{\theta}{2m}t^2\right)$, while (22.6) goes over into

$$V(z)=\frac{1}{\theta}\int_{-\infty}^{\infty}K(|z-z'|)\,e^{\gamma z'-V(z')}\,dz'. \qquad (22.6')$$

If we should explicitly introduce a normalization condition of the type $f=\int^{\infty}e^{\gamma z'-V(z')}dz'$, then (22.6) and (22.6') would only acquire the multiplier f before the integral operator.

Abstracting from the fact that in (22.6) z depends on t, then this equation upon the transformation $\varphi(z) = V(z) - \gamma z$ leads to an equation describing a system of particles in an external field of force (γz playing the role of potential function). Observing further that $V(z)$, $\eta(z)$ depend on the coordinates and the time in the combination $x+\frac{\gamma\theta}{2m}t^2$, we see that this fictitious field of force is in fact the result of the presence in the system of the possibility of self-accelerating motions. The solution for (t, x) has two possible forms, $\eta(t, x)=\eta\left(x+\frac{\gamma\theta}{2m}t^2\right)$ and $\eta(t, x)=\eta\left(x-\frac{\gamma\theta}{2m}t^2\right)$, and has an obvious sense: if the distribution of the particles of the system is known at some moment of time, then the spatial form of distribution will not change with the time, moving with constant acceleration $\gamma\frac{\theta}{2m}$ in one of two opposite directions.

We now take up the question as to the presence of even one essential solution for equation (22.6). Knowing one solution, of course, enables us to find other solutions as well, if we use the method of branching. Since for $\gamma = 0$, we arrive at the equation studied in Secs. 17 and 18, we take as the parameter of the expansion and endeavor to satisfy the equation of the formal expansion

$$V(z)=\varphi_0(z)+\gamma\varphi_1(z)+\gamma^2\varphi_2(z)+\dots \tag{22.7}$$

The successive approximations $\varphi_n(z)$ are defined by the system of equations that is easily obtained, observing that

$$V^{(n)}(z)|_{\gamma=0}=n!\varphi_n(z).$$

Designating $L\{\zeta(z)\}=\zeta(z)+\frac{1}{\theta}\int_{-\infty}^{\infty}K(|z-z'|)e^{-\varphi_0(z')}\zeta(z')dz'$, we have:

$$\left.\begin{aligned}
\varphi_0(z)&=\frac{1}{\theta}\int_{-\infty}^{\infty}K(|z-z'|)e^{-\varphi_0(z')}dz',\\
L\{\varphi_1(z)\}&=\frac{1}{\theta}\int_{-\infty}^{\infty}K(|z-z'|)e^{-\varphi_0(z')}z'\,dz',\\
L\{\varphi_2(z)\}&=\frac{1}{\theta 2!}\int_{-\infty}^{\infty}K(|z-z'|)e^{-\varphi_0(z')}\{z'-\varphi_1(z')\}^2dz',
\end{aligned}\right| \tag{22.8}$$

It follows from the first equation in (22.8) that $\varphi_0(z)=\varphi_0$ (φ_0 being constant) satisfies the equation. Two cases, however, should be distinguished:

a) If $K>0$, then φ_0 is determined for all values of θ;

b) If $K<0$, then φ_0 is determined only for sufficiently high temperatures $\theta \geqslant eK$, where $K=\int_{-\infty}^{\infty}K(|x-x'|)\,dx'$.

Observing further that

$$\int_{-\infty}^{\infty} K(|z-z'|)\, z'^n dz' = \int_{0}^{\infty} K(|\xi|)\{(z+\xi)^n+(z-\xi)^n\}\, d\xi =$$

$$= z^n K + c_n^2 K_2 z^{n-2} + \ldots + c_n^{2s} K_{2s} z^{n-2s} + \ldots + \begin{cases} K_n\ (n=2m) \\ c_n' K_{n-1} z\ (n=2m-1), \end{cases}$$

$$\left(K_n = 2\int_{0}^{\infty} K(|\xi|)\, \xi^n d\xi;\quad K_{2m+1}=0\right),\quad (n=2m),$$

we reduce the second equation of (22.8) to the form $L\{\varphi_1(z)\} = \varphi_0 z$, from which, setting $\varphi_1(z) = A_1 z$, we have:

$$\varphi_1(z) = \frac{\varphi_0}{1+\varphi_0} z.$$

The subsequent equations of the systems are solved in the same fashion, and give solutions for $\varphi_n(z)$ in the form of polynomials of the corresponding degree in z, for example,

$$\varphi_2(z) = \frac{1}{2!}\frac{\varphi_0}{(1+\varphi_0)^3}\left\{z^2 + \frac{K_2}{K}\frac{1}{1+\varphi_0}\right\},$$

$$\varphi_3(z) = \frac{1}{3!}\frac{\varphi_0(1-2\varphi_0)}{(1+\varphi_0)^5} z^3 + \frac{1}{2!}\frac{\varphi_0(1-3\varphi_0)}{(1+\varphi_0)^6}\frac{K_2}{K} z,$$

$$\varphi_4(z) = \frac{1}{4!}\frac{\varphi_0(1-6\varphi_0+4\varphi_0^2-4\varphi_0^3)}{(1+\varphi_0)^7} z^4 +$$

$$+\frac{1}{4!}\frac{\varphi_0(6-52\varphi_0^3+62\varphi_0^2-24\varphi_0^3)}{(1+\varphi_0)^8}\frac{K_2}{K} z^2 +$$

$$+\frac{1}{4!}\frac{\varphi_0}{1+\varphi_0}\left\{\frac{K_2^2}{K^2}\cdot\frac{4(\varphi_0^2-4\varphi_0)}{(1+\varphi_0)^8} + \frac{(1-6\varphi_0+4\varphi_0^2-4\varphi_0^3)}{(1+\varphi_0)^7}\frac{K_4}{K}\right\},$$

. . . .

It is extremely difficult to evaluate the successive approximations obtained and, at the same time, investigate the convergence of the series. However, assuming convergence of the series at least for small γ and large φ_0 (low temperatures), the nature of the distribution of particles may be evaluated at a certain moment of time (for simplicity's sake, t = 0) at the origin of coordinates. Setting $\rho(x) = e^{-\varphi(x)}$, , where

$$\rho(x) = \int_{-\infty}^{\infty} f(0, x, \xi)\, d\xi, \quad \text{and} \quad \varphi(x) = V(x) - \gamma x,$$, and confining

ourselves to the first term of the expansions, we have:

$$\varphi(0)=\varphi_0,\ \frac{d\varphi}{dx}\Big|_{x=0}=-\frac{\gamma}{1+\varphi_0},\ \frac{d^2\varphi}{dx^2}\Big|_{x=0}=\frac{\gamma^2}{(1+\varphi_0)^2}.$$

Thus, in the neighborhood of $x=0$ $\varphi(x)$ increases, and correspondingly ρ (x) decreases.

We were unable to investigate the asymptotic behavior of ρ (x) as the absolute value of $|x|\to\infty$. The magnitude $\gamma=\frac{1}{r_0}$ is closely connected with a characteristic length r_0 defining the size of the region of space where ρ (x) is different from a constant value (for $\gamma=0$ $r_0\to\infty$, , while ρ (x) coincides with a uniform distribution). Finding other solutions comes down to solving the equation

$$v(z)+\frac{1}{\theta_0}\int_{-\infty}^{\infty}K(|z-z'|)\,e^{-V_0(z')}\,v(z')\,dz'=$$
$$=\varepsilon\int_{-\infty}^{\infty}K(z,z')\,e^{-V_0(z')}\{1-v(z')\}\,dz'+$$
$$+\left(\frac{1}{\theta_0}+\varepsilon\right)\sum(-1)^n\frac{1}{n!}\int_{-\infty}^{\infty}K(z,z')\,e^{-V_0(z')}\,v^n(z')\,dz'$$
$$\left(V_0(z)=V(z)\Big|_{\theta=\theta_0},\ v(z)=V(z)-V_0(z),\ \frac{1}{\theta}=\frac{1}{\theta_0}+\varepsilon\right).$$

The solution discussed in this section was found by V. A. Yakovlev. At first sight, this solution appears to have paradoxical properties, indicating the possibility of the self-acceleration of the system under the action of merely internal forces. However, it should be kept in mind that this solution holds true only for a condition where the condition of closure formulated above does not apply. It therefore does not contradict the laws of conservation (see Chapter II), but presents a special nature of connection of the dynamic system with the surrounding medium, which in the theory being set forth does not reduce solely to the introduction of external forces (see in this connection, Point 8, Sec. 7, Chapter I).

PART II

APPLICATIONS OF THE THEORY

CHAPTER I

THE THEORY OF CRYSTALS

Section 23. Introduction

The Born theory of crystals, in which statistical and quantum-mechanics methods are used, has led to important results. Explanation of the general laws of the dynamics of the crystal lattice and of the character of thermal motion at low temperatures, solution of the problem of heat capacity of crystalline bodies, establishment of the existence of two branches in the oscillatory spectrum of crystals and their applications, are all only a part of the positive results of Born's theory.

However, up to the present, the solution of the problem of crystals on the basis of various physical theories has suffered from the following defects in fundamental principles.

1. The very fact of the existence of the crystalline structure with atoms localized in the vicinity of the points of the lattice has not been derived from the theory, but has been postulated. And yet, the experimental conditions to which the ensemble of particles being studied is subjected, do not necessarily lead to strict spatial localization of each atom around the corresponding point of the lattice.

2. It would seem that the process of crystallization, the process of the appearance of a periodic structure in the space distribution of the probability density, should be at the focal point of the theory. How does the transition into the crystalline state take place? It is, above all, a theory of the crystalline state that should reply to this question. However, at this point, it turned out, all the existing theories encounter difficulties.

3. It is likewise impossible to go forward in the theory of the crystalline state within the framework of previous theories for high temperatures as well, since the introduction of nonlinear oscillations of the crystal lattice can never bring us nearer to understanding the nature of fusion (every theory of perturbations is confined only to a region near to the unperturbed state).

4. Finally, we cannot overlook the factor in the existing methods of solving many-particle problems, which even in the gaseous state would predetermine the answer to the question as to what the structure of the lattice should be, what its period should be for the subsequent crystallization.

The ideas at the basis of the new approach to the theory of crystals are the following:

1. Spatially periodic distribution is a particular state of the motion of particles. The states of motion of particles in a collective are extremely diversified; one of them is spatial periodicity in the distribution of particles. From this point of view, a crystal is not a postulated construction, but a state of motion of the particles.

2. The fruitfulness of this point of view is apparent immediately in the possibility of a uniform solution of problems at both high and low temperatures, since the motion remains qualitatively the same in both these cases, merely the character of this motion changing, that is, its periodic structure.

3. An essentially new fact is that the determinate state of motion of each particle of the aggregate is only feebly dependent on the total number of particles. Even the problem of two particles in the theory being set forth entails essential features of the collective. Accordingly, in the theory the problem of two bodies has not only classical solutions but also new solutions, among which are periodic distribution as well. In this sense, it is possible to speak of a crystal of two atoms. It is not essential in principle that there be large aggregates of particles present.

Section 24. Discontinuous Formation of Period Structure and Criteria of Crystallization

We consider the equation for two particles in the simplest form (17.1):

$$V(r)=a\int_{-\infty}^{\infty}K(|r-r'|)\,e^{-\frac{V(r')}{\vartheta}}\,dr'$$

for (17.5):

$$V(r)-V(0)=\int_{-\infty}^{\infty}K(|r-r'|)\,\rho(0)\,(e^{-\frac{V(r')-V(0)}{\vartheta}}-1)\,dr'.$$

This equation was studied in detial in Sec. 17. It was shown there that the potential V(r) is connected with the particle density ρ (r) by the relation

$$\rho(r)=\rho(0)\,e^{-\frac{V(r)-V(0)}{\theta}}=ae^{-\frac{V(r)}{\theta}}.$$

We can introduce into equation (17.1) the parameter λ :

$$\lambda=\frac{\rho(0)}{\theta}$$

and

$$\varphi(r)=\frac{V(r)-V(0)}{\theta}.$$

Then equation (17.1) is written as follows:

$$\varphi(r)=\lambda\int_{-\infty}^{\infty}K(|r-r'|)\,(e^{-\varphi(r')}-1)\,dr.$$

This equation has the trivial solution

$$V(r)\equiv V(0)=\text{const.},$$

corresponding to uniform distribution of particles in space.

We now inquire into the existence and the conditions for the formation of spatially periodic solutions for the potential V(r) or, what amounts to the same, for the density of the distribution of particles in space. We here confine oursélves to the case of small amplitudes of deviation of the value of the potential from the constant magnitude

$$|\varphi(r)|\ll 1$$

and linear equation (17.5). We then obtain a linear integral equation belonging to the class of singular equations:

$$\varphi(r)+\lambda\int_{-\infty}^{\infty}K(|r-r'|)\,\varphi(r')\,dr'=0. \tag{24.1}$$

We shall seek (in addition to the trivial solution) the simplest spatially periodic solution of the type

$$\varphi(r)=\sum_{k}a_k\,e^{ikr};\quad k_x=k_y=k_z=\frac{2\pi}{d} \tag{24.2}$$

(where d is the period),

corresponding to a crystal of simple cubic structure. Substitution of this expression gives:

$$e^{ikr} = -\lambda e^{ikr} \int_{-\infty}^{+\infty} K(|r-r'|)\, e^{-ik(r-r')}\, dr'.$$

The integral that appears here is a magnitude independent of r, and consequently, function (24.2) is a solution if

$$1 = \lambda \int_{-\infty}^{+\infty} K^*(|r-r'|)\, e^{-ik(r-r')}\, dr', \tag{24.3}$$

where $K^+ = -K$.

Introducing the value of a parameter $\lambda = \lambda_0$, corresponding to the case $k_x = k_y = k_z = 0$ for $d = \infty$, in other words, the case of the absence of the crystalline state, we may write:

$$\lambda = \frac{1}{4\pi \int_0^\infty K^*(\rho) \frac{\sin k\rho}{k\rho} \rho^2\, d\rho} \geqslant \lambda_0 = \frac{1}{4\pi \int_0^\infty K^*(\rho)\, \rho^2\, d\rho}. \tag{24.4}$$

Now this is a basic criterion for the existence of the crystalline state. It contains the condition for the process of the formation of the crystal from the homogeneous phase, and makes it possible to determine the numerical value of the period.

If $\lambda < \lambda_0$, then there are no periodic solutions. The conditions for the beginning of crystallization may be written as follows:

$$4\pi \frac{\rho(0)}{\theta_{\text{кр}}} \int_0^\infty K^*(\rho)\, \rho^2\, d\rho = 1. \tag{24.5}$$

Condition (24.5) can only be satisfied when the integral is positive, that is, when the forces of attraction predominate over the forces of repulsion.

This criterion does not contain any concepts analogous to "spheres of action;" it is of integral nature. Here, the forces are automatically taken into consideration over the entire interval of their action.

It is essential to note the following features of the method presented:

1. If the root of equation (24.3) (value k) is unique, we have a purely periodic distribution. But, it may be (for a given character of the forces of interaction) that the equation gives several roots for k, which in the general case may be incommensurable. In this case, the solutions are of the class of almost periodic functions. In this special case, the crystal will have more complicated properties than the merely periodic phenomenon.

2. The treatment that has been given, based on linear approximation, does not furnish the possibility of defining the amplitudes of the change of potentials and the dependence of these amplitudes on the temperature and density. It is necessary here to inroduce a nonlinear theory.

3. The criterion of crystallization is approximate, since in introducing it no account was taken of the collective functional interactions, expressed by a functional sum.

In order to introduce these interactions in the treatment, we consider equation (17.7), obtained above, which possesses the form we require:

$$\varphi(r)=\sum_{n}\lambda_n\int_{-\infty}^{\infty}\cdots\int_{-\infty}^{\infty}K_n(r,r_1,\ldots,r_n)\{e^{\varphi(r_1)}\ldots e^{\varphi(r_n)}-1\}\,dr_1\ldots dr_n.$$

Linearizing this equation, we have:

$$\varphi(r)-\sum_{n}\lambda_n\int_{-\infty}^{\infty}\cdots\int_{-\infty}^{\infty}K_n(r,r_1,\ldots,r_n)\{\sum_{i}^{n}\varphi(r_i)\}\,dr_1\ldots dr_n, \qquad (24.6)$$

in which case, since the kernels are assumed to depend only on the modulus of difference of the arguments, we, aiming at a solution of the type exp ikr, in a completely analogous way, easily find the generalized criterion of crystallization:

$$1=-4\pi\int_{0}^{\infty}\sum_{n}\lambda_n\sum_{i}^{n}K_{n_i}\frac{\sin k\varrho}{k\varrho}\varrho^2\,d\varrho, \qquad (24.7)$$

where

$$K_{n_i}=\underbrace{\int\cdots\int}_{(n-1)}K_n(r,r_1,\ldots,r_n)\,dr_1\ldots dr_{i-1}\,dr_{i+1}\ldots dr_n.$$

Thus, calculation of the functional interactions has not in principle altered the basic result obtained above: the sudden appearance of a periodic distribution of density at a certain temperature. Only the external form of the crystallization criterion has changed. The period is now found from the more complicated equation (24.7).

The crystallization criterion (24.3) may be given more simply, but therefore, of course, in more approximate form for the case of forces of interaction decreasing sufficiently rapidly with the distance. Then in the integral in (24.1), the only essential factor is the region around the point r' = r.

We therefore expand φ(r') in a Taylor series in the neighborhood of point r' = r:

$$\varphi(r') = \varphi(r) + \sum_{i=1}^{3} (x - x_i) \frac{\partial}{\partial x_i} \varphi(r) + \frac{1}{2} \sum_{\substack{i=1 \\ j=1}}^{3} (x - x_i)(x - x_j) \frac{\partial^2}{\partial x_i \partial x_j} \varphi(r) + \ldots$$

Then

$$\int_{-\infty}^{\infty} K(|r - r'|)\varphi(r')\,dr' = \sum_{s=0}^{\infty} K_{2s} \nabla^{2s} \varphi, \qquad (24.8)$$

where the coefficients of the expansion may be termed "moments" of the energy of interaction:

$$\left. \begin{aligned} k_0 &= \int_{-\infty}^{\infty} K(|r - r'|)\,dr', \\ k_2 &= \int_{-\infty}^{\infty} K(|r - r'|)(x - x')^2\,dr', \\ &\ldots\ldots\ldots\ldots\ldots\ldots \end{aligned} \right\} \qquad (24.9)$$

Confining ourselves to only the first terms of the expansion, we have instead of (24.1) the differential equation

$$\Delta\varphi + \varkappa^2 \varphi = 0, \quad \varkappa^2 = \frac{1 + \lambda k_0}{\lambda k_2}, \qquad (24.10)$$

which, despite the approximation that has been made, still correctly represents the discontinuous nature of the new solutions to the initial integral equation that appear here.

Periodic solutions arise only on the condition: $\varkappa^2 > 0$. We observe that the criterion obtained from this for the beginning of crystallization coincides with that found above.

The spatial periods are defined by the formula:

$$\Lambda^2 = \frac{(2\pi)^2}{\varkappa^2} = (2\pi)^2 \frac{\lambda k_2}{1 + \lambda k_0}. \tag{24.11}$$

For sufficiently large λ (i.e., low temperatures), the magnitude of the periods is independent of the temperature and is determined by the radius of action of the molecular forces

$$\Lambda^2 = (2\pi)^2 \frac{k_2}{k_0}. \tag{24.12}$$

Section 25. Nonlinear Theory

It was emphasized previously that linear theory cannot describe the process of the gradual change in the character of crystalline structure as the temperature decreases. It only defines the conditions for the abrupt appearance of such a structure and makes it possible to establish the magnitude of the spatial period and its dependence on the forces of interaction, the density and the temperature of the medium. The size of the amplitude and the dependence between the phases of the individual harmonics remain unclear. In addition, of course, there is the question as to the reality of the crystallization criterion found, in the more general nonlinear theory.

In order to include nonlinear effects in the theory, we must exactly solve the equation

$$\varphi(r) = \lambda \int_{-\infty}^{\infty} K(|r - r'|)\, e^{-\varphi(r')}\, dr', \quad \lambda = \frac{a}{\theta}, \quad \varphi(r) = \frac{V(r)}{\theta}. \tag{25.1}$$

This is of the type of equations the study of which, as has been said, began with Liapunov's classical work, continued in the mathematical work of Schmidt, Lichtenstein and Nazarov⁺. The

⁺ For the literature see Sec. 18.

mathematical apparatus they developed makes it possible to find the individual branches of the solutions and their behavior upon continuous variation of parameter λ, if any kind of exact solution is known for a given value λ_0.

We formulate the problem of branching precisely as follows: in equation (25.1) to find the value for the variable parameter λ beginning with which a periodic solution branches off from the solution corresponding to uniform density; to show that this solution is exact; and to define variation of the amplitude of the periodic solutions, upon continuous variation of the temperature entering into λ.

We consider this problem. Under the condition

$$\int_{-\infty}^{\infty} K(|r-r'|)\,dr' = K \quad (|K| < \infty)$$

equation (25.1) has an obvious solution corresponding to the uniform distribution of the particles in space

$$\varphi(r) = \frac{V(r)}{\theta} = C.$$

We obtain for the magnitude C, the equation

$$C = \lambda_0 e^{-C} K, \quad \text{where} \quad \lambda_0 = \frac{a}{\theta}$$

is the corresponding characteristic number, but in turn

$$\rho(r) = ae^{-\frac{V(r)}{\theta}} = ae^{-C} = \text{const.}$$

We may normalize the density to unity $\rho = 1$, and hence we obtain an equation in the form

$$\varphi = \frac{K}{\theta}.$$

To elucidate the behavior of the solution in the neighborhood of λ_0, we set $\lambda = \lambda_0 + \varepsilon$ and $\varphi(r) = C + u(r)$. Substituting these in (25.1), we may, expanding exp as a series and taking into consideration that by definition C satisfies the equation at $\lambda = \lambda_0$, obtain the following equation, for solution:

$$u(r) - \lambda_0^* \int_{-\infty}^{\infty} K(|r - r'|)\, u(r')\, dr' = \tau K - \tau \int_{-\infty}^{\infty} K(|r - r'|)\, u(r')\, dr' +$$

$$+ (\lambda_0^* + \tau) \sum_{n=2}^{\infty} (-1)^n \frac{1}{n!} \int_{-\infty}^{\infty} K(|r - r'|)\, u^n(r')\, dr' \qquad (25.2)$$

$$(\lambda_0^* = \lambda_0 e^{-C}, \ \tau = \varepsilon e^{-C}).$$

We note that the linear equation not containing the small parameter τ has the form of (24.1), which we discussed above:

$$K\{\zeta(r)\} \equiv \zeta(r) + \lambda_0^* \int_{-\infty}^{\infty} K(|r - r'|)\, \zeta(r')\, dr' = 0; \qquad (25.3)$$

we saw that it had a periodic solution on the condition

$$\lambda_0^* \sigma(|k|) = -1 \qquad (25.4)$$

(criterion of crystallization), where

$$\sigma(|k|) = 4\pi \int_0^{\infty} K(\rho) \frac{\sin k\rho}{k\rho} \rho^2\, d\rho.$$

Since λ_0^+ is $>$ 0, then $|\underline{k}|$ must satisfy the condition

$$\sigma(|k|) < 0.$$

We solve nonlinear equation (25.2) by the method of successive approximations, setting

$$u(r) = \tau^{\frac{1}{n}} f_1(r) + \tau^{\frac{2}{n}} f_2(r) + \tau^{\frac{3}{n}} f_3(r) + \dots , \qquad (25.5)$$

where n is an integer (the case of branching solutions, n being the possible number of branches.

It may be seen by direct calculation that none of the cases except n = 2 gives any but a trivial solution, or are insoluble altogether. We therefore consider the case n = 2.

Substituting expansion (25.5) in (25.2) leads to the following infinite sequence of equations:

$$\left.\begin{aligned} &K\{f_1(r)\}=0,\\ &K\{f_2(r)\}=\int_{-\infty}^{\infty}K(|r-r'|)\left\{1+\lambda_0^*\frac{f_1^2(r')}{2!}\right\}dr',\\ &K\{f_3(r)\}=\int_{-\infty}^{\infty}K(|r-r'|)\Big\{-f_1(r')+\\ &\qquad\qquad+\lambda_0^*\left[f_1(r')f_2(r')-\frac{1}{3!}f_1^3(r')\right]\Big\}dr',\\ &\dots\dots\dots\dots\dots\dots\dots\dots\dots, \end{aligned}\right\} \quad (25.6)$$

where $K\{f_i\}$ has the form (25.3). We solve this sytem. Obviously, we have (see Sec. 24),

$$f_1(r)=C_1\sin kr,$$

where C_1 is an arbitrary constant, and the value of k is determined by relation (25.4). The subsequent equation should define constant C_1.

We substitute this solution in the right side of the second equation of (25.6) and, taking into consideration that

$$\int_{-\infty}^{\infty}K(|r-r'|)\begin{Bmatrix}\sin nkr'\\ \cos nkr'\end{Bmatrix}dr'=\sigma(n|k|)\begin{Bmatrix}\sin nkr\\ \cos nkr,\end{Bmatrix} \quad (25.7)$$

we obtain:

$$f_2(r)=a+b\cos 2kr+C_2\sin kr, \quad (25.8)$$

where

$$a=K\frac{1+\lambda_0^*\frac{C_1^2}{4}}{1+\lambda_0^*K},\quad b=\frac{\sigma(2|k|)}{\sigma(|k|)-\sigma(2|k|)}\,\frac{C_1^2}{4}. \quad (25.8')$$

Solution (25.8) holds good under the obvious condition that the denominator in (25.8') does not vanish

$$\sigma(|\mathbf{k}|)-\sigma(2|\mathbf{k}|)\neq 0.$$

The third term in (25.8) C_2 sin kr is a solution of a homogeneous equation with the new arbitrary constant C_2.

Accordingly, in the second approximation, constant C_1 is not defined. It is therefore necessary to consider the equation of the third approximation.

The equation for f_3 contains on the right side, the term A sin kr, with a magnitude k. Sin kr satisfy a homogeneous equation. Hence, as the equation for f_3 is solvable only under the condition

$$A=\frac{C_1^3}{4}\left\{\frac{\lambda_0^* K}{1+\lambda_0^* K}-\frac{1}{2}\,\frac{\sigma(2|\mathbf{k}|)}{\sigma(|\mathbf{k}|)-\sigma(2|\mathbf{k}|)}-\frac{1}{2}\right\}-\frac{\frac{1}{\lambda_0^*}}{1+\lambda_0 K}=0, \tag{25.9}$$

$$(\sigma(|\mathbf{k}|)-\sigma(3|\mathbf{k}|)\neq 0).$$

This equation defines the value of C_1. If the sign of the coefficient of C_1^2 is not different from the sign of the second number, then it is sufficient to consider $\lambda=\lambda_0-\varepsilon$, and a real value for C_1 will be assured.

The second condition for the solvability of the equation for f_3 is:

$$\sigma(|\mathbf{k}|)-\sigma(3|\mathbf{k}|)\neq 0.$$

In this way we obtain the following expression for f_3:

$$f_3(r)=A_3\sin 3kr+C_3\sin kr,$$

where C_3 sin kr is the solution of a homogeneous equation with a new arbitrary constant. From the condition for the solvability of the fourth equation, we obtain $C_2 = 0$.

We see that the following approximations do not furnish anything beyond periodic solutions with a short period, since the first part will always be a periodic function. This follows directly from (25.7). As for the homogeneous equation, it is alike in all the approximations.

Continuing the process of systematic solution of system (25.6), we obtain:

$$\left.\begin{aligned} f_{2n}(r) &= \sum_{m=0}^{n} B_m \cos 2mkr + C_{2n} \sin kr, \\ f_{2n-1}(r) &= \sum_{m=1}^{n} A_{2m+1}^{(2\lambda-1)} \sin(2m+1)kr + C_{2n-1} \sin kr \end{aligned}\right\} \qquad (25.10)$$

under the condition:

$$\sigma(|k|) - \sigma(n|k|) \neq 0.$$

Assuming that all the $C_m\ (m < 2n-1)$ are determined, we show that C_{2n-1}, C_{2n} are uniquely defined by the solvability condition for equations (25.6) for f_{2n+1} and f_{2n+2}. The calculations employed for determining C_{2n-1} may be easily applied to prove $C_{2n} = 0$ ($n = 2, 3, \ldots$).

It is obvious that $A_{2m+1}^{(2n-1)}, f_{2n-2}^{(2)}, \ldots, f_1$ does not depend on C_{2n-1}.

Further, we have:

$$K\{f_{2n}(r)\} = \int_{-\infty}^{\infty} K(|r-r'|)\{Q_{2n}(f_{2n-2}, f_{2n-3}, \ldots, f_1) + \lambda_0^* [f_1(r') f_{2n-1}(r') + R_{2n}(f_{2n-2}, \ldots, f_1)]\}\, dr';$$

if we compare this expression with (25.6), then the meaning of Q_{2n} and R_{2n} will be obvious.

From this it is not hard to determine that

$$f_{2n}(r) = C_{2n-1}\left\{\frac{C_1}{2}\frac{\lambda_0^* K}{1+\lambda_0^* K} + \frac{C_1}{2}\frac{\sigma(2|k|)}{\sigma(|k|) - \sigma(2|k|)}\cos 2kr\right\} +$$

+ (terms not containing C_{2n-1}).

In a completely analogous manner

$$K\{f_{2n}(r)\} = \int_{-\infty}^{\infty} K(|r-r'|)\Big\{-f_{2n-1}(r') + \lambda_0^*\Big[[f_1 f_{2n} +$$

$$+ f_2 f_{2n-1} - \frac{1}{2} f_1^2 f_{2n-1}\Big] + (\text{terms not containing } C_{2n-1})\Big\} dr'.$$

The solvability solution for the last equation may be written, after some simple calculation, as:

$$C_{2n-1} = \frac{1}{N_0} P^{(1)}_{2n+1}\{f_{2n-1}, f_{2n-3}, \dots, f_1\}. \qquad (25.11)$$

Here $N_0 = N_0 = \frac{2}{1+\lambda_0 K}$, and $P^{(1)}_{2n+1}$ represents an expression consisting of coefficients of lower approximations and definable by a choice of those of them that are present with sin kr, and of combination $f_i^{(\alpha)} f_j^{(\beta)} \dots f_r^{(\gamma)}$, not containing C_{2n-1}. Consequently, the conditions for solvability of (25.11) are linear with respect to C_{2n-1}.

In this way, we have obtained a formal solution for u (r) in the form:

$$u(r) = \sum_{m=0}^{\infty} \tau^{\frac{2m+1}{2}} \Big\{ \sum_{l=1}^{m} A^{(2m+1)}_{2l+1} \sin(2l+1)kr + C_{2m+1} \sin kr \Big\} +$$

$$+ \sum_{m=1}^{\infty} \tau^m \Big\{ \sum B^{(2m)}_{2l} \cos 2m\, kr + C_{2m} \sin kr \Big\}. \quad (25.12)$$

The series obtained, on condition of being convergent, represent a periodic function with a period defined by the vector k.

As we see, the apparatus proposed makes it possible to define not only the condition for the appearance of a periodic structure and the magnitude of the period, but also gives the way of systematically determining the amplitude of the several harmonics in Fourier series (25.12).

In order to give a solution in final form, it is still necessary to determine the magnitude of the expansion parameter τ. To this end, we use the normalization condition, choosing as the field of integration a volume that is a multiple of an elementary cell of periodicity. We have:

$$\frac{1}{V_0}\int_{V_0} ae^{-c}\, e^{-u(r)}\, dr = 1,$$

setting $$a = a_0(1 + \tau\mu_1 + \tau^2\mu_2 + \tau^3\mu_3 + \dots)$$

and expanding $e^{-u(r)}$ in a series in $u(\underline{r})$, we obtain a sequence of equations :

$$a_0 e^{-c} = 1,$$

$$\frac{1}{V_0}\int_{V_0} (-C_1 \sin kr)\, dr = 0,$$

$$\frac{1}{V_0}\int_{V_0} \left\{-f_2(r) + \mu_1 + \frac{1}{2} f_1(r)\right\} dr = 0,$$

$$\dots\dots\dots\dots\dots\dots$$

It is not hard to see that this system uniquely defines all the μ_i. Since, further

$$\tau = \varepsilon e^{-c} = e^{-c}\left(\frac{a}{\theta} - \frac{a_0}{\theta_0}\right),$$

then substitution in this of the value for a gives us an equation determining $\tau = \tau(\theta, \theta_0, \mu_i)$, that is

$$\tau = \left[\frac{1}{\theta} - \frac{1}{\theta_0} + \frac{1}{\theta}(\mu_1\tau + \mu_2\tau^2 + \dots)\right].$$

For $\theta \sim \theta_0$

$$\tau = \frac{1}{\theta} - \frac{1}{\theta_0}.$$

Section 26. Investigation of Convergence of the Expansion u ($\underline{r}$)

It is necessary to show that the series obtained (25.12) converges. This question was studied in collaboration with Yakovlev. The essential aspect is that the initial equation is singular (inside domain of integration), and therefore the general theorems of the theory of branching cannot immediately be transferred to it. In showing the convergence in (25.12), we therefore justify the entire course of the argument hitherto.

We make use of Nazarov's method of proof for nonsingular nonlinear integral equations, introducing the modifications necessary for each case.

We consider the function

$$S(z) = (\lambda_0^* + \nu^2)(e^z - 1 - z) + \nu^2(1 + z) \tag{26.1}$$

or, what comes to the same thing,

$$S(z) = (\lambda_0^* + \nu^2)\left(\frac{z^2}{2!} + \frac{z^3}{3!} + \frac{z^4}{4!} + \dots\right) + \nu^2(1 + z), \tag{26.2}$$

where λ_0+, $\nu^2 = \tau$ has the same meaning as before.

We set $z(\nu) = \nu x_0 + \nu^2(x_1^2 + y_1^2) + \nu^3(x_2 + y_2) + \dots$ and insert this series in (26.2). We obtain:

$$S(z) = \nu^2 S_2 + \nu^3 S_3 + \nu^4 S_4 + \dots,$$

where

$$S_2 = 1 + \lambda_0^* \frac{x_0^2}{2!}, \quad S_3 = x_0 + \lambda_0^* \left[x_0(x_1 + y_1) + \frac{x_0^3}{3!}\right], \dots \tag{26.3}$$

Comparing S_2, S_3 ... with the expressions under the integral sign in (25.6), we see that the corresponding expressions coincide if we perform the change

$$f_i(r) \quad \text{for} \quad x_{i-1}+y_{i-1}$$

and change the indices. Accordingly, if $x_0 \geqslant |C_1 \sin kr|$, $x_i + -y_i \geqslant |f_i(r)|$,
then S_n will be majorant for the corresponding expressions of (25.6). For the purpose of forming the entire group of $\{x_i\}, \{y_i\}$, satisfying these equation, we set $x_0 = C_1$.

Now considering the sequence

$$\sigma(|k|)-\sigma(0)|, \quad |\sigma(|k|)-\sigma(2|k|), \ldots, \ |\sigma(|k|)-\sigma(n|k|)|, \ldots$$

and noting that $\lim \sigma(n|k|)=0$, we assume that there exists an $\eta>0$, such that

$$|\sigma(|k|)-\sigma(n|k|)>\eta \tag{26.4}$$

for any $n = 0, 1, 2, \ldots$.

Introducing the designation $f_{2n}(r)=P_{2n}(r), f_{2n+1}(r)=P_{2n}+C_{2n+1}\sin kr$, it is easy to see that

$$|f_{2n}|<\frac{S_{2n}}{\eta}, \quad |P_{2n+1}|<\frac{S_{2n+1}}{\eta}. \tag{26.5}$$

defining $y_n=\frac{S_{n+1}}{\eta}$, we get:

$$y_{2n-1}>|f_{2n}|, \ y_{2n}>|P_{2n+1}|,$$

where constants x_{2n+1} ($n = 0, 1, 2, \ldots$) are arbitrary, subject only to the condition $x_{2n+1} \gg 0$. We shall later make use of this freedom of choice.

We now define x_{2n} in such a way that $x_{2n}>|C_{2n+1}\sin kr|$
We consider the solvability equation for (25.11) and evaluate the right side of this equation, that is, $P^{(1)}_{2n+1}$. We write two successive equations from (25.6) in somewhat different form

$$L\{f_{2n}\} = \int\limits_{-\infty}^{\infty} K(|r-r'|)\{Q_{2n}(f_{2n-2}, \ldots, f_1) + \\ + \lambda_0^* M_{2n}(f_{2n-1}, \ldots, f_1)\}\, dr',$$

$$L\{f_{2n+1}\} = \int\limits_{-\infty} K(|r-r'|)\{Q_{2n-1}(f_{2n-1}, \ldots, f_1) + \\ + \lambda_0^* M_{2n+1}(f_{2n}, \ldots, f_1)\}\, dr'.$$

Apart from the signs and coefficients, the expressions $f_1(Q_{2n+1} + \lambda_0 M_{2n})$, consist of the same terms as the expression

$$Q_{2n+1} + \lambda_0 M_{2n+1} \backsim -f_{2n-1} + \lambda_0^*[f_1 f_{2n} + f_2 f_{2n-1}] + f_1(Q_{2n} + \lambda_0^* M_{2n}),$$

with the exception of a group consisting of three terms, that is $Q_{2n+1} + \lambda_0^* M_{2n+1}$ where the sign denotes correspondence in the sense indicated above.

For each member in M_{2n}, Q_{2n}, having the form

$$\frac{r}{m!} f_i f_k \ldots f_l = \Phi_{2n}^{ik\ldots l} \begin{pmatrix} i+k+\ldots+l=2n \\ r \leqslant m \\ m \leqslant 2n \end{pmatrix}, \tag{26.6}$$

in the expressions M_{2n+1}, Q_{2n+1}, the corresponding term has the form

$$\frac{p}{(m+1)!} f_1 f_i f_k \ldots f_l = \Phi_{2n+1}^{1, i, k \ldots l} \begin{pmatrix} 1+i+k+\ldots+l=2n+1 \\ p \leqslant m+1 \\ m \leqslant 2n+1 \end{pmatrix}. \tag{26.7}$$

Inasmuch as

$$\frac{r}{m!} \geqslant \frac{p}{(m+1)!}, \quad \text{TO} \quad |f_1 \Phi_{2n}^{ik\ldots l}| \geqslant |\Phi_{3n+1}^{1, i, k \ldots l}|. \tag{26.7'}$$

As we know, $P^{(1)}_{2n+1}$ is obtained from $Q_{2n+1} + {}_0 M_{2n+1}$ by excluding the coefficients in sin <u>kr</u> that do not contain C_{2n-1}, that is, from the expression $-P^{(2)}_{2n-1} + \lambda_0^* [f_1 f^0_{2n} + f_2 P_{2n-1}] +$ the other terms not containing C_{2n-1} (f^0_{2n} denotes the part of f_{2n} without terms with C_{2n-1}).

In this expression, replacing all the quantities by their greatest values, and taking into consideration, on the basis of (26.7') that the majorant function $f_2(Q_{2n} + \lambda_0 M_{2n})$ is still more majorant for the terms not containing C_{2n-1}, we have:

$$|P^{(1)}_{2n+1}| < y_{2n-2} + \lambda_0^* \left(x_0 \frac{S_{2n}}{\gamma_1} + \frac{S_2}{\eta} y_{2n-2}\right) + x_0 S_{2n},$$

all

$$|P^{(1)}_{2n+1}| < \alpha S_{2n} + \beta y_{2n-2}.$$

Taking into consideration relationship (25.11), we obtain:

$$|C_{2n-1}||N_0| < \alpha S_{2n} + \beta y_{2n-2},$$

$$\alpha = x_0 \left(1 + \frac{\lambda_0^*}{\eta}\right), \quad \beta = 1 + \lambda_0^* \frac{S_3}{\gamma_1}.$$

Further defining x_{2n-2} by the equation

$$x_{2n-2}|N_0| = \alpha S_{2n} + \beta y_{2n-2},$$

we have

$$x_{2n-2} > |C_{2n-1}|. \tag{26.8}$$

Since x_{2n-1} is arbitrary, we assume in general

$$\left.\begin{aligned} x_n|N_0| &= \alpha S_{n+2} + \beta y_n, \\ y_n &= \frac{S_{n+1}}{\gamma_1} \quad (n = 1, 2, \ldots). \end{aligned}\right\} \tag{26.9}$$

Hence, x_n, y_n, defined by this system, satisfy the condition

$x_n + y_n > |f_{n+1}(r)|$. But it is still necessary to show that the systems (26.9), considered along with (26.3), actually define x_n, y_n as functions from x_0. We prove the solvability of this system, and to this end introduce series:

$$\left.\begin{aligned} x &= \nu^2 x_1 + \nu^3 x_2 + \nu^4 x_3 + \ldots, \\ y &= \nu^2 y_1 + \nu^3 y_2 + \nu^4 y_3 + \ldots \end{aligned}\right\} \tag{26.10}$$

Multiplying the first of equations (26.9) by ν^{n+2}, and the second to the ν^{n+1}, and summating each separately, according to n, we obtain:

$$\nu x |N_0| = \alpha S(\nu, x, y) + \beta \nu y - \alpha \nu^2 S_2,$$
$$y = \frac{1}{\eta} S(\nu, x, y).$$

Setting $x = \nu X$, $y = \nu Y$ and canceling ν^2 out of the first equation and ν from the second, we have:

$$\Phi_1(\nu, X, Y) = 0,$$
$$\Phi_2(\nu, X, Y) = 0,$$

where

$$\Phi_1(\nu, X, Y) = -|N_0| X + (\lambda_0^* + \nu^2)\alpha \left\{\frac{e^{\nu(x_0+X+Y)} - 1 - \nu(x_0+X+Y)}{\nu^2}\right\} +$$
$$+ \alpha[1 + \nu(x_0 + X + Y)] + \beta Y - \alpha S_2,$$
$$\Phi_2(\nu, X, Y) = -Y + (\lambda_0^* + \nu^2)\frac{1}{\eta}\left\{\frac{e^{\nu(x_0+X+Y)} - 1 - \nu(x_0 + X + Y)}{\nu^2}\right\} +$$
$$+ \nu[1 + \nu(x_0 + X + Y)].$$

Since

$$\Phi_1\Big|_{\substack{\nu=0\\X=0\\Y=0}} = 0, \qquad \Phi_2\Big|_{\substack{\nu=0\\X=0\\Y=0}} = 0$$

and

$$\frac{D(\Phi_1, \Phi_2)}{D(X, Y)} = \begin{vmatrix} -|N_0| + \lambda_0^* \alpha x_0, & \beta + \lambda_0^* \alpha x_0 \\ 0, & -1 \end{vmatrix} \neq 0, \cdot$$

then equations $\Phi_1(\nu, X, Y) = 0, \ \Phi_2(\nu, X, Y) = 0$ implicitly define $X = X(\nu), \ Y = Y(\nu)$ in the neighborhood of $= 0$.

Consequently, there also exist functions x, y defined by (26.10), that is series representing them converge at least over a finite interval in the neighborhood of $\nu = 0$.

Thus, the series of $z(\nu) = \nu x_0 + \nu^2 (x_1^2 + y_1^2) + \nu^3 (x_2 + y_2) + \cdots$ likewise converges, and since it is majorant for the series u (r), the group is completed.

Section 27. Role of Collective Interactions

Up to the present, we have considered the potential of the field in which each particle moves, in the form

$$V(r) = \int_{-\infty}^{\infty} K(|r - r'|)\rho(r')\,dr'.$$

We now consider the more complicated case of N uniform particles (see Sec. 8)

$$V(r) = \sum_{n=1}^{N-1} \int_{-\infty}^{\infty} \cdots \int_{-\infty}^{\infty} K_n(r, r_1, \ldots, r_n)\rho(r_1)\ldots\rho(r_n)\,dr_1\ldots dr_n \tag{27.1}$$

and pose the question as to the existence of an exact periodical solution in this case as well.

This question is of particular interest inasmuch as, as we have seen above, introduction of collective interactions into the Gibbs' method leads to specific difficulties (see Sec. 5).

On the basis of the results obtained in Sec. 17, the problem consists in the solution of equation (17.4)

$$\varphi(r) = \sum_{n=1}^{N-1} \lambda_n \int_{-\infty}^{\infty} \cdots \int_{-\infty}^{\infty} K_n(r, r_1, \ldots, r_n) \exp\{-\sum_{j=1}^{n} \varphi(r_j)\}\,dr_1\ldots dr_n \tag{27.2}$$

$$\left(\varphi(r) = \frac{V(r)}{\theta}, \quad \rho(r) = ae^{-\varphi(r)}, \quad \lambda_n = \frac{a^n}{\theta}\right).$$

Assuming that

$$K_n(r, r_1, r_2, \ldots, r_n) = \mathcal{K}_n(|r-r_1|, \; |r_2-r_3|, \ldots),$$
$$\left|\int_{-\infty}^{\infty} K_n(r, r_1, \ldots, r_n)\, dr_1 \ldots dr_n\right| < +\infty,$$

we have the obvious solution corresponding to the case of uniform density:

$$\left.\begin{aligned} \varphi_0(r) &= C, \\ C &= \sum_{n=1}^{N-1} \lambda_n \cdot K_n e^{-nC}, \end{aligned}\right\} \tag{27.3}$$

where

$$K_n = \int_{-\infty}^{+\infty} \ldots \int K_n(r_1, \ldots, r_n)\, dr_1 \ldots dr_n.$$

Setting $\theta = \theta_0$, we seek a solution to (27.2) that differs from a constant for $\theta \neq \theta_0$, setting:

$$\Phi(r) = C + \varphi(r),$$
$$\lambda_n = \lambda_{n_0} + \tau,$$

where τ is defined by the relation

$$\frac{a}{\theta} = \frac{a_0}{\theta_0} + \tau$$
$$\left(\lambda_{n_0} = \frac{a_0}{\theta_0}\right). \tag{27.4}$$

We obtain the equation

$$\Delta\{u(r)\} = \sum_{l=1}^{N-1} \tau e^{-lc} K_l -$$
$$- \sum_{l=1}^{N-1} \tau e^{-lc} \int_{-\infty}^{\infty} \ldots \int K_l(r, r_1, \ldots, r_l) \left\{\sum_{j=1}^{l} u(r_j)\right\} dr_1 \ldots dr_l +$$
$$+ \sum_{l=1}^{N-1} (\lambda_l + \tau) e^{-lc} \int_{-\infty}^{\infty} \ldots \int K_l(r, r_1, \ldots, r_l) \times$$
$$\times \left[\sum_{n=2}^{\infty} (-1)^n \frac{1}{n!} \left\{\sum_{j=1}^{l} u(r_j)\right\}^n\right] dr_1 \ldots dr_l, \tag{27.5}$$

where

$$\Lambda\{u(r)\} = u(r) + \\ + \sum_{l=1}^{N-1} \lambda_l e^{-lc} \int_{-\infty}^{+\infty} \cdots \int K_l(r, r_1, \ldots, r_l) \left\{ \sum_{j=1}^{l} u(r_j) \right\} dr_1 \ldots dr_l.$$

This equation is thus far exact. We now apply the apparatus of the theory of branching and we assume

$$\varphi(r) = \tau^{\frac{1}{2}} \varphi_1(r) + \tau \varphi_2(r) + \tau^{\frac{3}{2}} \varphi_3(r) + \ldots, \tag{27.6}$$

in which case we obtain a sequence of equations for determining $\phi_i(\underline{r})$.

The first of these equations has the form

$$\Lambda\{f_1(r)\} = 0. \tag{27.7}$$

A solution of it will be constituted by $f_1(r) = C_1 \sin \underline{k_0 \xi}$ on condition that $|\underline{k_0}|$ satisfies the equation

$$-1 = \sum_{l=1}^{N-1} \lambda_l e^{-lc} \sum_{j=1}^{l} \sigma_{0\,1\ldots l}^{(j)}(|\mathbf{k}_0|), \tag{27.8}$$

The second equation of the sequence, after substitution of the first solution in its right side leads to the form:

$$\sigma_l^{(j)}(|\mathbf{k}_0|) = \int_{-\infty}^{\infty} \cdots \int K_l(r, r_{(1)}, \ldots, r_{(l)}^{(l)}) e^{-i\mathbf{k}_0(r - r_{(j)})} dr_{(1)} \ldots dr_{(l)};$$

$$k_0 = |\mathbf{k}_0|.$$

$$\Delta\{f_2(r)\} = \sum_{l=1}^{N-1} e^{-lc} K_l + \sum_{l=1}^{N-1} \lambda_l e^{-lc} \frac{C_1^2}{2!} \int \underset{-\infty}{\overset{\infty}{\cdots}} \int K_l(\boldsymbol{r}, \boldsymbol{r}_{(1)}, \ldots, \boldsymbol{r}_{(l)}) \times$$

$$\times \left\{ \sum_{j=1}^{l} \sin^2 \boldsymbol{k}_0 \boldsymbol{r}_{(j)} \right\} d\boldsymbol{r}_{(1)} \ldots d\boldsymbol{r}_{(l)} +$$

$$+ \sum_{l=1}^{N-1} \lambda_l e^{-lc} C_1^2 \sum_{1 \leqslant s < t \leqslant l} \int \underset{-\infty}{\overset{\infty}{\cdots}} \int K_l(\boldsymbol{r}, \boldsymbol{r}_{(1)}, \ldots, \boldsymbol{r}) \times$$

$$\times \sin \boldsymbol{k}_0 \boldsymbol{r}_{(s)} \sin \boldsymbol{k}_0 \boldsymbol{r}_{(t)} \, d\boldsymbol{r}_{(1)} \ldots d\boldsymbol{r}_{(l)}.$$

Further designating

$$\sigma_l^{(s, t, \ldots, r)}(mk_0, pk_0, \ldots, qk_0) = \int \underset{-\infty}{\overset{\infty}{\cdots}} \int K_l(\boldsymbol{r}, \ldots, \boldsymbol{r}_{(l)}) \times$$

$$\times e^{im \boldsymbol{k}_0 (\boldsymbol{r} - \boldsymbol{r}_{(s)}) + ip \boldsymbol{k}_0 (\boldsymbol{r} - \boldsymbol{r}_{(t)}) + \ldots + iq \boldsymbol{k}_0 (\boldsymbol{r} - \boldsymbol{r})_{(l)}} d\boldsymbol{r}_{(1)} \ldots d\boldsymbol{r}_{(l)},$$

we have the obvious formulas

$$\int \underset{-\infty}{\overset{\infty}{\cdots}} \int K_l(\boldsymbol{r}, \boldsymbol{r}_{(1)}, \ldots, \boldsymbol{r}_{(l)}) \sin \boldsymbol{k}_0 \boldsymbol{r}_{(s)} \sin \boldsymbol{k}_0 \boldsymbol{r}_{(t)} \, d\boldsymbol{r}_{(1)} \ldots d\boldsymbol{r}_{(l)} =$$

$$= \frac{1}{2} \sigma_l^{(s, t)}(\boldsymbol{k}_0, \boldsymbol{k}_0) - \frac{1}{2} \sigma_l^{(s, t)}(\boldsymbol{k}_0, \boldsymbol{k}_0) \cos 2\boldsymbol{k}_0 \boldsymbol{r},$$

$$\int \underset{-\infty}{\overset{\infty}{\cdots}} \int K_l(\boldsymbol{r}, \boldsymbol{r}_{(1)}, \ldots, \boldsymbol{r}_{(l)}) \sin^2 \boldsymbol{k}_0 \boldsymbol{r}_{(s)} \, d\boldsymbol{r}_{(1)} \ldots d\boldsymbol{r}_{(s)} \ldots d\boldsymbol{r}_{(l)} =$$

$$= \frac{1}{2} K_l - \frac{1}{2} \sigma_l^{(s)}(2\boldsymbol{k}_0) \cos 2\boldsymbol{k}_0 \boldsymbol{r}.$$

Making use of these relationships, we write the equation for f2(r) in the form

$$\Delta\{f_2(r)\} = \sum_{l=1}^{N-1} e^{-lc} K_l +$$

$$+ \sum_{l=1}^{N-1} \lambda_l e^{-lc} \frac{C_1^2}{2} \left\{ \sum_{1 \leqslant s < t \leqslant l} \sigma_l^{(s, t)}(k_0, k_0) + \frac{1}{2} l K e_l \right\} -$$

$$- \cos 2\boldsymbol{k}_0 \boldsymbol{r} \sum_{l=1}^{N-1} \lambda_l e^{-lc} \frac{C_1^2}{2} \left\{ \sum_{1 \leqslant s < t \leqslant l} \sigma_l^{(s, t)}(k_0, k_0) + \frac{1}{2} \sum_{1 \leqslant s \leqslant l} \sigma_l^{(s)}(2k_0) \right\} \qquad (27.9)$$

It is not hard to verify the correctness of the following relationships:

$$\Lambda\{A\} = A\,(1 + \sum_{l=1}^{N-1} \lambda_l e^{-lc} lK_l) \quad \text{(A being a constant)}$$

$$\Lambda\{A \cos 2\boldsymbol{k}_0\boldsymbol{r}\} = A\,(1 + \sum_{l=1}^{N-1} \lambda_l e^{-lc} \sum_{j=1}^{l} \sigma_l^{(j)}(2k_0))\cos 2\boldsymbol{k}_0\boldsymbol{r}.$$

We find from this a solution for $f_2(r)$

(27.10)

$$f_2(\boldsymbol{r}) = A_2 + B_2 \cos 2\boldsymbol{k}_0\boldsymbol{r}$$

under the condition

$$1 + \sum_{l=1}^{N-1} \lambda_l e^{-lc} \sum_{j=1}^{l} \sigma_l^{(j)}(2k_0) \neq 0,$$

where A_2 and B_2 are equally determined constants.

Inserting $f_1(\underline{r})$ and $f_2(\underline{r})$ in the succeeding equations of the system, we obtain:

$$\begin{aligned}\Lambda\{(f_3\boldsymbol{r})\} = \sin \boldsymbol{k}_0\boldsymbol{r} \sum_{l=1}^{N-1} \Big\{ -c_l e^{-lc} \sum_{j=1}^{l} \sigma_l^{(j)}(k_0) + \lambda_l e^{-lc} \Big[lA_2 \sum_{j=1}^{l} \sigma_l^{(j)}(k_0) - \\ - B_2 \Big(\sum_{1<j<s<l} \frac{1}{2} \sigma_l^{(j,s)}(k_0,\ 2k_0) + \frac{1}{2} \sum_{j=1}^{l} \sigma_l^{(j)}(k_0) \Big) - \\ - \frac{1}{3!} C_1^2 \Big(\sum_{\substack{1<s,t,j<l \\ s\neq t\neq j}} \sigma_l^{(s,t,j)}(k_0,\ k_0,\ k_0) + \frac{9}{4} \sum_{j=1}^{l} \sigma_l^{(j)}(k_0) + \\ + \frac{3}{4} \sum_{1<j<s<l} \sigma_l^{(j,s)}(k_0,\ 2k_0) \Big) \Big] \Big\} + B_3^0 \sin 3\boldsymbol{k}_0\boldsymbol{r}.\end{aligned} \qquad (27.11)$$

The equation under consideration is solvable only under the conditions that the coefficient in sin $\underline{k_0 r}$ vanishes and

$$1+\sum_{l=1}^{N-1}\lambda_l e^{-lc}\sum_{j=1}^{l}\sigma_l^{(j)}(3k_0)\neq 0. \tag{27.12}$$

The condition that the coefficient in sin $\underline{k_0 r}$ vanish, gives an equation for determining C_1. This equation has the form

$$\sum_{l=1}^{N-1}\Big\{\sum_{j=1}^{l} e^{-lc}\sigma_l^{(j)}(k_0)+A_2\sum_{j=1}^{l}\lambda_l e^{-lc}e\sigma^{(j)}e(k_0)-$$
$$-\frac{1}{2}B_2\lambda_l e^{-lc}\Big(\sum_{1\leqslant j<s\leqslant l}\sigma_l^{(j,s)}(k_0,\ 2k_0)+\sum_{j=1}^{l}\sigma_l^{(j)}(k_0)\Big)-$$
$$-\frac{C_1^2}{6}\lambda_l e^{-lc}\Big(\sum_{1\leqslant s\neq j\neq t<l}\sigma_l^{(j,s,t)}(k_0,k_0,k_0)+$$
$$+\frac{3}{4}\sum_{1\leqslant j<s\leqslant l}\sigma_l^{(j,s)}(k_0,\ 2k_0)+\frac{9}{4}\sum_{j=1}^{l}\sigma_l^{(j)}(k_0)\Big)\Big\}=0. \tag{27.13}$$

Here A_2 and B_2 contain C_1^2. The solvability of the equation with respect to real values of C_1 may be shown in the same way as in Sec. 25, that is, if the equation gives an imaginary value for C_1, then an expansion is performed for $\lambda^+ = \lambda - \tau$. The subsequent construction of the solutions goes forward in the same way as in Sec. 26. We add to each particular solution of the system the general solution of the corresponding homogeneous equation C_n sin $\underline{k_0 r}$, and we define C_n from the following approximations. In general, $C_{2m} = 0$, and C_{2n-1} are defined from the linear equation of solvability for $f_{2n+1}(\underline{r})$. The linearity of these equations follows from the fact that C_{2n-1} is introduced into $f_{2n-1}(\underline{r})$, while the equations for $f_{2n}(\underline{r})$ and $f_{2n+1}(\underline{r})$ do not contain terms of the type $f_{2n-1}^3(r),\ f_{2n-1}(r)f_{2n}(r),\ f_{2n}^3(\underline{r})$, so that the powers of τ corresponding to these terms are greater than $\frac{2n+1}{2}$ (the power of τ , for which the equation for $f_{2n+1}(\underline{r})$ is selected).

We can thus construct a series for $u(\underline{r})$ that, on the condition of convergence, represents a periodic function. There are no difficulties in proving the convergence of the series: it is sufficient to require that the conditions be satisfied

$$\left|1+\sum_{l=1}^{N-1} \lambda_l e^{-lc} \sum_{j=1}^{l} \sigma_l^{(j)}(nk_0)\right| > \rho \qquad (\rho > 0) \tag{27.14}$$

and to take a third function $S(z)$ as in Sec. 26 as the initial function.

We further introduce the normalization condition:

$\frac{1}{V_0}\int\limits_{V_0} fe^{-c-u(r)}\,dr = 1$ (V_0 being a volume which is a multiple of the cube of periodicity) and, setting $f = f_0(1+\tau\mu_1+\tau^2\mu_2+\tau^3\mu_3+\ldots)$, we define all the constants f_0, μ_i. Thereby the crystallization equation (27.8) is reduced to the form

$$\theta_0 = -\sum_{n=1}^{N-1}\sum_{j=1}^{n} \sigma_n^{(j)}(k_0). \tag{27.15}$$

The parameter of expansion τ is defined by the equation

$$\tau = \left[\frac{1}{\theta} - \frac{1}{\theta_0} + \frac{1}{\theta}(\tau\mu_1 + \tau^2\mu_2 + \ldots)\right]\exp\left\{-\frac{\sum\limits_{n=1}^{N-1} K_n}{\theta_0}\right\} \tag{27.16}$$

In this way, a periodic solution for a system of particles is likewise realized for more complex laws of interaction as well.

Section 28. "Lattice" Solution for $\theta = 0$

In the preceding paragraphs we have sought a solution for the initial equation in the class of spatially periodic functions, assuming that $\theta \neq 0$. As has been mentioned, the point $\theta = 0$, represents a singular value of this parameter in the initial equation. The previous methods of solution are therefore inapplicable here.

To obtain a limiting solution, for $\theta \to 0$, of the equation considered above (17.1)

$$V(r)=\int_{-\infty}^{\infty} K(|r-r'|)\,a(\theta)\,e^{-\frac{V(r')}{\theta}}\,dr', \tag{28.1}$$

we apply Laplace's asymptotic formula[+]. For the three-dimensional case, this formula has the following form. Let there be, for $k \to \infty$, $h(x, y, z)$ a bounded, doubly-differentiable function and $\Phi(x, y, z)$ a bounded function. Further, let:

$$h(x, y, z) > 0,$$
$$(a-\eta \leqslant x \leqslant a+\eta;\quad b-\eta \leqslant y \leqslant b+\eta;\quad c-\eta \leqslant z \leqslant c+\eta),$$
$$h_{x_i}(a, b, c)=0,\quad h_{x_i x_i}(a, b, c)<0\quad (i=x, y, z),$$

where the index x_i designates the derivative along the i-coordinate. Further, $h(x, c_2, c_3)$ $h(c_1, y, c_3)$ and (c_1, c_2, z) do not decrease in the respective intervals

$$(a \leqslant x \leqslant a+\alpha),\quad (b \leqslant y \leqslant b+\beta),\quad (c \leqslant z \leqslant c+\gamma).$$

and do not increase in the intervals

$$(a-\alpha \leqslant x \leqslant a),\quad (b-\beta \leqslant y \leqslant b),\quad (c-\gamma \leqslant z \leqslant c)$$

Then

$$\int_{a-\alpha}^{a+\alpha}\int_{b-\beta}^{b+\beta}\int_{c-\gamma}^{c+\gamma} \Phi(x, y, z)\,e^{kh(x, y, z)}\,dx\,dy\,dz \sim 8\Phi(a, b, c)\,e^{kh(a, b, c)} \times$$
$$\times\left(-\frac{\pi}{2kh_{xx}(a, b, c)}\right)^{\frac{1}{2}}\left(-\frac{\pi}{2kh_{yy}(a, b, c)}\right)^{\frac{1}{2}}\left(-\frac{\pi}{2kh_{zz}(a, b, c)}\right)^{\frac{1}{2}}. \tag{28.2}$$

In our case k corresponds to $\frac{1}{\theta}$, $h(x, y, z)$ corresponds to

[+] See, for example, Widder, The Laplace transformation, p. 277-280, 194 - (one dimension).

$V(\underline{r})$ and Φ (x, y, z) corresponds to $K(|\underline{r}-\underline{r}'|)$. We set

$$c=nd,\ \alpha=\beta=\gamma=\frac{d}{2}(l,\ m,\ n=0\pm1,\ \ldots),$$ that is,

we assume that point (a, b, c) is at a point of the cubic crystal lattice period d (we consider an infinite crystal). We integrate over the volume of a cube with edge d, with at the center of this cube a lattice point $\underline{r}_s$ = (ld, md, nd) (s indicating, for brevity, the group of numbers l, m, n). Then we impose the following conditions on the function $V(\underline{r})$ at the point $\underline{r}_s$:

$$V(r_s)<0,\ V_{x_i}(r_s)=0,\ V_{x_ix_i}(r_s)>0\ \ (x_i=x,\ y,\ z). \tag{28.3}$$

We shall test the possibility of this assumption after having obtained the solution. We observe that the condition $V(\underline{r}) < 0$ is realized for the case of attractive forces; the conditions imposed on the derivatives define the lattice points as points corresponding to a minimum potential energy.

Applying Laplace's formula to equation (28.1) indicates $\theta \longrightarrow 0$, we obtain by summation for all values of l, m, n from $-\infty$ to $+\infty$

$$V(r)=a(\theta)\sum_s K(|r-r_s|)\,e^{-\frac{V(r_s)}{\theta}}\frac{(2\pi\theta)^{\frac{3}{2}}}{(V_{xx}V_{yy}V_{zz})^{\frac{1}{2}}}. \tag{28.4}$$

Requiring that

$$V(r_s)=\text{const.},\ V_{xx}(r_s)=V_{yy}(r_s)=V_{zz}(r_s)=\text{const.}$$

(for any s), (28.5)

we obtain:

$$V(r)\rightarrow\sum_s K(|r-r_s|). \tag{28.6}$$

On the other hand, if the maximums $\rho(\underline{r})$ are situated near the lattice points, it is natural to assume

$$V(r_s)\rightarrow a(\theta)e^{-\frac{V(r_s)}{\theta}}\left(\frac{2\pi\theta}{V_{xx}(r_s)}\right)^{\frac{3}{2}}\sum_s K(|r-r_s|). \tag{28.7}$$

Equating (28.6) and (28.7), we find:

$$a(\theta) = e^{-\frac{V(r_s)}{\theta}} \left(\frac{V_{xx}(r_s)}{2\pi\theta} \right)^{\frac{3}{2}}. \qquad (28.8)$$

Hence, the asymptotic form of $V(\underline{r})$ for $\theta \to 0$ is a periodic function (28.7) with an arbitrary period d.

We find the expression for density

$$\rho(r) = a(\theta) e^{-\frac{V(r)}{\theta}}.$$

To this end we expand $V(\underline{r})$ in this formula in a series in the neighborhood of each of the points $\underline{r}_s$ in powers of $(\underline{r} - \underline{r}_s)$ and confine ourselves to the quadratic terms. By virtue of (28.3) and (28.5), we have:

$$V_s(r) = V(r_s) + (r - r_s)^2 \frac{V_{xx}(r_s)}{2}. \qquad (28.9)$$

Substituting (28.9) in the expression for $\rho(\underline{r})$, taking (28.7) into consideration and introducing the designation

$$2\frac{\theta}{V_{xx}} = \alpha^2, \qquad (28.10)$$

we obtain for ρ_s:

$$\rho_s(r) = (\alpha\pi^{\frac{1}{2}})^{-3} e^{-\frac{[(x-ld)^2+(y-md)^2+(z-nd)^2]}{\alpha^2}}. \qquad (28.11)$$

This is the value of the density around a lattice point $\underline{r}_s$ = (ld, md, nd).

The complete density is obtained by summation of $\rho_s(\underline{r})$ for all the lattice points:

$$\rho(x, y, z) = (\alpha\pi^{\frac{1}{2}})^{-3} \sum_s e^{-\frac{[(x-ld)^2+(y-md)^2+(z-nd)^2]}{\alpha^2}}. \qquad (28.12)$$

We consider the normalization question. Let there be in d^3, the volume of a cube with its center at a lattice point (ld, md, nd) even a single particle; in that case, the integral with respect to ρ over this volume should be equal to unity. It is clear from (28.10) that this condition is satisfied if we consider that

$$\int_{l,d-\frac{d}{2}}^{l,d+\frac{d}{2}} \sum e^{-\frac{(x-ld)^2}{\alpha^2}} dx = \int_{-\infty}^{\infty} e^{-\frac{(x-ld)^2}{\alpha^2}} dx = \alpha\pi^{\frac{1}{2}}.$$

We see from (28.11) that for $\alpha \longrightarrow 0$, the expression for the density is

$$\rho(x, y, z) \to \text{const.} \sum_{s} \delta(x-ld)\,\delta(y-md)\,\delta(z-nd). \tag{28.13}$$

Hence, for $\theta = 0$, the solution is given by formula (28.13).

This solution is a lattice of point particles. In conclusion, we point out that the initial solution as to the course of the potential and its first two derivatives is satisfied if the conditions

$$\sum_{s} K'(|r-r_s|)=0 \qquad \text{and} \qquad \sum_{s} K''(|r-r_s|)>0$$

are satisfied for all the lattice points.

Section 29. Intermediate Temperatures

It is important, in the theory of the crystalline state, to explain the nature of the dependence of the size of the periods on the changes in temperatures.

For values of the temperatures and densities satisfying the crystallization criterion, a periodic solution branches off from the solution corresponding to the uniform distribution. Following one of these branches, we consider the question as to the behavior of the periodic solution along the branch. Do there exist along this course parts of new branchings? What is the character of these new solutions?

To solve this problem we should approach from the side of low temperatures as well. As we have explained if $\theta \longrightarrow 0$, the asymptotic form for expressing the potential and density is purely periodic, of a "lattice" structure. According to the results of Sec. 28, we can say that the effects connected with an increase in temperature on a "lattice" solution are comprised in a "spreading" of the maximums of density around the lattice points, without, however, destroying the periodicity of the distribution. The question arises as to whether, despite this, there may not be new solutions, which do not correspond to the purely periodical ones? The existence of such solutions would indicate the appearance of saltatory changes in the lattice structure, and might give the key to understanding the nature of superconductivity.

We consider the initial equation, which we now write in the form

$$V(r) = \int_{-\infty}^{\infty} K(|r - r'|)\, a(\theta)\, e^{-\frac{V(r')}{\theta}}\, dr', \tag{29.1}$$

$$\rho(r) = a(\theta) \exp\left\{-\int_{-\infty}^{\infty} \frac{K(|r - r'|)\, \rho(r')\, dr'}{\theta}\right\} \tag{29.2}$$

Let ρ_0 and V_0 be exact periodic solutions of this equation found in Sec. 25, We seek an equation for a function of small deviations from the periodic solution. We set

$$V = V_0 + u, \quad \rho = \rho_0 + \rho_1, \tag{29.3}$$

where

$$u \ll V_0, \quad \rho_1 \ll \rho_0.$$

Confining ourselves to the terms linear in u and ρ_1, we obtain the two following equations:

$$u(r) + \frac{1}{\theta} \int_{-\infty}^{\infty} K(|r - r'|)\, \rho_0(r')\, u(r')\, dr' = 0, \tag{29.4}$$

$$\rho_1(r) + \frac{\rho_0(r)}{\theta} \int_{-\infty} K(|r - r'|)\, \rho_1(r')\, dr' = 0. \tag{29.5}$$

It has been taken into account here that $\rho_0(r) = a(\theta) e^{-\frac{V_0(r)}{\theta}}$.

The following relationship exists between u and ρ_1

$$\rho_1 = \rho_0 \frac{u}{\theta}, \tag{29.6}$$

which is obtained by varying equation $\rho = ae^{-\frac{V}{\theta}}$. By means of this equation, we are convinced of the equivalence of forms (29.4) and (29.5).

It is an essential point that a linear equation such as we saw in Sec. 18 determines the criterion for the appearance of branching and the character of the new solutions arising.

We rewrite equation (29.4) in the form

$$\psi(r) = \lambda \int_{-\infty}^{\infty} K(|r - r'|) e^{\varphi_0(r', \lambda)} \psi(r')\, dr', \tag{29.7}$$

where

$$\lambda = -\frac{a}{\theta} \sigma(0); \quad \psi = u(r); \quad \varphi_0(r, \lambda) \doteq -\frac{V_0(r, \lambda)}{\theta},$$

K is now normalized.

Integral equation (29.7) belongs to a type that has been little studied. Parameter enters into it in complex form, by means of the function $\varphi_0(\underline{r},\ )$.

But, on the basis of physical considerations, we may consider as the most probable case the particular case in which the periodic structure over a certain temperature interval does not depend greatly on λ, so that this dependence may be ignored. In this particular case, the equation is solved exactly, and gives us an idea as to the nature of the solutions that arise. We establish the basic properties of equation (29.7).

1. We show that any periodic functions possessing the same period as φ_0 are solutions of (29.7).

For, we rewrite equation (29.7) thus:

$$\psi(r) - \lambda \int_{-\infty}^{\infty} K(|r'|) \rho(r - r') \psi(r - r')\, dr' = 0, \tag{29.8}$$

where now

$$(\rho(r) = e^{\varphi_0(r)}),$$

from which we see that along with a function $\psi(\underline{r})$, the function $\psi(\underline{r}) = \psi(\underline{r} + \underline{n}d)$ (where $\underline{n}$ is an integral vector) likewise satisfies this equation, d being the period of function $\varphi_0(\underline{r})$.

2. In addition to periodic solutions with a period d, integral equation (29.8) likewise has other solutions of the type

$$\psi(r, k) = e^{\frac{2\pi ikr}{d}} u(r, k), \tag{29.9}$$

where $\underline{k}$ is an arbitrary real vector, and $u(\underline{r}, \underline{k})$ is a periodic function of $\underline{r}$ with period $\underline{d}$.

In point of fact, substituting of (29.9) in (29.8) gives

$$u(r, k) - \lambda \int_{-\infty}^{\infty} K(|r'|) e^{\frac{2\pi ikr'}{d}} \rho(r - r') u(r - r') dr' = 0, \tag{29..0}$$

that is, an equation of the same type as (29.8) except for a different kernel, which, however, 'is likewise dependent on the difference of arguments. Consequently, equation (29.10) likewise has periodic solutions with the period d.

3. The spectrum of eigenvalues of λ for a given $\underline{k}$ [in the system of functions (29.9)] is discrete. We show this by reducing the equation in question to an equation of the Fredholm type; we note that functions (29.9) satisfy the condition

$$\psi(r + nd, k) = e^{i2\pi kn} u(r), \tag{29.11}$$

where $\underline{n}$ is an integral vector.

By means of this transformation, integral equation (29.7) is reduced to an integral equation with a finite interval of integration.

We represent the integral appearing in (29.7) in the form of a sum

$$\int_{-\infty}^{\infty} K(|r - r'|) \rho(r') \psi(r') dr' = \sum_{n} \int_{d^3} K(|r - r'|) \rho(r') \psi(r') dr',$$

where integration on the right side extends over the elementary cell d^3 with number (n_x, n_y, n_z).

Making use of the condition of periodicity for $\rho(\underline{r})$ and condition (29.11), we have:

$$\int_{d^3} K(|r-r'|)\rho(r')\psi(r')\,dr' = \int_{d^3} K(|r-nd-r'|)\rho(r')\,e^{2\pi ikn}\psi(r')\,dr'.$$

Consequently, the initial equation is brought into the form

$$\psi(r)-\lambda\int_{d^3} G(|r-r'|,\ k)\rho(r')\psi(r')\,dr', \qquad (29.12)$$

where

$$G(|r-r'|,\ k)=\sum_n e^{ikn}K(r-r'-nd). \qquad (29.13)$$

The equation obtained belongs to the ordinary type of Fredholm integral equations, and, consequently, for each real value of $\underline{k}$, there is an infinite discrete spectrum of real eigenvalues $\lambda_1(k),\ \lambda_2(k),\ldots,\ \lambda_n(k),\ \ldots$. We call the region of variation $\lambda_i = \lambda_i(k)$ a zone. In this way, the state of the particle may be described by the number of the zone. Within each zone λ depends continuously on $\underline{k}$. We proceed to find this dependence.

Let there be an expansion into a Fourier series

$$\rho(r)=e^{\varphi_1(r)}=\sum \rho_h e^{2\pi i(hr)};\quad h=h_1b_1+h_2b_2+h_3b_3;$$

where h_1, h_2, h_3 are whole numbers; $\underline{b}_1$, $\underline{b}_2$, $\underline{b}_3$ are unit vectors of the axes of the "inverse lattice."

We represent $\psi(\underline{r})$ by the series

$$\psi(r)=\sum_h a_h e^{i(k+2\pi h,\ r)},$$

and substitution of these expressions in (29.12) gives

$$a_g = \lambda\sigma(|k + 2\pi g|)\sum_h a_{g-h}\rho_h \qquad (29.14)$$

(g = $\frac{2\pi m}{d}$, m being an integral vector).

The condition for the solvability of the infinite system of equations (29.14) is the equivalence to zero of the determinant

$$\Delta(\lambda, k) =$$

$$\begin{vmatrix} \cdot & \cdot & \cdot & \cdot & \cdot \\ \ldots & \{\lambda\sigma(|k|)\}^{-1} & \rho_{1,0,0} & \rho_{0,1,0} & \ldots \\ \ldots & \rho_{-1,0,0} & \{\lambda\sigma|k+2\pi b|\}^{-1} & \rho_{-1,1,0} & \ldots \\ \ldots & \rho_{0,-1,0} & \rho_{1,-1,0} & \{\lambda\sigma|k+2\pi b|\}^{-1} & \ldots \\ \cdot & \cdot & \cdot & \cdot & \cdot \end{vmatrix} = 0. \qquad (29.15)$$

By solving this equation, we obtain λ, corresponding to the wave vecot k.

We give the final result: the equation

$$\psi(r) - \lambda \int_{-\infty}^{\infty} K(|r - r'|)\rho(r')\psi(r')\,dr' = 0.$$

$$\rho(r + d) = \rho(r)$$

has a banded spectrum of eigenvalues of λ, with a denumerable number of bands $\lambda_1(k)$, $\lambda_2(k)$, ... , $\lambda_n(k)$ ...... and with eigenfunctions of the type

$$\psi_n(r, k) = e^{\frac{i2\pi kr}{d}} u_n(r, k); \quad u_n(r + d, k) = u_n(r, k),$$

where k is an arbitrary real vector. Within each band λ depends continuously on k.

It may therefore be said that a type of crystals is possible in which the purely periodic structure is systematically replaced by an "almost" periodic structure under continuous temperature change. In this connection, even an infinitely extended crystal is not a purely periodical phenomenon, but comprises organically within it irregularities of structure. The sequential arising and disappearing of these irregularities with continuous temperature changes takes place in a saltatory way.

It is essential to stress that we obtained this result only in a particular case, under the condition that the initial periodic solution depends only slightly on the temperature. The exact solution of this problem is still attended with great difficulties.

It is of great interest to explain how the irregularity arises in the field of low temperatures. Is there not here a certain minimum temperature after which an irregularity arises that exists along with the periodic structure? This question was discussed by Pereleshin[+].

We start from equation (29.1). We are interested in a solution of equation (29.5) that is close to a "lattice" solution. We shall mean by $\rho_0(\underline{r})$ a solution of type (28.12). Introducing kernel $K'(|\underline{r}-\underline{r}'|) = -K^+(|\underline{r}-\underline{r}'|)$ and designating ρ_1 by ρ, we write equation (29.5) in the form

$$\rho(r) - \frac{\rho_0(r)}{\theta} \int_{-\infty}^{\infty} K^*(|r-r'|)\rho(r')\,dr' = 0. \tag{29.16}$$

Since equation (29.16) is equivalent to equation (29.4), it has a solution of the same type. Consequently, equation (29.16) has periodic solutions with a period equal to the period d of function ρ_0 and, in addition, solutions

$$\rho(r) = e^{ikr}u(r), \tag{29.17}$$

where $\underline{k}$ is a real vector and $u(\underline{r})$ a periodic function with the period d.

Our problem is to specify the conditions under which solutions of type (29.17) arise with $D = \frac{2\pi}{k} \gg d$, where D is a new period, imposed on the fundamental period of the lattice d. To be able to obtain some evaluations, we pass from integral equation (29.16) to a differential equation. For this purpose, we expand the portion under the integral $\rho(\underline{r}')$ in (29.16) into a Taylor series around point $\underline{r}$ in powers of $(\underline{r}'-\underline{r})$. Since $K^*(|r-r'|)$ is an even function of its argument, then

$$\int_{-\infty}^{\infty} K^*(|r-r'|)\rho(r')\,dr' = \sum_{n=0}^{\infty} K^*_{2s}\nabla^{2s}\rho =$$
$$= K^*_0 + K^*_2\nabla^2\rho + K^*_4\nabla^4\rho + \ldots, \tag{29.18}$$

[+] A. Pereleshin, Zh. ETF 18, 449 (1947).

where

$$K^*_{2n} = \frac{1}{2n!} \int_{-\infty} K^* (|r - r'|) (|r' - r|)^{2n} dr'.$$

Substituting (29.18) in (29.16), we have:

$$K^*_2 \Delta\rho + K^*_4 \Delta^2 \rho \ldots + \left(K^*_0 - \frac{\theta}{\rho_0}\right)\rho = 0.$$

Ignoring terms containing derivatives of the fourth and higher orders, we obtain an equation which we shall study further

$$\Delta\rho + \frac{1}{K_2}\left(K^*_0 - \frac{\theta}{\rho_0(r)}\right)\rho(r) = 0. \qquad (29.19)$$

Our ignoring the terms with higher derivatives may be justified in the following way. If we approximate the function $K^+ (|\underline{r} - \underline{r}'|)$, which decreases rapidly with the distance, as follows:

$$K^* (|r - r'|) = \begin{cases} V_0 & \text{for } 0 < |r - r'| < R, \\ 0 & \text{for } |r - r'| > R, \end{cases}$$

where R is "the radius of action of the molecular forces," then

$$K^*_{2n} = \frac{2\pi V_0 R^{2n+3}}{(2n!)\, 2n+3},$$

from which

$$\frac{K^*_{2n}}{K^*_0} = \frac{10 R^{2n-2}}{2n!\,(2n+3)}.$$

If we set $\rho(r) \sim e^{ikr}$, then $|\nabla^{2n}\rho| = |k|^{2n}\rho$.

In view of this, we obtain for the order of magnitude of the ratio of $\frac{K_{2n}^*}{K_2^*}\nabla^{2n}\rho$ to $\nabla^2\rho$,

$$\frac{10R^{2n-2}|k|^{2n-2}}{2n!(2n+3)}\sim\left(\frac{R}{D}\right)^{2n-3}$$

The radius of action of the molecular forces is equal in order of magnitude to the lattice period d. In this way, we see that the ignoring that has been done is justified only for the case of periods $D \gg d$, but it is precisely periods of this kind that interest us. Upon excluding the terms in question, we can "lose" only periods $D \ll d$.

The temperature at which the fundamental periodicity of the crystal is violated is the temperature at which solutions of type (29.17) arise for equation (29.16).

We now consider equation (29.19). Since we consider only a simple cubic lattice, equation (29.19) is symmetrical with respect to variables x, y, z [it is obvious from (28.12) that the "lattice" solution $\rho_0(x, y, z)$ is symmetrical with respect to x, y, z]. For simplicity's sake, we therefore shall confine ourselves to studying equation (29.19) on the presupposition that the variations of functions $\rho(\underline{r})$ and $\rho_0(\underline{r})$ take place only upon variations of x, while these functions do not change as y and z vary; in other words, we conduct an averaging of functions $\rho(x, y, z)$ and $\rho_0(x, y, z)$ for y and z. We designate the function averaged in this way by $\rho(x)$ and $\rho_0(x)$. (Because of the equality of status of variables x, y, z, we would obtain the same results if the averaging were done for x and z for x and y).

On the basis of (28.12), we have:

$$\rho_0(x)=\frac{1}{d^2\alpha\sqrt{\pi}}\sum_l e^{-\frac{(x-ld)^2}{\alpha^2}} \qquad (29.20)$$

We find $\rho_0(x)_{min}=\rho_0\left(kd+\frac{d}{2}\right)$ (which we shall need later),

$$\rho_0(x)_{min}=\frac{1}{d^2\alpha\sqrt{\pi}}\sum_l e^{-\frac{d^2}{4\alpha^2}(2(k-l)+1)^2}$$

or

$$\rho_0(x)_{min}=\frac{2}{d^2\alpha\sqrt{\pi}}e^{-\frac{d^2}{4\alpha^2}}+\frac{2}{d^2\alpha\sqrt{\pi}}\sum_{m-1}^{\infty}e^{-\frac{d^2}{4\alpha^2}(2m+1)^2} \qquad (29.21)$$

After averaging for y and z, instead of equation (29.19), we obtain the following equation:

$$\frac{d^2\rho(x)}{dx^2} + \frac{1}{K_2^*}\left(K_0^* - \frac{\theta}{\rho_0(x)}\right)\rho(x) = 0. \qquad (29.22)$$

Equation (29.22) is an equation of the second order with a periodic coefficient (since $\rho_0(x)$ is a "lattice" solution averaged for y and z, having a period equal to the lattice period d). We have here what is called Hill's equation.

We introduce the function ρ_0. For simplicity's sake, we replace it by the following cuve:

$$\rho_0(x) = \rho_a \text{ при } 0 < x < a, \quad \rho_0(x) = \rho_b, \text{ при } -b > x < 0, \qquad (29.23)$$

where a + b = d, $\rho_0(x)$ is a periodic function with the period d, a is the "node width," b the "internodal width," ρ_a the "density at the node," ρ_b "internodal density." The quantities a, b, ρ_a, ρ_b are functions of the temperature θ. The relationship a ρ_a + b ρ_b = 1 exists between them. For $\theta \to 0$, $a \to 0$, $b \to d$, $\rho_a \to +\infty$, $\rho_b \to 0$.

We shall consider the interval (- b, a) of the length d. Substituting (29.23) in (29.22), we obtain two different equations for the integrals (- b, 0) and (0, a):

$$\frac{d^2\rho}{dx^2} + \beta^2\rho = 0, \quad 0 < x < a, \quad \beta^2 = \frac{1}{K_2}\left(K_0 - \frac{\theta}{\rho_a}\right), \qquad (29.24)$$

$$\frac{d^2\rho}{dx^2} - \gamma^2\rho = 0, \quad -b < x < 0, \quad \gamma^2 = \frac{1}{K_2}\left(\frac{\theta}{\rho_b} - K_0\right). \qquad (29.25)$$

Solution of equations (29.24) and (29.25) are had in the form

$$\rho(x) = e^{ikx}u(x), \quad u(x+d) = u(x)$$

and we obtain

$$u_1(x) = Ae^{i(-k+\beta)x} + Be^{-i(k+\beta)x}, \qquad (29.26)$$

$$u_2(x) = Ce^{(-ik+\gamma)x} + De^{-(ik+\gamma)x}. \qquad (29.27)$$

Coefficients A, B, C and D are determined from the continuity conditions for functions u_1 (x) and u_2 (x) and their derivatives at point x = 0, and the periodicity conditions at point x = - b and x = a.

The condition for the solvability of the system that we obtain of four homogeneous equations in four unknowns A, B, C, D is the equivalent to zero of the determinant of the system. This condition has the form

$$\cos kd = \cos \beta a \cos \gamma_1 b - \frac{\beta^2 + \gamma_1^2}{2\beta\gamma_1} \sin \beta a \sin \gamma_1 b. \qquad (29.28)$$

If $\gamma^2 < 0$, then introducing $\gamma_1^2 = -\gamma^2$, we obtain in an analogous way, the following condition for solvability:

$$\cos kd = \cos \beta a \operatorname{ch} \gamma b - \frac{\beta^2 - \gamma^2}{2\beta\gamma} \sin \beta a \operatorname{sh} \gamma b. \qquad (29.29)$$

In order to obtain values for k from (29.28) and (29.29), we must take into account the dependence of a, b, β, γ, γ_1 on the temperature.

We consider the case $\theta \to 0$. To confirm the sing of γ^2, we must evaluate $\lim \frac{\theta}{\rho_b}$ as $\theta \to 0$. For ρ_b, we employ $\rho_0(x)_{min}$ from (29.21), which for low temperatures is approximately equal to

$$\rho_0(x)_{min} \sim \frac{2}{d^2 \alpha \sqrt{\pi}} e^{-\frac{d^2}{4\alpha}}.$$

It is obvious from this that $\lim \frac{\theta}{\rho_b} = +\infty.$

Thus, in the case $\theta \to 0$, we must apply equation (29.28), since $\gamma^2 > 0$. For $\theta \to 0$, $\beta^2 \to \frac{K_0}{K_2} \sim \frac{1}{R^2} \sim \frac{1}{d^2}$, that is, $\beta a - \frac{a}{d} \to 0$ (since as $\theta \to 0$, a $\to 0$); therefore cosine β a may be as an approximation replacedby unity and $\frac{\sin \beta a}{\beta}$ by a, after which we have:

$$\cos kd \sim \operatorname{ch} \gamma b - \frac{\beta^2 - \gamma^2}{2\gamma} a \operatorname{sh} \gamma b.$$

For $\theta \to 0$, $\gamma^2 \to \infty$, that is, $\gamma^2 \gg \beta^2$.
Hence

$$\cos kd \sim \operatorname{ch} \gamma b + \frac{\gamma a}{2} a \operatorname{sh} \gamma b,$$

that is, we see that the right side is greater than unity; it tends to $+\infty$ as $\theta \to 0$. Consequently, as $\theta \to 0$, there are no solutions of the type $e^{ikx}u(x)$ with real k's.

On the other hand, if for low enough temperatures the quantity γ^2 is negative (it changes sign upon further decrease in temperature), then approximately, replacing $\cos \beta a$ in (29.29) by unity and $\frac{\sin \beta a}{\beta}$ by a, we obtain:

$$\cos kd \sim \cos \gamma_1 b - \frac{\beta^2 + \gamma_1^2}{2\gamma_1} a \sin \gamma_1 b \sim \cos \gamma_1 b,$$

that is, in this case there are real values for k. Hence, even from crude evaluations like this, it will be seen that for $\gamma^2 < 0$, real k can exist, but not for $\gamma^2 > 0$, or for $\gamma^2 > \beta^2$.

We get more precise evaluations. Considering kd, β a, γ b, γ_1b to be small as compared with unity, we expand the right and left sides of (29.28) and (29.29) as series, and confine ourselves to the quadratic terms. From (29.28), we obtain:

$$k = \left[\frac{\beta^2 a}{d} - \frac{\gamma^2 b}{d}\right]^{\frac{1}{2}}, \tag{29.30}$$

and from (29.29)

$$k = \left[\frac{\beta^2 a}{d} + \frac{\gamma_1^2 b}{d}\right]^{\frac{1}{2}}. \tag{29.31}$$

From these formulas we find that in order for real k's to exist, it is necessary that the condition be satisfied

$\gamma^2 < \frac{a\beta^2}{b}$. For $\gamma^2 = 0$, we have for k, $k = \beta\left(\frac{a}{d}\right)^{\frac{1}{2}}$

or approximately (since at low temperatures $\beta \sim \frac{1}{R} \sim \frac{1}{d}$),

$$kd \sim \left(\frac{a}{d}\right)^{\frac{1}{2}}, \quad \frac{\lambda}{d} \sim \left(\frac{d}{a}\right)^{\frac{1}{2}} \gg 1. \tag{29.32}$$

The condition for the existence of real k's may be written down in another way as well (employing (29.24) and 29.25)):

$$\theta \cdot \left(\frac{b}{\rho_b} + \frac{a}{\rho_a}\right) < K_0 \cdot d, \tag{29.33}$$

and this inequality may be written approximately in a simpler form, if we consider that at low temperatures $\frac{a}{\rho_a} \ll \frac{b}{\rho_b}$ and $b \sim d$. We then have:

$$\frac{\theta}{\rho_b} < K_0. \tag{29.34}$$

Relation (29.34) replaces our previous condition $\gamma^2 < 0$.

Since at low temperatures, $\rho_b \sim c\theta^{\frac{1}{2}} e^{-\frac{\text{const.}}{\theta}}$, then from conditions (29.33) and (29.34) it follows that solutions of the type $e^{ikx} u(x)$ with real k exist at a temperature greater than a certain minimum temperature θ_{min}, which is defined by the equation

$$\theta_{min} \frac{b(\theta_{min})}{\rho_b(\theta_{min})} + \frac{a(\theta_{min})}{\rho_a(\theta_{min})} = K_0 d, \tag{29.35}$$

or approximately

$$\frac{\theta_{min}}{\rho_b(\theta_{min})} = K_0. \tag{29.36}$$

We evaluate the order of magnitude of θ_{min}. To this end, we introduce the following approximative relations:

$K_0 = \frac{4\pi}{3} d^3 \theta$ ($\theta_{пл}$ the fusion temperature):

$\alpha^2 = \frac{d^2}{2} \frac{\theta}{\theta_{пл}}$, since $x^2 = \frac{\alpha^2}{2} = \left(\frac{d}{2}\right)^2 \frac{\theta}{\theta_{пл}}$

Setting

$$\rho = \frac{2}{d^2 \sqrt{\pi}} e^{-\frac{d^2}{4\rho^2}}$$

and substituting all these in (29.36), we have:

$$z^{\frac{3}{2}} e^{-z} = \frac{3}{32\sqrt{\pi}},$$

where

$$z = \frac{\theta_{пл}}{2\theta_{min}}.$$

The value for z obtained in this way is about equal to 5.5, from which $\theta_{min} \sim 0.1\ \theta_{пл}$. It follows from this that after a certain temperature, the periodic density distribution in the crystal is suddenly destroyed.

In this way we see that it is possible that a minimum temperature exists at which almost periodic solutions arise. Evaluation gives a reasonable order of magnitude for θ_{min}. It would appear that this result is a weighty argument for the rationality of our approach to the approach to the problem of superconductivity from the point of view of the new theory of crystals.

Section 30. Crystallization of a Binary Mixture

Yakovlev considered the conditions for the formation of a crystalline structure in a system consisting of interacting particles of various sorts. He conducted such an investigation for a system consisting of two kinds of particles ($N_1 \gg 1$, $N_2 \gg 1$) with masses m_1, m_2 and kernels for the energy of the integral interactions of the form:

$$K_{ij}(\boldsymbol{r}, \boldsymbol{r}') = K_{ij}(|\boldsymbol{r} - \boldsymbol{r}'|) \quad (i, j = 1, 2).$$

Generalization of the theory to the case of a system with many components does not present any special difficulties.

For a system consisting of particles of a different "kind" the equations appear as follows:

$$\boldsymbol{v}\nabla_r f_1(\boldsymbol{r}, \boldsymbol{v}) + \frac{1}{m_1} \nabla_r V_1(\boldsymbol{r}) \nabla_v f_1(\boldsymbol{r}, \boldsymbol{v}) = 0,$$

$$\boldsymbol{v}\nabla_r f_2(\boldsymbol{r}, \boldsymbol{v}) + \frac{1}{m_2} \nabla_r V_2(\boldsymbol{r}) \nabla_v f_2(\boldsymbol{r}, \boldsymbol{v}) = 0,$$

Here we set $\frac{\partial f_1}{\partial t} = \frac{\partial f_2}{\partial t} = 0$ (stationary case), so that

$$V_1(r) = (N_1 - 1)\int\limits_{-\infty}^{\infty} K_{11}(r, r') f_1(r', v')\, dv'\, dr' +$$

$$+ N_2 \int\limits_{-\infty}^{\infty} K_{12}(r, r') f_2(r', v')\, dv'\, dr',$$

$$V_2(r) = (N_2 - 1)\int\limits_{-\infty}^{\infty} K_{22}(r, r') f_2(r', v')\, dv'\, dr' +$$

$$+ N_1 \int\limits_{-\infty}^{\infty} K_{12}(r, r') f_1(r, v)\, dr'\, dv'$$

(omitting functional terms of higher order).

By substituting

$$f_1 = \varphi_1(r)\, \psi_{01}\, e^{-\frac{v^2}{2\theta} m_1}, \quad f_2 = \varphi_2(r)\, \psi_{02}\, e^{-\frac{v^2}{2\theta} m_2}$$

(where ψ_{01}, ψ_{02} are defined by the obvious normalization conditions in velocities space) the equations in question lead to a system of nonlinear integral equations

$$\left.\begin{aligned} U_1(r) &= \lambda_1 \int\limits_{-\infty}^{\infty} K_{11}(|r - r'|)\, e^{-U_1(r')}\, dr' + \\ &\quad + \lambda_2 \int\limits_{-\infty}^{\infty} K_{12}(|r - r'|)\, e^{-U_2(r')}\, dr', \\ U_2(r) &= \lambda_2 \int\limits_{-\infty}^{\infty} K_{22}(|r - r'|)\, e^{-U_2(r')}\, dr' + \\ &\quad + \lambda_1 \int\limits_{-\infty}^{\infty} K_{12}(|r - r'|)\, e^{-U_1(r')}\, dr'. \end{aligned}\right\} \quad (30.1)$$

Here we introduce the designations:

$$\varphi_1(r) = f_1 e^{-U_1(r)}, \quad U_1(r) = \frac{V_1(r)}{\theta}, \quad \lambda_1 = \frac{f_1}{\theta},$$

$$\varphi_2(r) = f_2 e^{-U_2(r)}, \quad U_2(r) = \frac{V_2(r)}{\theta}, \quad \lambda_2 = \frac{f_2}{\theta}.$$

We add to system (30.1) the normalization condition determining the constant f_1, f_2

$$\frac{1}{V_0}\int_{V_0} (f_1 e^{-U_1(r)} + f_2 e^{-U_2(r)})\, dr = 1. \qquad (30.2)$$

In the case of a periodic solution, we take V_0 as a multiple of the volume of the cell of periodicity, while for a uniform distribution V_0 may be taken at will. We shall look for periodical solutions of $U_i(\underline{r})$, $\varphi_i(\underline{r})$ as branches of the obvious solutions of system (30.1) $U_i(\underline{r}) = c_i^0$ (assuming that this phrase

$$K_{ij} = \int_{-\infty}^{\infty} K_{ij}(|r - r'|)\, dr'$$ has a meaning).

We have:

$$c_1^0 = \lambda e^{-c_1^0} K_{11} + \lambda_2 e^{-c_2^0} K_{12},$$

$$c_2^0 = \lambda e^{-c_1^0} K_{12} + \lambda_2 e^{-c_2^0} K_{22},$$

$$1 = f_{10} e^{-c_1^0} + f_{20} e^{-c_2^0}.$$

Further denoting $f_{10} e^{-c_1^0} = \alpha,\ f_{20} e^{-c_2^0} = \beta$, we have

$$c_1^0 = \frac{\alpha K_{11} + \beta K_{12}}{\theta}, \quad c_2^0 = \frac{\alpha K_{12} + \beta K_{22}}{\theta}, \quad \alpha + \beta = 1.$$

This solution depends on two arbitrary parameters θ, α. Setting a value $\theta = \theta_0$, we seek a solution different from the constant one, in the form:

$$\left.\begin{aligned} \lambda_1^* = \lambda_1 + \tau,\ \lambda_2^* = \lambda_2 + \eta, \\ U_1(r) = c_1^0 + u_1(r), \\ U_2(r) = c_2^0 + u_2(r). \end{aligned}\right\} \tag{30.3}$$

Inserting (30.3) in (30.1), we transform the equations to the following form

$$\begin{aligned} & L_1\{u_1(r),\ u_2(r)\} = \\ & = \int_{-\infty}^{\infty} K_{11}(r, r') e^{-c_1^0} \left\{ \tau - \tau u_1(r') + (\lambda_1 + \tau) \sum_{n=2}^{\infty} (-1)^n \frac{1}{n!} u_1^n(r') \right\} dr' + \\ & \qquad + \lambda_2 \int_{-\infty}^{\infty} K_{12}(r, r') e^{-c_2^0} \left\{ \sum_{n=2}^{\infty} (-1)^n \frac{1}{n!} u_2^n(r') \right\} dr' + \\ & + \eta \int_{-\infty}^{\infty} K_{12}(r, r') e^{-c_2^0} \left\{ 1 - u_2(r') + \sum_{n=2}^{\infty} (-1)^n \frac{1}{n!} u_2^n(r') \right\} dr', \\ & L_2\{u_1(r),\ u_2(r)\} = \\ & = \int_{-\infty}^{\infty} K_{12}(r, r') e^{-c_1^0} \left\{ \tau - \tau u_1(r') + (\lambda_1 + \tau) \sum_{n=2}^{\infty} (-1)^n \frac{1}{n!} u_1^n(r') \right\} dr' + \\ & \qquad + \lambda_2 \int_{-\infty}^{\infty} K_{22}(r, r') e^{-c_2^0} \left\{ \sum_{n=2}^{\infty} (-1) \frac{1}{n!} u_2^n(r') \right\} dr' + \\ & \qquad + \eta \int_{-\infty}^{\infty} K_{22}(r, r') e^{-c_2^0} \left\{ 1 - u_2(r') + \sum_{n=2}^{\infty} (-1)^n \frac{1}{n!} u_2^n(r') \right\} dr', \end{aligned} \tag{30.4}$$

where

$$L_1\{u_1, u_2\} \equiv$$
$$\equiv u_1(r) + \frac{\alpha}{\theta_0}\int_{-\infty}^{\infty} K_{11}(r, r')\, u_1(r')\, dr' + \frac{\beta}{\theta_0}\int_{-\infty}^{\infty} K_{12}(r, r')\, u_2(r')\, dr',$$

$$L_2\{u_1, u_2\} \equiv$$
$$\equiv u_2(r) + \frac{\beta}{\theta_0}\int_{-\infty}^{\infty} K_{22}(r, r')\, u_2(r')\, dr' + \frac{\alpha}{\theta_0}\int_{-\infty}^{\infty} K_{12}(r, r')\, u_1(r')\, dr'.$$

We consider the system of linear homogeneous equations of the Fredholm type .

$$L_1\{\zeta_1(r), \zeta_2(r)\} = 0,$$
$$L_2\{\zeta_1(r), \zeta_2(r)\} = 0.$$

This system has nontrivial bounded solutions:

$$\zeta_1(r) = c_1 \sin kr + c_2 \cos kr,$$
$$\zeta_2(r) = Ac_1 \sin kr + Ac_2 \cos kr,$$

where c_1, c_2 are arbitrary constants,

$$A = -\frac{\alpha}{\theta_0}\,\frac{\sigma_{12}(k)}{1 + \frac{\beta}{\theta_0}\sigma_{22}(k)}$$

on the condition

$$\operatorname{Det}\{\sigma_{ij}(k)\} = \begin{vmatrix} 1 + \frac{\sigma_{11}(k)}{\theta_0}\alpha, & \frac{\sigma_{12}(k)}{\theta_0}\beta \\ \frac{\sigma_{12}(k)}{\theta_0}\alpha, & 1 + \frac{\sigma_{22}(k)}{\theta_0}\beta \end{vmatrix} = 0, \tag{30.5}$$

$$\sigma_{ij}(k) = \int_{-\infty}^{\infty} K_{ij}(r, r') \exp\{ik(r - r')\}\, dr'; \quad k = |k| .$$

We can satisfy equation (30.5) using any θ_0. Assuming that θ_0 is chosen in conformity with (30.5), we shall solve system (30.4). We represent the solutions of this system in the form of the following power expansions:

$$
\left.\begin{aligned}
u_1(r) &= \tau^{\frac{1}{2}} f_1(r) + \tau f_2(r) + \tau^{\frac{3}{2}} f_3(r) + \ldots, \\
u_2(r) &= \tau^{\frac{1}{2}} \chi_1(r) + \tau \chi_2(r) + \tau^{\frac{3}{2}} \chi_3(r) + \ldots, \\
\eta &= \tau\mu_1 + \tau^2\mu_2 + \tau^3\mu_3 + \ldots, \\
f_1 &= f_{10}(1 + \tau\xi_1^{(1)} + \tau^2\xi_2^{(1)} + \tau^3\xi_3^{(1)} + \ldots), \\
f_2 &= f_{20}(1 + \tau\xi_1^{(2)} + \tau^2\xi_2^{(2)} + \tau^3\xi_3^{(2)} + \ldots).
\end{aligned}\right\} \qquad (30.6)
$$

The constant $\xi_i^{(1)}, \xi_i^{(2)}$ are determined from the normalization conditions and μ_i, $f_i(\underline{r})$, $\chi_i(\underline{r})$ are independent of this condition. We shall therefore first solve system (30.4) and only then determine $f_i^{(1)}$, $\xi_i^{(2)}$. Substitution of the first three expansions of (30.6) in (30.4) leads, as before, to an infinite sequence of systems of integral equations. The first of them has the form:

$$
L_1\{f_1(r), \chi_1(r)\} = 0,
$$
$$
L_2\{f_1(r), \chi_1(r)\} = 0.
$$

On the basis of (30.3), we have:

$$
f_1(r) = c_1 \sin kr,
$$
$$
\chi_1(r) = Ac_1 \sin kr
$$

(c_1 being an arbitrary constant).

The second system has the form:

$$
L_1\{f_2, \chi_2\} = \int_{-\infty}^{\infty} K_{11}(r, r') e^{-c_1^0}\left(1 + \lambda_1 \frac{f_1^2(r')}{2!}\right) dr' + \int_{-\infty}^{\infty} K_{12}(r, r') e^{-c_2^0}\left(\lambda_2 \frac{\chi_1^2(r')}{2!} + \mu_1\right) dr',
$$

$$
L_2\{f_2, \chi_2\} = \int_{-\infty}^{\infty} K_{12}(r, r') e^{-c_1^0}\left(1 + \lambda_1 \frac{f_1^2(r')}{2!}\right) dr' + \int_{-\infty}^{\infty} K_{22}(r, r') e^{-c_2^0}\left(\lambda_2 \frac{\chi_1^2(r')}{2!} + \mu_1\right) dr'.
$$

Substitution in the right side of this system of expressions for $f_1(r)$, $\chi_1(r)$ and elementary transformations give:

$$f_2(r) = A_2^{(1)} + B_2^{(1)} c_1^2 \cos 2kr, \quad \text{where} \quad A_2^{(1)} = A_{2,1}^{(1)} + A_{2,2}^{(1)} c_1^2 + A_{2,3}^{(1)} \mu_1,$$
$$\chi_2(r) = A_2^{(2)} + B_2^{(2)} c_1^2 \cos 2kr, \qquad A_2^{(2)} = A_{2,1}^{(2)} + A_{2,2}^{(2)} c_1^2 + A_{2,3}^{(2)} \mu_1,$$

$A_{ij}^{(k)}$ being independent of c_1, μ_1. It is essential to note that the given expressions for $f_2(r)$, $\chi_2(r)$ apply only under the conditions that $\operatorname{Det}\{K_{ij}\} \neq 0$ and $\operatorname{Det}\{\sigma_{ij}(2k)\} \neq 0$,

where the designations of the determinants indicate that they are formed from (30.4) by simple replacement of the corresponding elements. Under these conditions, the second system of equations is solvable for any value of c_1 and μ_1.

We consider the third system of the sequence:

$$L_1\{f_3, \chi_3\} = \int_{-\infty}^{\infty} K_{11}(r, r') e^{-c_1^0}\left\{-f_2(r') + \lambda_1\left(f_1 f_2 - \frac{1}{3!} f_1^3\right)\right\} dr' +$$
$$+ \int_{-\infty}^{\infty} K_{12}(r, r') e^{-c_2^0}\left\{\lambda_2\left(\chi_1\chi_2 - \frac{1}{3!}\chi_1^3\right) - \mu_1\chi_1\right\} dr'.$$
$$L_2\{f_3, \chi_3\} \int_{-\infty}^{\infty} K_{12}(r, r') e^{-c_1^0}\left\{-f_1(r') + \lambda_1\left(f_1 f_2 - \frac{1}{3!} f_1^3\right)\right\} dr' +$$
$$+ \int_{-\infty}^{\infty} K_{22}(r, r') e^{-c_2^0}\left\{\lambda_2\left(\chi_1\chi_2 - \frac{1}{3!}\chi_1^3\right) - \mu_1\chi_1\right\} dr'.$$

Simple transformations of the right side give expressions for them $A_{3,0}^{(i)} \sin kr + B_{3,0}^{(i)} \sin 3kr \quad (i = 1, 2)$, respectively.

The system is only solvable on the condition that

$$A_{3,0}^{(1)} = A_{3,0}^{(2)} = 0,$$

or in expanded form:

$$\left.\begin{aligned}&\sigma_{11}(k)e^{-c_1^0}\left\{-1+\lambda_1\left(A_2^{(1)}-\frac{1}{2}B_2^{(1)}c_1^2-\frac{c_1^2}{8}\right)\right\}+\\&\quad+\sigma_{12}(k)e^{-c_2^0}A\left\{\lambda_2\left(A_2^{(2)}-\frac{1}{2}B_2^{(2)}c_1^2-\frac{A^2}{8}c_1^2\right)-\mu_1\right\}=0,\\&\sigma_{12}(k)e^{-c_1^0}\left\{\quad 1+\lambda_1\left(A_2^{(1)}-\frac{B_2^{(1)}c_1^2}{2}-\frac{c_1^2}{8}\right)\right\}+\\&\quad+\sigma_{22}(k)e^{-c_1^0}A\left\{\lambda_2\left(A_2^{(2)}-\frac{B_2^{(2)}c_1^2}{2}-\frac{A^2c^2}{8}\right)-\mu_1\right\}=0.\end{aligned}\right\} \tag{30.7}$$

Equation (30.7) is a system in two unknowns c_1 and μ_1. The condition of consistency is the requirement

$$\sigma_{11}(k)\sigma_{22}(k)-\sigma_{12}^2(k)\neq 0. \tag{30.8}$$

From this, we have:

$$-1+\lambda_1\left(A_2^{(1)}-\frac{B_2^{(1)}c_1^2}{2}-\frac{c_1^2}{8}\right)=0,$$

$$-\mu_1+\lambda_2\left(A_2^{(2)}-\frac{B_2^{(2)}c_1^2}{2}-\frac{A^2c_1^2}{8}\right)=0.$$

Further, making use of the expression for $A_2^{(i)}$, $B_2^{(i)}$, we rewrite this system in the form

$$\left.\begin{aligned}&c_1^2\left(A_{2,2}^{(1)}-\frac{B_2^{(1)}}{2}-\frac{1}{8}\right)+\mu_1A_{2,3}^{(1)}=\frac{1}{\lambda_1}-A_{2,1}^{(1)},\\&c_1^2\left(A_{2,2}^{(2)}-\frac{B_2^{(2)}}{2}-\frac{A^2}{8}\right)+\mu_1\left(A_{2,3}^{(2)}-\frac{1}{\lambda_2}\right)=A_{2,1}^{(2)}.\end{aligned}\right\} \tag{30.9}$$

Here all the coefficients $A_{ij}^{(k)}$, $B_2^{(i)}$ are independent of c_1 and μ_1. Assuming that the determinant of this system differs from zero, it is easy to determine c_1^2 and μ_1. If it should prove in this that c_1^2 is negative, then we need merely change the sign of τ; upon this, only the signs of $A_{2,1}^{(1)}$, $A_{2,1}^{(2)}$ change signs, and likewise the entire right side of the first equation of (30.9). If, further,

$\mathrm{Det}\{s_{ij}(3k)\} \neq 0$, then the general solution will be:

$$f_3(r) = A_3^{(1)} \sin 3kr + c_3 \sin kr,$$
$$\chi_3(r) = A_3^{(2)} \sin 3kr + Ac_3 \sin kr.$$

Construction of the following approximations is done in the usual manner. To each particular solution for $f_{2n-1}(r)$ $\chi_{2n-1}(r)$ we add the solution of a homogeneous equation with a new arbitrary constant c_{2n-1}. The following equations for $f_{2n}(r)$, $\chi_{2n}(r)$ still contain the constant μ_n and are always solvable. The equations for $f_{2n+1}(r)$, $\chi_{2n+1}(r)$ contain on the right side only the first powers of μ_n, c_{2n-1} (no new constant μ_{n+1} is introduced), and the condition of solvability gives two equations, which are sufficient to determine them.

We seek an explicit expression for the solvability conditions, to which end, we consider two successive equations

$$\left.\begin{aligned}
L_1\{f_{2n}, \chi_{2n}\} &= \int_{-\infty}^{\infty} K_{11}(r, r') e^{-c_1^0} \lambda_1 f_1 f_{2n-1}\, dr' + \\
&+ \int_{-\infty}^{\infty} K_{12}(r, r') e^{-c_2^0} (\mu_n + \lambda_2 \chi_1 \chi_{2n-1})\, dr' + \\
&+ \{\text{чл., не сод. } c_{2-1}, \mu_n\}, \\
L_1\{f_{2n+1}, \chi_{2n+1}\} &= \int_{-\infty}^{\infty} K_{11}(r, r') e^{-c_1^0} \Big\{-f_{2n-1}(r') + \\
&+ \lambda_1 \Big(f_1 f_{2n-1} + f_2 f_{2n+1} - \frac{1}{2} f_1^2 f_{2n-1}\Big)\Big\} dr' + \\
+ \int_{-\infty}^{\infty} K_{12}(r, r') e^{-c_1^0} &\Big\{-\mu_1 \chi_{2n-1} - \mu_n \chi_1 + \lambda_2 \Big(\chi_1 \chi_{2n} + \\
&+ \chi_2 \chi_{2n-1} - \frac{1}{2} \chi_1^2 \chi_{2n-1}\Big)\Big\} dr' + \{\text{чл., не сод. } c_{2n-1}, \mu_n\}.
\end{aligned}\right\} \tag{30.10}$$

We have not written down the equations of the systems here, since by virtue of (30.8), they give the same solvability equations. Taking into account the fact that

$$f_{2n-1}(r) = P_{2n-1}^{(1)}(r) + c_{2n+1} \sin kr,$$
$$\chi_{2n-1}(r) = P_{2n-1}^{(2)}(r) + Ac_{2n-1} \sin kr,$$

where $P_{2n-1}^{(i)}(r)$ do not depend on c_{2n-1}, μ_n, we have:

$$f_{2n}(r) = \frac{c_1 c_{2n-1}}{2} A^{(1)}_{2n,1} + \frac{c_1 c_{2n-1}}{2} A^{(1)}_{2n,8} \cos 2kr + \mu_n A^{(1)}_{2n,2} + \\ + \{\text{чл., не сод. } c_{2n-1}, \mu_n\},$$

$$\chi_{2n}(r) = \frac{c_1 c_{2n-1}}{2} A^{(2)}_{2n,1} + \frac{c_1 c_{2n-1}}{2} A^{(2)}_{2n,8} \cos 2kr + \mu_n A^{(2)}_{2n,2} + \\ + \{\text{чл., не сод. } c_{2n-1}, \mu_n\}.$$

Since, as we can easily see, $A^{(i)}_{2n,2} = A^{(i)}_{2,8}$, $A^{(i)}_{2n,8} = 2B^{(i)}_2$,

then after some simple calculations from the equation for $f_{2n+1}(r)$, $\chi_{2n+1}(r)$

we obtain the condition for solvability:

$$\left.\begin{aligned} & c_{2n-1} c_1^3 \left(A^{(1)}_{2,3} - \frac{B^{(1)}_2}{2} - \frac{1}{8}\right) + \mu_n c_1 A^{(1)}_{2,8} = \\ & \qquad = R^{(1)}_{2n+1}\{\mu_{n-1}, \mu_1, c_{2n-3}, \ldots, c_1, \ldots\}, \\ & c_{2n-1} c_1^2 \left(A^{(2)}_{2,2} - \frac{B^{(2)}_2}{2} - \frac{A_2}{8}\right) + \mu_n c_1 \left(A^{(2)}_{2,8} - \frac{1}{\lambda_2}\right) = \\ & \qquad = K^{(2)}_{2n+1}\{\mu^{(2)}_{2n+1}, \ldots, \mu_1, c_{2n-3}, \ldots, c_1 \ldots\}. \end{aligned}\right\} \quad (30.11)$$

Equations (30.11) are always consistent with respect to c_{2n-1}, μ_n, since the determinant of the system differs from zero by virtue of (30.9). Thus conditions (30.9), (30.11) provide the determination of all the c_{2n-1}, μ_n, and therefore the construction of a formal solution for $U_1(\underline{r})$, $U_2(\underline{r})$, . In addition to these conditions, we have still another necessary condition

$$\operatorname{Det}\{\sigma_{ij}(nk)\} \neq 0 \quad (n = 0, 2, 3, \ldots). \qquad (30.12)$$

Finally, we obtain the following periodic solutions:

$$U_1(r) = \sum_{m=0}^{\infty} \tau^{\frac{2m+1}{2}} \left(\sum_{l=1}^{m} A^{(1)} \sin(2l+1)\, kr + c_{2m+1} \sin kr \right) +$$
$$+ \sum_{m=1}^{\infty} \tau^{m} \sum_{l=0}^{m} B_{m,l}^{(1)} \cos 2lkr,$$
$$U_2(r) = \sum_{m=0}^{\infty} \tau^{\frac{2m+1}{2}} \left(\sum_{l=1}^{m} A^{(2)} \sin(2l+1)\, kr + Ac_{2m+1} \sin kr \right) +$$
$$+ \sum_{m=1}^{\infty} \tau^{m} \sum_{l=0}^{m} B_{m,l}^{(2)} \cos 2lkr.$$

We now determine the constant $\xi_i^{(1)}, \xi_i^{(2)}$, entering into expansion for f_1, f_2 from (30.6). The normalization condition (30.2) takes on the form:

$$\frac{1}{V_0} \int\limits_{(V_0)} \{ f_{10} e^{-c_1^0} (1 + \tau\xi_1^{(1)} + \tau^2 \xi_2^{(1)} + \ldots)\, e^{U_1(r)} +$$
$$+ f_{20} e^{-c_2^0} (1 + \tau\xi_1^{(2)} + \tau^2 \xi_2^{(2)} + \ldots)\, e^{-U_2(r)} \}\, dr = 1,$$

or

$$\frac{1}{V_0} \int\limits_{(V_0)} \{ \alpha (1 + \tau\xi_1^{(1)} + \tau^2 \xi_2^{(1)} + \ldots)(1 - U_1(r) + \ldots) +$$
$$+ \beta (1 + \tau\xi_1^{(2)} + \tau^2 \xi_2^{(2)} + \ldots)(1 - U_2(r) + \ldots) \}\, dr = 1.$$

Inserting in the expression under the integral for $U_1(\underline{r})$, $U_2(\underline{r})$ their expansions (30.6) with the already defined functions $f_1(\underline{r})$, $\chi_i(\underline{r})$, we obtain the sequence of equations:

$$\frac{1}{V_0}\int_{(V_0)}\{-f_1(r)\alpha-\chi_1(r)\beta\}\,dr=0,$$

$$\frac{1}{V_0}\int_{(V_0)}\left\{\alpha\left(\xi_1^{(1)}-f_2(r)+\frac{1}{2}f_1^2(r)+\beta\left(\xi_1^{(2)}-\chi_2(r)+\frac{1}{2}\chi_1^2(r)\right)\right\}dr=0.$$

. . . .

All the odd equations are satisfied identically; from the even equations, we obtain expressions for $\alpha\xi_i^{(1)}+\beta\xi_i^{(2)}$, , in particular,

$$\alpha\xi_1^{(1)}+\beta\xi_1^{(2)}=\alpha\left(A_2^{(1)}-\frac{c_1^2}{4}\right)+\beta\left(A_2^{(2)}-\frac{Ac_1^2}{4}\right).$$

By definition $\tau=\lambda_1^*-\lambda_1=\frac{f_1}{\theta}-\frac{f_{10}}{\theta_0}$, $\gamma_i=\frac{f_2}{\theta}-\frac{f_{20}}{\theta_0}$ and consequently,

$$\left.\begin{aligned}&\tau=f_{10}\left\{\frac{1}{\theta}-\frac{1}{\theta_0}+\frac{1}{\theta}(\tau\xi_1^{(1)}+\tau^2\xi_2^{(1)}+\tau^3\xi_3^{(1)}+\ldots)\right\},\\&\tau\mu_1+\tau^2\mu_2+\tau^3\mu_3+\ldots=\\&=f_{20}\left\{\frac{1}{\theta}-\frac{1}{\theta_0}+\frac{1}{\theta}(\tau\xi_1^{(2)}+\tau^2\xi_2^{(2)}+\tau^3\xi_3^{(2)}+\ldots)\right\}.\end{aligned}\right\}\tag{30.13}$$

Multiplying the first of equations (30.13) by $e^{-c_1^0}$, , and the second by $e^{-c_2^0}$ and adding, we have:

$$\begin{aligned}&\tau e^{-c_1^0}+e^{-c_2^0}(\tau\mu_1+\tau^2\mu_2+\ldots)=\\&\qquad=\frac{1}{\theta}-\frac{1}{\theta_0}+\frac{1}{\theta}\{\tau(\alpha\xi_1^{(1)}+\beta\xi_1^{(2)})+\tau^2(\alpha\xi_2^{(1)}+\beta\xi_2^{(2)})+\ldots\}.\end{aligned}$$

Replacing $\alpha\xi_i^{(1)}+\beta\xi_i^{(2)}$ by their values obtained from the normalization condition, we obtain an equation defining τ as a function of other known magnitudes. For $\theta\approx\theta_0$

$$\tau=\left(\frac{1}{\theta}-\frac{1}{\theta_0}\right)\frac{1}{e^{-c_1^0}+\mu_1e^{-c_2^0}}.\tag{30.14}$$

Further, we multiply the first equation of (30.13) by f_{20} and the second by f_{10} and subtract the first from the second, obtaining:

$$\tau(f_{10}\mu_1 - f_{20}) - \tau^2 f_{10}\mu_2 + \tau^3 f_{10}\mu_3 + \ldots =$$
$$= \frac{f_{10}f_{20}}{\theta}\{\tau(\xi_1^{(2)} - \xi_1^{(1)}) + \tau^2(\xi_2^{(2)} - \xi_2^{(1)}) + \ldots\}$$

From which we find

$$\left.\begin{array}{l} \frac{f_{10}f_{20}}{\theta}(\xi_1^{(2)} - \xi_1^{(1)}) = f_{10}\mu_1 - f_{20}, \\ \frac{f_{10}f_{20}}{\theta}(\xi_2^{(2)} - \xi_2^{(1)}) = f_{10}\mu_2, \\ \ldots\ldots\ldots\ldots \\ \ldots\ldots\ldots\ldots \end{array}\right\} \quad (30.15)$$

Equations (30.15) together with the equations for $\alpha\xi_i^{(1)} + \beta\xi_i^{(2)}$ define all the constants $\xi_i^{(1)}, \xi_i^{(2)}$ in particular. In this way, the conception of the solution (30.6) is completely fulfilled.

We now go to equation (30.5), which defines the parameters of crystallization, and we write it in the form

$$\theta_0^2 + \theta_0(\alpha\sigma_{11}(k_0) + \beta\sigma_{22}(k_0)) + \alpha\beta(\sigma_{11}(\boldsymbol{k}_0)\sigma_{22}(\boldsymbol{k}_0) - \sigma_{12}^2(k_0)) = 0 \quad (30.16)$$
$$k_0 = |\boldsymbol{k}_0|.$$

θ_0 designates KT_0, where T_0 is the temperature of crystallization, $\underline{k_0}$ the wave vector of the lattice, α, β – the concentrations of the components.

The character of the curve $\theta_0 = \theta_0(k_0\alpha)$ is determined by the sign of the expression $\Delta = \sigma_{12}^2(k_0) - \left(\frac{\sigma_{11}(k_0) + \sigma_{22}(k_0)}{2}\right)^2$. Correspondingly, in the interval $0 \leqslant \alpha \leqslant 1$, $\theta_0(k_0\alpha)$ (k_0 being fixed) is represented by segments of curves of the second order; two segments, $\Delta = 0$, $\Delta < 0$, and one if $\Delta > 0$.

The result obtained gives us reason to assume that these curves correspond to Tamman's well-known crystallization curves for binary mixtures with unlimited mutual solvability, both in the liquid and the solid states.

Section 31. Comparison With Experiment for Case of Argon

It is important to keep in mind that the theory set forth requires a knowledge of the law of the action of forces over the entire interval (the energy of interaction comes under the integral sign, integration being performed over the entire interval from zero to ∞). At present the experimental data do not give this law for the entire interval in question. Furthermore, a role is played in the theory by collective interactions, which are definable by kernels K_1 ..., K_{N-1}, which in general are introduced first and with respect to which there are, as yet, no experimental data. Of course, it is not a defect of the theory that new factors enter into it, for it has hitherto been unknown and has not been investigated experimentally.

In view of these circumstances it is impossible to pretend to quantitative precision for the comparison with experimental data. The value of the theory consists rather in obtaining qualitatively new results and in their generality, which is only slightly dependent on the concrete form of the forces of interaction. Tokareva[+] considered the quantitative consequences of the linear theory of crystallization, and compared them with the experimental data for argon. Crystals of the noble gases are the simplest objects for comparing theory and experiment, since a theory is valid for them that does not take into consideration the internal degrees of freedom of the particle.

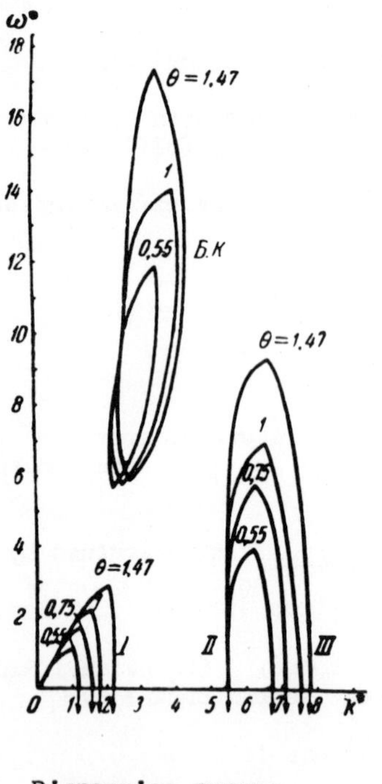

Fig. 2. Dispersion curves for crystals.

The point of departure was the criterion for the appearance in a system of many particles of solutions for the deviation of the density ρ from the stationary value ρ_0 of the wave type

$$\rho - \rho_0 = a \exp(i\omega t - ikr). \quad (31.1)$$

This criterion, as we saw, was:

$$\frac{4\pi N}{\theta} \int_0^\infty K(\rho) \frac{\sin k\rho}{k\rho} \rho^2 d\rho = -\frac{1}{2\int_0^\infty x e^{-x^2} \cos vx\, dx}, \quad (31.2)$$

where $v = \frac{2\omega}{kc}$, $\theta = kT$

[+] Vestnik MGU, No. 3-4; Uch. zap. Fizmata MOPI (1950).

It includes the crystallization condition as a particular case for $\omega = 0$.

Tokareva shows the following relationship for the potential function of interaction:

$$K(\rho) = a\left(1 - e^{-\frac{K_{12}}{a}}\right), \qquad (31.3)$$

where a is a parameter to be determined from experiments, K_{12} the energy of interaction of two isolated particles. In approximation (31.3), some account is taken as well of the effects of the collective interactions, since it is only for $a > K_{12}$ that the formula coincides with the corresponding formula for "paired" interactions. The conclusion may be drawn from indirect experimental data that $a \sim \theta$. For the potential energy of interaction, there was taken the Lenard-Jones potential

$$K_{12} = 4u\left\{\left(\frac{r_0}{r}\right)^{12} - \left(\frac{r_0}{r}\right)^{6}\right\}, \qquad (31.4)$$

where u and r_0 are two parameters of this interaction. The minimum of K_{12} is equal to u, and occurs for $r = \sqrt[6]{2}\, r_0$, the potential K_{12} is equal to zero for $r = r_0$, positive for $r < r_0$ and negative for $r > r_0$:

For argon $r_0 = 3{,}408$ Å and $u = 165{,}0 \cdot 10^{-16}$ erg;

For neon $r_0 = 2{,}734$ Å and $u = 48{,}81 \cdot 10^{-16}$ erg;

For helium $r_0 = 2{,}56$ Å and $u = 14{,}04 \cdot 10^{-16}$ erg.

In nondimensional units, the equation being investigated (the dependence of ω^+ and k^+) is:

$$\frac{1}{\int\limits_0^\infty z\left\{\exp\left[-4\theta^*\left(z^{-12} - z^{-6}\right)\right] - 1\right\}\frac{\sin k^* z}{k^*}\,dz} = \beta\int\limits_0^\infty x e^{-x^2}\cos vx\,dx \quad (31.5) \qquad (31.5)$$

$$\left(\omega^* = \frac{\omega}{c r_0},\quad k^* = \frac{k}{\frac{1}{r_0}},\quad c = \sqrt{\frac{2\theta}{m}},\quad v = \frac{2\omega^*}{k^*},\quad \theta^* = \frac{u}{\theta},\quad \beta = 8\pi N r_0^3\right).$$

TABLE 1

	- 2.5	- 2	- 1.47	- 1	- 0.85	- 0.754	- 0.695	-0.55	- 0.40
k	I(k)	I(k)	I(k)	I(k)	I(k)	I(k)	I(k)	I(k)	I(k)
0.00	0.199	0.487	1.019	1.182	1.577	1.968	2.269	3.710	5.597
0.50	0.225	0.379	0.692	1.389	1.880	2.346	2.822	4.892	16.313
1.00	0.285	0.503	0.710	2.188	3.131	4.200	4.990	12.345	-48.309
1.50	0.433	0.791	1.746	5.731	10.504	21.413	52.356	-22.573	-9.901
2.00	1.052	2.049	6.498	-19.379	-12.034	-9.852	-8.977	-7.490	-6.527
2.50	10.526	-7.600	-6.116	-5.181	-5.333	-5.415	-5.435	-5.669	-5.910
3.00	-1.597	-2.099	-3.071	-4.212	-4.606	-4.933	-5.139	-5.738	-6.464
3.50	-1.138	-1.704	-2.885	-4.560	-5.266	-5.827	-6.161	-7.174	-8.467
4.00	-1.215	-2.119	-5.593	-6.640	-7.886	-8.873	-9.533	-11.468	-14.265
4.50	-1.782	-3.353	-9.560	-17.545	-18.726	-21.791	-24.038	-30.030	-38.610
5.00	-3.413	-13.869	-58.479	104.187	90.090	86.207	85.470	99.009	100.000
5.50	8.695	12.391	12.484	16.181	18.518	20.325	21.336	25.063	30.120
6.00	3.311	4.233	7.379	11.933	13.986	15.823	16.639	19.646	23.981
6.50	2.304	4.152	6.720	12.740	15.037	16.778	17.790	22.124	26.316
7.00	3.134	6.173	13.106	18.484	21.930	24.038	26.188	30.864	36.765
7.50	4.629	11.709	19.436	48.165	54.054	59.880	62.893	71.428	-
8.00	12.345	27.933	92.592	-102.041	-142.857	-169.491	-192.308	-294.117	-
8.50	-8.196	-11.534	-35.842	-33.333	-41.152	-46.296	-51.813	-66.667	-
9.00	-0.902	-11.197	-25.445	-22.026	-30.581	-34.602	-37.594	-46.089	-

The results of the numerical integration of system (31.5) are introduced into Fig. 2.

The values of the function in the crystallization formula:

$$I(k^*) = \left(\int_0^\infty z \exp\left[-4\theta^* (z^{-12} - z^{-6}) - 1\right] \frac{\sin k^* z}{k^*}\, dz \right)^{-1}, \qquad (31.6)$$

obtained by numerical integration are given in Table 1.

Figure 3 gives the dependence of the length of the period on the temperatures.

Upon consideration of Fig. 2, we come to the following conclusions.

First conclusion. The existence of periods. Fig. 2 shows that for $\omega = 0$ (the stationary case), there are several points at which the curve $\omega = \omega(k)$ crosses the k axis. The points of intersection with the axis of abscissas give the numerical values of the space periods $\left(\frac{1}{k^+}\right)$. In the temperature interval

$$0{,}4 \leqslant \theta^* \leqslant 2{,}25 \text{ *)}$$

there are three periods, of which the first (I) in essence depends on temperature, the second (II) does not depend on the temperature and is equal to $d^* = \frac{2\pi}{k_2^*}$,

where $k_2^* \approx 5{,}45;\ d^* = \frac{d}{r_0}$,

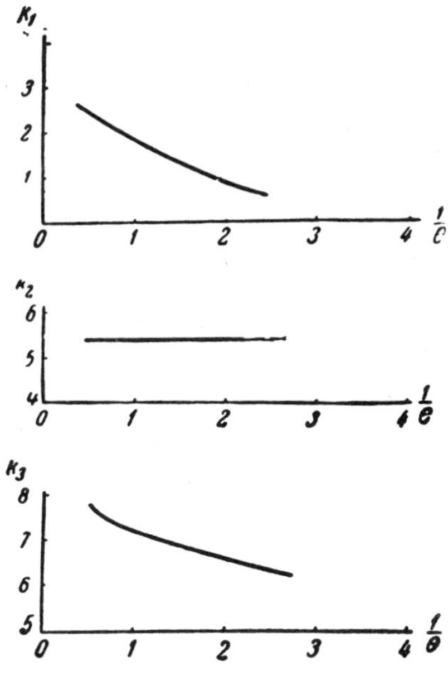

Fig. 3. Dependence of length of periods on temperature.

and the third (III) depends on the temperature but to a lesser degree than the first. The formation of these periods takes place consecutively and abruptly as the temperature drops continuously.

---

* The temperature of crystallization for argon corresponds to $\theta^+ = 1.47$; for neon $\theta^+ = 1.48$.

In this way, it is possible to speak of a fine structure of crystallization.

Second conclusion. Apart from the existence of periods and their dependence on the temperature, the "new crystal" is characterized by a definite periodic structure, which does not coincide with the stationary value of the periods (for $\omega = 0$). For at a definite temperature, a new loop is formed on the dispersion curve, this loop being of the type of an ellipse, and its area increasing as the temperature decreases (observable on Fig. 2: Б.К.). Since this oval does not cut the axis, it constitutes periodic solutions for the density

$$\{\sim \exp i\,(\omega^* t^* \pm k^* x^*)\} \qquad \omega^* \neq 0, \quad k^* \neq 0,$$

characteristic for which is the presence of a phase velocity of translation that is not equal to zero, $v = \frac{\omega^*}{k^*} \neq 0$; hence, this type of periodic structure is realized only in motion (we shall designate this class of periodic solutions as "shifting periods").

Third conclusion. The spectrum of frequencies (dependence of ω on k) has a banded structure. There are forbidden regions. As the temperature decreases, these regions shrink and the separate bands join. The envelope of the ovals is a straight line, which must represent the ordinary curve of dispersion: ω = const. k. At a sufficiently low temperature, the forbidden zones degenerate into discretely situated points on this straight line.

Fourth conclusion. The phase transition curve. To the various periods there correspond different critical temperatures at which they appear.

If the observable temperature of fusion is identified with the critical temperature at which the fundamental period (not dependent on the temperature) appears or disappears, we obtain the phase transition curve on a plane $\left(\frac{1}{\theta}, \frac{\beta}{2}\right)$ (Fig. 4).

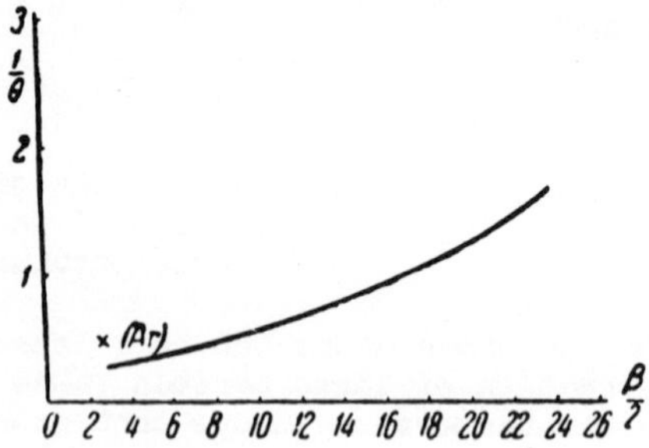

Fig. 4. Gas-crystal phase transition curve.

The existence of this curve indicates the possibility of crystallization not at a single value for the temperature and density, but at all points of the curve. There are no periodic solutions above this curve. The crystalline state is realized for temperature and density values lying below the curve.

This curve is defined by the formula (equal sign)

$$\min \frac{1}{\int\limits_0^\infty z\{\exp[-4\theta^*(z^{-12}-z^{-6})]-1\} \frac{\sin k^* z}{k^*}\, dz} \begin{cases} < & \text{There are periodic solutions} \\ = & 4\pi N r_0^3 \\ > & \text{There are no periodic solutions} \end{cases} \qquad (31.7)$$

where $\frac{2\pi}{k^+}$ is the fundamental period.

The experimental values of these quantities are known to us from the literature only for argon. The difference between the theoretical and experimental values is 15 - 20% for argon.

It is hard to expect a better agreement, if we take into account the approximate nature of the formula for the kernel K.

Passage from theory to experiment. From the experimental point of view, the calculation that has been made raises the primary question as to the confirmation of the qualitatively new conclusions of the theory:

1. The existence of periods and their gradual appearance with decreasing temperature.

2. The existence of forbidden zones in the oscillation spectrum of the crystal near the temperature of fusion.

CHAPTER II

VIBRATION PROPERTIES OF ELECTRON PLASMA*

The physical concept of plasma has not been defined with sufficient clarity in the literature. Basically, the singular oscillatory properties of an aggregate of charged particles have led to the need of distinguishing this aggregate, the plasma, from gas.

In the theory being set forth, the properties of plasma are precisely defined by the initial postulates and equations. The following are characteristic of the theory:

1. The absence of conceptions of collisions between electrons. The concept of cross-section, which is at the basis of the notion of the kinetic theory of gases, cannot be introduced for electrons. We have seen that dispersion at a Coulomb center leads to a divergent integral for the complete diameter, which indicates the impossibility of exact introduction of this concept.

2. The interactions between electrons (and nuclei) are assumed to be collective, which is especially characteristic for the Coulomb laws of forces. In this chapter, we shall consider the oscillatory properties of plasma.

Section 32. Vibration Properties for a Given Initial Distribution

In Sec. 20, we considered solutions of Cauchy's problem for arbitrary forces and arbitrary initial distribution function. Our task now is to make the results there obtained concrete, especially for a Coulomb law of forces.

Let us consider a mixture of two kinds of particles: electrons and positively charged particles, which are near equilibrium. The initial equations, as we saw in Sec. 19, permit solutions corresponding to a uniform distribution in space of both kinds of particles, with a Maxwell distribution of velocities. This requires that in the case of uniform distribution the total charge of any assigned volume should be equal to zero.

We assume that at some given initial moment of time the distribution function for the electron is different from the distribution function corresponding to a uniform distribution. The question arises as to the change in this perturbation in the

---

* See the author's monograph "Theory of Vibration Properties of Electron Gas and Its Applications," Uch. zapiski MGU 75, No. 2 (1945).

course of time. The following simplifications are admissible under this statement of the problem:

1. Linearization of the initial equations, if it is presupposed that the amplitudes of the perturbations are small as compared to the value of the f-function in the unperturbed state.

2. In view of the presence of high frequencies of vibration, the translation of heavy ions may be ignored, and they may be considered as distributed with uniform density independently of the distribution of the electrons.

Equations (19.3) or (20.1) may be taken as the initial equations

$$\frac{\partial\varphi}{\partial t} + \boldsymbol{v}\,\mathrm{grad}_r\varphi = $$
$$= \frac{1}{m}\,\mathrm{grad}_v f_0 \cdot \mathrm{grad}_r \int_{-\infty}^{\infty} K(|\boldsymbol{r}-\boldsymbol{r}'|) \int_{-\infty}^{\infty} \varphi(\boldsymbol{r}', \boldsymbol{v}', t)\, d\boldsymbol{v}' \cdot d\boldsymbol{r}', \qquad (32.1)$$

where K is now the kernel of the Coulomb interaction, $f_0(v^2)$ is the unperturbed function of distribution of the electron.

The solution of this equation for the deviation of the density of electrons from the uniform distribution

$\rho(\boldsymbol{r}, t) = \int_{-\infty}^{\infty} \varphi(\boldsymbol{r}, \boldsymbol{v}, t)\, d\boldsymbol{v}$ led to a Volterra integral equation of the second class

$$q_k(t) = F_k(t) + \int_{t_0}^{t} G_k(t-\tau)\, q_k(\tau)\, d\tau, \qquad (32.2)$$

where $q_k(t)$ is the Fourier amplitude in the expansion

$$\rho(\boldsymbol{r}, t) = \sum_k q_k(t)\, e^{ikr}, \qquad (32.3)$$

$F_k(t)$ a function defined by the initial perturbation [formula (20.6)], and $G_k(t)$ a form of the unperturbed distribution function [formula (20.7)].

The solution of (32.2) is given by integral (20.9)

$$q_k(t) = \frac{1}{2\pi i} \int_{\sigma_0 - i\infty}^{\sigma_0 + i\infty} e^{pt} \frac{F_k(p)}{1 - G_k(p)} \, dp. \tag{32.4}$$

As was mentioned in Sec. 20, the situation of the poles of the function under the integral sign defines the properties of the solution. The eqution for the poles (dispersion equation) is

$$-\frac{i\sigma(k)}{m} \int_{-\infty}^{\infty} \frac{k\nabla_v f_0}{p + ikv} \, dv = 1. \tag{32.5}$$

It connects the values of p and k in solutions of the type

$$\exp(pt + ikr).$$

For the Coulomb law of forces

$$\sigma(k) = 4\pi \int_0^\infty K(\rho)\rho^2 \frac{\sin k\rho}{k\rho} \, d\rho =$$

$$= \lim_{\varkappa \to 0} 4\pi \int_0^\infty \frac{e^2}{\rho} e^{-\varkappa\rho} \rho^2 \frac{\sin k\rho}{k\rho} \, d\rho = 4\pi \frac{e^2}{k^2} > 0. \tag{32.6}$$

We saw in Sec. 20, that for any law of the forces the condition $\sigma(k) > 0$ leads to an oscillatory change in the perturbations of density in the course of time. This result is quantitatively made more precise by the fact that for a given wave number k, the magnitude p has a complex value.

We consider the dispersion equation (32.5) for the case of Coulomb forces. Without destroying the generality, we shall choose a system of coordinates with an axis x directed along k. Then equation (32.5) will have the form

$$\frac{4\pi e^2}{k} \int_{-\infty}^{\infty} \frac{\xi \frac{\partial \Phi_0}{\partial \xi}}{k\xi - \omega} \, d\xi \, d\eta \, d\zeta = 1, \qquad \text{where} \qquad p = -i\omega. \tag{32.7}$$

Function $\Phi_0(\xi, \eta, \zeta)$ entering into the dispersion equation is a distribution function for the stationary state. In this section, we consider an electron gas characterized in the stationary state by the Maxwell distribution function

$$\Phi_0(\varepsilon) = N\left(\frac{m}{2\pi\theta}\right)^{\frac{3}{2}} e^{-\frac{\varepsilon}{\theta}}, \tag{32.8}$$

where $\varepsilon = \frac{m}{2}(\xi^2 + \eta^2 + \zeta^2)$. Substituting (32.8) in (32.7) and integrating with respect to η and ζ , we obtain:

$$-\frac{4\pi e^2 N}{k}\sqrt{\frac{m}{2\pi(\theta)^3}}\int_{-\infty}^{\infty}\frac{\xi e^{-\frac{m\xi^2}{2\theta}}}{k\xi - \omega}\, d\xi = 1. \tag{32.9}$$

Here it is convenient to introduce nondimensional magnitudes.

We take as the unit of frequency the familiar characteristic frequency of an electron gas

$$\omega_0 = \sqrt{4\pi N\frac{e^2}{m}}.$$

We take as the unit of the wave number, the reciprocal of the Debye distance D

$$D = \frac{1}{\varkappa}, \quad \varkappa = \sqrt{\frac{4\pi e^2 N}{\theta}}.$$

Then the unit of velocity will be

$$\frac{\omega_0}{\varkappa} = \sqrt{\frac{\theta}{m}}.$$

We write

$$\omega^* = \frac{\omega}{\omega_0}, \quad k^* = \frac{k}{\varkappa}, \quad \xi^* = \frac{\xi\varkappa}{\omega_0} = \xi\sqrt{\frac{m}{\theta}} = x.$$

Introducing all this into (32.9), we obtain the dispersion equation in the form

$$\int_{-\infty}^{\infty} \frac{xe^{-\frac{x^2}{2}}\,dx}{x - \frac{\omega^*}{k^*}} = -\sqrt{2\pi}\cdot k^{*2}. \tag{32.10}$$

The fact that the quantities e, m, N, θ defining the electron plasma do not appear explicitly in the formula obtained, but only through chosen dimensional units, shows that in the system under consideration, the time interval defined by the frequency ω_0 and the space interval defined by a Debye distance are natural measurements, intrinsic to the system under consideration.

This is also indicated by the circumstance that the Debye formula for static polarization is automatically included in the dispersion formula obtained, and is a particular case of it. For, setting $\omega = 0$ in (32.10), we obtain

$$k^{*2} = -1, \tag{32.11}$$

that is, $k = \pm i\kappa$. Consequently, the spatial dependence of solutions (32.4) will in this case be determined by the terms $e^{\pm \kappa r}$. We are interested in the way in which the inhomogeneity of density is eliminated that arises in an electron gas, and we shall therefore consider <u>k</u> as given in the dispersion equation.

We limit ourselves to considering the case that is most interesting physically, when the macroscopic inhomogeneity arising in an electron plasma is great as compared to the Debye distance, that is,

$$|k^*| = \left|\frac{k}{\varkappa}\right| \cong 2\pi\frac{d}{D} < 1, \qquad \text{where} \qquad k = \frac{2\pi}{D}.$$

In this case, we represent the expression under the integral in the form of a power series in k:

$$\frac{1}{x - \frac{\omega^*}{k^*}} = -\frac{k^*}{\omega^*}\left[1 + \frac{x}{\omega^*}k^* + \left(\frac{x}{\omega^*}\right)^2 k^{*2} + \ldots\right], \tag{32.12}$$

which here, too, uniformly converted in the interval $|x| < \frac{\omega^*}{k^*}$.

Substituting (32.12) and (32.10) and for the time confining ourselves to the first term that does not vanish, we have:

$$-\frac{k^*}{\omega^*}\int x^2 e^{-\frac{1}{2}x^2}\,dx = -\sqrt{2\pi}k^{*2},$$

where the integration may extend only under the interval

$|x| < \frac{\omega^*}{k^*}$; however, since $\frac{\omega^*}{k^*} > 1$, then without any noticeable error, we may consider the limits as infinite; then we have:

$$\omega^{*2} = 1,$$

or, for ω^+:

$$-\left(\frac{k^*}{\omega^*}\right)^2\sqrt{2\pi} = -\sqrt{2\pi}\,k^{*2} \tag{32.13}$$

from which, consequently, we have the constant value for the cyclical frequency

$$\omega = \pm\omega_0 = \pm\sqrt{\frac{4\pi Ne^2}{m}}. \tag{32.14}$$

Thus, in the approximation under consideration, the dispersion law is such that the cyclical frequency ω does not depend on the wave number and is equal to a constant ω_0, which characterizes a plasma. This indicates the presence of an anomalously great dispersion, of such a nature that the magnitude of the group velocity is equal to zero. This means that in this approximation, there is no propagation of the perturbations; the macroscopic inhomogeneity produced does not diffuse in space, as in an ordinary gas, but oscillates with a frequency ω_0.

The law of dispersion also defines the dependence of the solutions on time. We have for the variation of the electron density ρ (x, y, z, t):

$$\rho(x, y, z, t) = \int_{-\infty}^{\infty} e^{i\omega_0 t - ikr}\, c(k)\,dk, \tag{32.15}$$

where amplitudes c(k) are defined by the initial condition for

$$\rho(x, y, z, 0) = \int_{-\infty}^{\infty} e^{ikr} c(k)\, dk. \tag{32.16}$$

Consequently, from (32.15) and (32.16)

$$\rho(x, y, z, t) = \rho(x, y, z, 0)\, e^{i\omega_0 t}, \tag{32.17}$$

which coincides with the formula obtained from elementary considerations.

We pass to the next approximation, the calculation of the first two nonvanishing terms in expansion (32.12); we have:

$$\left\{\left(\frac{k^*}{\omega^*}\right)^2 \int x^2 e^{-\frac{1}{2}x^2} dx + \frac{k^*}{\omega^*} \int x^4 e^{-\frac{1}{2}x^2} dx\right\} = \sqrt{2\pi}\, k^{*2}.$$

Here, too, we may without any notable error extend integration to infinity. We get:

$$\left\{\left(\frac{k^*}{\omega^*}\right)^2 \sqrt{2\pi} + \left(\frac{k^*}{\omega^*}\right)^4 3\sqrt{2\pi}\right\} = \sqrt{2\pi} \cdot k^{*2}. \tag{32.18}$$

Consequently, we have the following equation to define ω as a function of k:

$$\frac{1}{\omega^{*2}} + \frac{3k^{*2}}{\omega^{*4}} = k^{*2},$$

or

$$\omega^{*4} - \omega^{*2} - 3k^{*2} = 0. \tag{32.19}$$

Since the initial formula (32.18) is valid only in the case $|k^*| < 1$ (of the second order), the solution of equation (32.19) must be represented in the form of a series in k, and we must limit ourselves to the first two nonvanishing terms. We thus have:

$$\omega^{*2} = \frac{1}{2} \pm \frac{1}{2}(1 + 12k^{*2})^{\frac{1}{2}} \approx \frac{1}{2} \pm \frac{1}{2}(1 + 6k^{*2}).$$

for a root with a plus sign, we have:

$$\omega^{*2} = 1 + 3k^{*2}, \tag{32.20}$$

that is, a further specification of formula (32.13). Passing to the usual units, we have:

$$\omega^2 = \omega_0^2 + \frac{3\theta}{m}k^2. \tag{32.21}$$

A root with a negative sign does not correspond to the initial assumption $\frac{\omega^*}{k^*} > 1$ and it should therefore be rejected.

Taking into consideration that k is small, we may likewise write

$$\omega = \omega_0 + \frac{3}{2}\frac{\theta}{\omega_0} \cdot k^2. \tag{32.22}$$

As has been shown, in the case ($\sigma(k) < 0$), the oscillations are damped. Approximating the Maxwell function in accordance with formula (20.11), we obtain a simple result (20.13):

$$\rho_k(r, t) = q_k(t)e^{ikr} = a_k e^{ikr - \gamma t} \cos \omega_0 t,$$

$$\omega_0^2 = \frac{\sigma(k) k^2 N}{m}, \quad \gamma = kc, c = \sqrt{\frac{2\theta}{m}}, \tag{32.23}$$

$\sigma(k) = \frac{4\pi e^2}{k^2}$ and hence, $\omega_0^2 = \frac{4\pi Ne^2}{m}$. Formula (32.22), and likewise the diffusion character of the oscillations (32.23) were results obtained in work over the period 1938 -1945 (see the monograph, "Theory of Vibration Properties of Electron Gas and Its Applications.").

In 1946, Landau gave further precision to results (32.23), involving the damping. He assumed that in contour interval (32.4), the function under the integral sign could be prolonged into the left half plane of the complex variable as well (the expression under the integral, generally speaking, is determined only in the right half of the plane, since Re(p) $>$ 0). Then, choosing the path of integration in a suitable way, as is shown in Fig. 5, the decrement of damping may be defined. We present his discussion here+ .

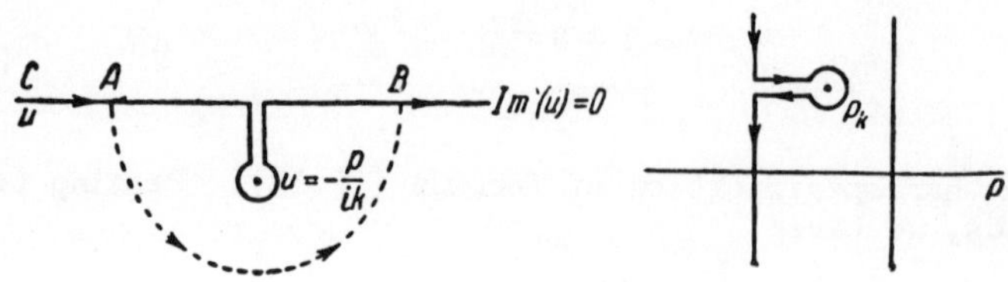

Fig. 5.

To calculate this decrement, Landau starts from the postulate (confirmed by the results) that as k $\longrightarrow$ 0, the real part of p_k likewise tends to zero (as we have seen, exponentially), while the imaginary part remains finite. Then point $u = \frac{p_k}{ik}$, which is enveloped by court C and small values of k, is situated at a finite distance from the imaginary axis and very close to the real axis (under the latter). He sets

$$p_k = -i\omega - \gamma;$$

the quantity $\gamma (0 < \gamma \ll \omega)$ is the damping decrement sought. He then chooses on the real axis a point A situated sufficiently close to $u = -\frac{p_k}{ik}$, in such a way that its distance to that point is still great as compared to $c\,|\mathrm{Im}(u)|$, he draws to this point the semicircle AB, embracing the singular point $u = -\frac{p_k}{ik}$ (dotted line in Fig. 5, a) and replaces a corresponding portion of the path of integration of C by it.

The integral over a rectilinear segment of path is real in the

+ L. Landau Zh.E.T.F. 16, No. 7 (1946)

boundary Re(p) = 0; in the approximation under consideration, it may be set equal to $\frac{4\pi ne^2}{mp^2}$. The integral over the semi-circle is equal to the residue with respect to the pole multiplied by πi (half of the complete circumference).

As a result, Landau obtained a dispersion equation in the form

$$-\frac{4\pi ne^2}{mp^2}+i\frac{4\pi^2 e^2}{mk^2}\frac{df_0\left(-\frac{p}{ik}\right)}{du}=1.$$

Setting here $p = -i\omega - \gamma$ as solving this equation by successive approximations, he obtained the following final expression for the damping decrement γ:

$$\gamma=\omega_0\sqrt{\frac{\pi}{\gamma}}\frac{1}{(ka)^2}e^{-\frac{1}{2}(ka)^2}. \qquad (32.24)$$

Thus, the damping decrement decreases exponentially as k increases.

The domain of applicability of formulas (32.22), (32.24) is defined by the condition $\gamma \ll \omega$. And this leads to an inequality, which may be written in the form

$$ka \ll 1.$$

Further, Landau considers the inverse limiting case of large k

$$ka \gg 1.$$

He sets again $p = -i\omega - \gamma$. As will be confirmed by the result, as $k \to \infty$ both the magnitudes ω and γ increase without limit, but in such a way that for large k, $\omega \ll \gamma$, while the particular $\frac{\omega}{k}, \frac{\gamma}{k}$ tend to zero and infinity, respectively. Then the pole $u = -\frac{p}{ik}$ is located relatively near the imaginary axis, but at a great distance from the real axis. Re(u) is small, and Im(u) is large. Since function f_0 increases exponentially for large imaginary values of u, then in the integral

$$\frac{4\pi ie^2}{km}\int_c \frac{df_0}{du}\cdot\frac{du}{(p+iku)}=1 \qquad (32.25)$$

it is sufficient to confine ourselves only to integration over a circle around the pole (Fig. 5, b), ignoring the integral along the real axis. In this way, we obtain equation (32.25) in the form

$$\frac{4\pi e^2}{mk^2} 2\pi i \frac{df_0\left(-\frac{p}{ik}\right)}{du} = 1$$

or, inserting the expression for $f_0(u) = m\sqrt{\frac{m}{2\pi\Theta}}e^{-\frac{mu^2}{2\Theta}}$,

$$\sqrt{2\pi}\frac{p}{\omega_0 (ka)^3} e^{\frac{p^2}{2\omega_0^2 (ka)^2}} = 1. \tag{32.26}$$

Taking the moduli from both sides of the equality and employing the postulated inequality $\gamma \gg \omega$, we can obtain:.

$$\xi e^{\frac{\xi^2}{2}} = \frac{1}{\sqrt{2\pi}}(ak)^2, \tag{32.27}$$

where

$$\xi = \frac{\gamma}{\omega_0 ka}.$$

Further, the phase multiplier of the expression on the left side of (32.27) is to the same approximation

$$\exp\left(\frac{i\gamma\omega}{\omega_0^2 a^2 k^2}\right).$$

Since the right side contains a positive real quantity, this multiplier must be equal to + 1. Hence, Landau finds

$$\frac{\gamma\omega}{\omega_0^2 a^2 k^2} = \pi$$

(equating here 3π, 5π, ..., we should obtain, as can be shown, not the closest root of equation (32.25) to the imaginary axis). Together with the definition of magnitude ξ this give

$$\omega = \pi\sqrt{\frac{\Theta}{m}}\frac{k}{\xi}, \gamma = \sqrt{\frac{\Theta}{m}}k\xi. \tag{32.28}$$

with function ξ (k), implicitly defined by equation (32.27), these formulas define the frequency and the damping decrement of the oscillations. The function ξ (k) is a function of k that increases very slowly (fundamentally, as the square root of the logarithm). The ratio $\frac{\gamma}{\omega}$ increases with k like ξ^2, that is, fundamentally like ln ka.

Section 33. Nondamping Waves

We saw in Sec. 21 of the first part that a principal property of Cauchy's problem (assigning the initial value of the function) is, in general, the separation of the initial moment of time. We repeat once again here that physically the setting apart of the initial moment of time involves, from our point of view, an interference from without in the system. But we are at the same time interested in other temporal processes as well, which characterize the natural development of the system, without any kind of important interference from without. The solution of the initial equations of the type of the half sum of the lagging and leading potentials discloses, as we saw in Sec. 21, temporal processes in which no moment of time is set off as special along the entire key axis. From the physical point of view, therefore, these processes naturally relate to a special class. Observation of these processes is, of course, not involved in any physical operation introducing a foreign element into the system.

We show the character of these processes as applied to the electron plasma. The initial linearized equation for deviation of the density from the uniform distribution (21.32) has the form

$$\rho(\boldsymbol{r}, t) = \int\limits_{-\infty}^{\infty} \frac{1}{m} \nabla_{\boldsymbol{v}} f_0 d\boldsymbol{v} \left\{ \frac{1}{2} \int\limits_{-\infty}^{t} d\tau \nabla_{\boldsymbol{r}} \int\limits_{-\infty}^{\infty} K(|\boldsymbol{r} - \boldsymbol{v}(t-\tau) - \boldsymbol{r}'|) \rho(\boldsymbol{r}', t) d\boldsymbol{r}' + \right.$$

$$\left. + \frac{1}{2} \int\limits_{+\infty}^{t} d\tau \nabla_{\boldsymbol{r}} \int\limits_{-\infty}^{\infty} K(|\boldsymbol{r} - \boldsymbol{v}(t-\tau) - \boldsymbol{r}'|) \rho(\boldsymbol{r}', t) d\boldsymbol{r}' \right\}. \tag{33.1}$$

Substitution of solutions of the type

$$\rho(\boldsymbol{r}, t) = \sum_{\boldsymbol{k}} q_{\boldsymbol{k}}(t) e^{i\boldsymbol{k}\boldsymbol{r}} \tag{33.2}$$

leads to the following equation for determing $q_k(t)$:

$$q_k(t) = \int_{-\infty}^{\infty} G_k(|t-\tau|)\, q_k(\tau)\, d\tau, \tag{33.3}$$

where

$$G_k(t-\tau) = \frac{\sigma(k)}{m} \int_{-\infty}^{\infty} e^{-ikv(t-\tau)} i(k\nabla_v) f_0 dv. \tag{33.4}$$

Equation (33.3) has a solution of the type

$$q_k(t) = a_k e^{i\omega t}.$$

Substitution gives the fundamental condition for solvability of this equation in the given class of functions of the following form:

$$\int_{0}^{\infty} G_k(t) \cos \omega t \, dt = 1 \tag{33.5}$$

or, as has been noted above,

$$\frac{\sigma(k)}{m} \int_{-\infty}^{\infty} \frac{\frac{\partial f_0}{\partial \xi} d\xi}{\xi - \frac{\omega}{k}} = 1; \quad f_0 = \int_{-\infty}^{\infty} f_0(\xi, \eta, \zeta)\, d\eta\, d\zeta, \tag{33.6}$$

taking the principal value of the integral. We make concrete condition (33.5) or (33.6) for the Coulomb law of forces and the Maxwell distribution function.

Then

$$\sigma(k) = 4\pi \int_{0}^{\infty} \frac{e^2}{\rho} \frac{\sin k\rho}{k\rho} \rho^2 d\rho = \frac{4\pi e^2}{k^2}, \tag{33.7}$$

$$G_k(t) = -\frac{2\sigma(k) N}{mc^2} \frac{1}{c^3} \pi^{-\frac{3}{2}} \int_{-\infty}^{\infty} e^{-ikvt} ikve^{-\frac{v^2}{c^2}} dv, \quad c^2 = \frac{2\theta}{m}. \tag{33.8}$$

the tabular integral is expressed by

$$\int_{-\infty}^{\infty} e^{-\frac{v^2}{c^2}} v \sin kvt\, dv = \frac{kc^3}{4} \sqrt{\pi}\, te^{-\frac{k^2c^2t^2}{4}},$$

and consequently,

$$G_k(t) = -\frac{\sigma(k)}{m} k^2 te^{-\frac{k^2c^2t^2}{4}}. \tag{33.9}$$

Equation (33.5) assumes the form

$$1 = -\omega_0^2 \int_0^{\infty} te^{-\frac{k^2c^2t^2}{4}} \cos \omega t\, dt; \quad \omega_0^2 = \frac{4\pi Ne^2}{m}. \tag{33.10}$$

We introduce the nondimensional unit

$$k^{*2} = \frac{k^2}{k_0^2}, \quad \omega^* = \frac{\omega}{\omega_0}, \quad k_0 = \frac{c}{\omega_0}$$

and we obtain the fundamental solvability criterion in the form of the formula

$$k^{*2} = -4 \int_0^{\infty} xe^{-x^2} \cos v^* x\, dx, \quad v^* = \frac{2\omega^*}{k^*}, \tag{33.11}$$

which gives the dependence of ω^+ on k^+, in implicit form. The right side contain the special function:

$$I(v^*) = -\int_0^{\infty} xe^{-x^2} \cos v^* x\, dx. \tag{33.12}$$

TABLE 2

		$I(\nu) = -\int_0^\infty xe^{-x^2} \cos \nu x \, dx$
1	0.	-0.450
2	1.0	-0.286
3	1.5708	-0.092
4	2.0	+0.043
5	2.5	+0.121
6	3.0	+0.142
7	4.0	+0.105
8	5.0	+0.059
9	6.0	+0.032
10	7.854	+0.017
11	10.0	+0.0103

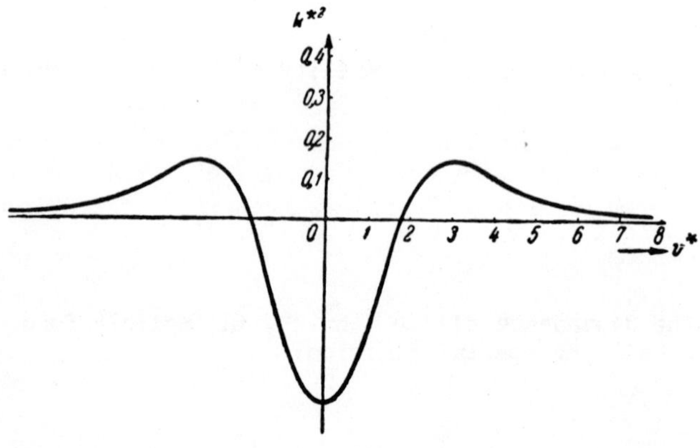

Fig. 6.

Table 2 gives the values for this function obtained by numerical integration; the graph of formula (33.11) is given in Fig. 6. The curves for the dependence of ω^{+} on k^{+}, likewise obtained by means of numerical integration, are given in Fig. 7.

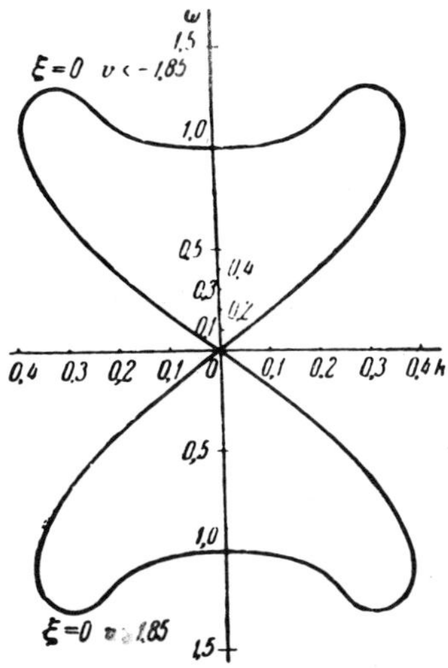

Fig. 7. Dispersion curve for electrons with Maxwell distribution function.

We note the fundamental characteristics of the dispersion curve.

1. The presence of maximum frequencies and maximum wave numbers leads to the fact that nondamping waves cannot be propagated in an electron plasma with frequencies as high as desired and with wave numbers as great as desired. At this point, we come up against a difference in principle from the ordinary equations of acoustics.

2. For small k we can speak of two branches, an acoustic branch when for $k \to 0$ and $\omega \to 0$ as well, and a second branch when for $k \to 0$, $\omega \to \omega_0$. In the field of small k, these two branches may be approximated by the following expressions:

$$\left.\begin{aligned} \omega^2 &= \omega_0^2 + c^2k^2, \\ \omega^2 &= c_1^2k^2, \end{aligned}\right\} \tag{33.13}$$

where

$$c = \sqrt{\frac{3\theta}{m}}, \quad c_1 \sim 1,3\sqrt{\frac{\theta}{m}}.$$

The size of the maximum frequency is determined by the value

$$\omega_{max} = 1.284\ \omega_0.$$

and the value of the maximum wave number by

$$k_{max} = 0.534\ k_0.$$

The dispersion equation longitudinal waves in an electron gas with any initial distribution function may be obtained at once, if we find the solution of the linearized equation

$$-\frac{\partial \varphi^{(l)}}{\partial t} = \operatorname{div}_r v\varphi^{(l)} + \operatorname{div}_v \frac{e}{m} e^{(l)} f_0,$$
$$\operatorname{div} e^{(l)} = 4\pi e \int \varphi^{(l)}\, dv \qquad (33.14)$$

in the wave form. For, setting

$$\varphi^{(l)}(r,\ v,\ t) = T(t)\, \varphi_0^{(l)}(r,\ v),$$
$$e^{(l)}(r,\ t) = T(t)\, e_0^{(l)}(r) \qquad (33.15)$$

and substituting in the given equations, we obtain:

$$\frac{\frac{dT}{dt}}{T} = i\omega; \quad T(t) = T(0)\, e^{i\omega t}. \qquad (33.16)$$

Thus, the dependence on the time is determined at once, while as yet ω is an undetermined constant of integration. To determine the amplitude, we have the system:

$$i\omega\varphi_0^{(l)} = \operatorname{div}_r v\varphi_0^{(l)} + \frac{e}{m} e_0^{(l)} \operatorname{grad}_v f_0,$$
$$\operatorname{div} e_0^{(l)} = 4\pi e \int \varphi_0^{(l)}\, dv, \qquad (33.17)$$
$$\left(\operatorname{div}_v \frac{e}{m} e_0^{(l)} f_0 \equiv \frac{e}{m} e_0^{(l)} \operatorname{grad}_v f_0\right),$$

and since by definition $\operatorname{rot} e^{(l)}=0$, the solutions must be sought in the form of expansions along plane longitudinal waves. For an individual wave, we have:

$$i(\omega-kv)\,g_k^{(l)}=-\frac{e}{m}\frac{1}{|k|}a_k(k\,\mathrm{grad}_v)f_0$$

$$-i|k|\,a_k=4\pi e\int g_k^{(l)}(v)\,dv, \tag{33.18}$$

where $\underline{k}_1$ is a unique vector in the direction of propagation. Substitution gives

$$\varphi_0^{(l)}(r,\ v)=g_k^{(l)}(v)\,e^{-ikr}, \tag{33.19}$$

$$e_0^{(l)}(r)=k_1a_ke^{-ikr},$$

from which we have the following equation for determing $g_k^{(1)}(v)$:

$$g_k^{(l)}(v)=-\frac{4\pi e^2}{m|k|}\frac{(k\nabla_v)f_0}{\omega-kv}\int g_k^{(l)}(v)\,dv. \tag{33.20}$$

This formula determines both the law of dispersion of longitudinal waves and the form of $g_k^{(1)}(v)$, the perturbed function of distribution for any t, that is, it gives the solution to the problem. See Sec. 20 for critical remarks as to the possibility of the vanishing of the denominator in this formula, and an answer to these observations. Integrating the above relationship on the right and the left with respect to the velocities, we obtain as the solvability condition a dispersion equation of the following type:

$$-\frac{4\pi e^2}{mk^2}\int\frac{(k\nabla_v)f_0(v)}{\omega-kv}dv=1 \tag{33.21}$$

or, assuming the x axis is directed along $\underline{k}$+:

$$-\frac{4\pi e^2}{mk}\int\frac{\frac{\partial\Phi_0}{\partial\xi}\,d\xi}{\omega-k\xi}=1, \tag{33.22}$$

where $\Phi_0(\xi)=\int\int f_0(\xi,\ \eta,\ \zeta)\,d\eta\,d\zeta.$

+ It is essential to note that as yet we have never assumed that ω and k are real. (see Sec. 35).

Integrating by parts and taking into account that at the boundaries

$$\Phi_0 \Big|_{\xi=-\infty}^{\xi=+\infty} = 0,$$

we may also represent the dispersion equation in the form

$$\frac{4\pi e^2}{m} \int \frac{\Phi_0(\xi)\, d\xi}{(\omega - k\xi)^2} = 1. \tag{33.23}$$

The form of $\Phi_0(\xi)$ may thus far be arbitrary. We consider some of the simplest cases:

1. $\Phi_0(\xi) = N\delta(\xi)$. All the particles have the same zero velocity. Then we have:

$$\frac{4\pi e^2}{m} N \frac{1}{\omega^2} = 1, \tag{33.24}$$

that is, a well-known result.

2. $$\Phi_0(\xi) = \frac{N}{2\xi_0} \left\{ \int_{-\infty}^{\xi} \delta(\xi + \xi_0)\, d\xi - \int_{-\infty}^{\xi} \delta(\xi - \xi_0)\, d\xi \right\},$$

that is, the distribution by velocities is given in the form of a rectangle with sides $-\xi_0$; $+\xi_0$, which gives

$$\frac{\partial \Phi_0}{\partial \xi} = \frac{N}{2\xi_0} \left\{ \delta(\xi + \xi_0) - \delta(\xi - \xi_0) \right\}. \tag{33.25}$$

The factor $2\xi_0$ in the denominator is introduced on the basis of the normalization requirement

$$\int_{-\xi_0}^{+\xi_0} \Phi_0(\xi)\, d\xi = N.$$

Using the general form of the dispersion equation, for example, (33.22), we have

$$-\frac{4\pi e^2 N}{2\xi_0 m k} \left\{ \frac{1}{\omega + k\xi_0} - \frac{1}{\omega - k\xi_0} \right\} = 1 \tag{33.26}$$

or

$$\frac{\omega^2}{\omega^2 - k^2\xi_0^2} = 1; \quad \omega^2 = \omega_0^2 + k^2\xi_0^2. \qquad (33.27)$$

3. For an electron beam with strictly fixed velocities ξ_0, we have:

$$\Phi_0(\xi) = N\delta(\xi - \xi_0),$$

which gives

$$\frac{\omega_0^2}{(\omega - k\xi_0)^2} = 1, \qquad (33.28)$$

or

$$\omega = \omega_0 + k\xi_0.$$

This relationship shows that for $\omega = 0$, a real value of k is possible, that is, a stationary periodic structure occurs, the period $\left|\frac{1}{k}\right|$ is defined by the formula $\frac{\xi_0}{\omega_0}$

4. Assuming that an electron beam has a distribution of the velocities concentrated around a value ξ_0 for the velocity in the form of a rectangle with sides $\xi_0 - \frac{\Delta\xi}{2}$; $\xi_0 + \frac{\Delta\xi}{2}$; we have:

$$\frac{\partial\Phi_0}{\partial\xi} = \frac{N}{\Delta\xi}\left\{\delta\left(\xi - \xi_0 + \frac{\Delta\xi}{2}\right) - \delta\left(\xi - \xi_0 - \frac{\Delta\xi}{2}\right)\right\},$$

which gives

$$\frac{\omega_0^2}{(\omega - k\xi_0)^2 - k^2\left(\frac{\Delta\xi}{2}\right)^2} = 1 \qquad (33.29)$$

or

$$\omega = k\xi_0 \pm \sqrt{\omega_0^2 + k^2\left(\frac{\Delta\xi}{2}\right)^2}.$$

Hence, in the case under consideration there are two phase velocities of propagation ($\frac{\omega}{k}$) depending on the presence of the two signs. In the more general case, the presence of several electron beams

$$\frac{\partial \Phi_0}{\partial \xi} = \sum_i \frac{N_i}{\Delta \xi_i} \left\{ \delta \left(\xi - \xi_i + \frac{\Delta \xi_i}{2} \right) - \delta \left(\xi - \xi_i - \frac{\Delta \xi_i}{2} \right) \right\}$$

we obtain:

$$\sum_i \frac{4\pi \frac{e^2}{m} N_i}{(\omega - k\xi_i)^2 - k^2 \left(\frac{\Delta \xi}{2} \right)^2} = 1; \tag{33.30}$$

neglecting the dispersion of velocities in each beam, we have:

$$\sum_i \frac{4\pi \frac{e^2}{m} N_i}{(\omega - k\xi_i)^2} = 1. \tag{33.31}$$

Recently, Haeff+, Pierce and others have applied the dispersion equation to a stream of electrons in vacuum tubes. The dispersion equation is obtained by them in the following way. When there are several electron beams, moving with different initial velocities, it is possible to base description of them on the equation of motion, Poisson's equation, and the equation of continuity:

$$\frac{\partial \boldsymbol{v}_i}{\partial t} + (\boldsymbol{v}_i \operatorname{grad}) \boldsymbol{v}_i = -\frac{1}{\rho_i} \operatorname{grad} V,$$

$$\Delta V = 4\pi e \sum_i \rho_i, \tag{33.32}$$

$$\operatorname{div} \rho_i \boldsymbol{v}_i = -\frac{\partial \rho_i}{\partial t}.$$

The connection between the beams is by means of the forces of electric interaction.

If we now assume that the space charge of the unperturbed motion of the beams is completely compensated (that is, introduce the analogue of the positive "phon" and linearize the initial system of equations on that presupposition, we can obtain:

+ A. Haeff, P.I.R.E. 37, 4 (1949) and Phys. Rev. 75, 1546 (1949).

$$\frac{\partial v_i}{\partial t} + (v_i^0 \operatorname{grad}) v_i = -\frac{1}{\rho_i} \operatorname{grad} V_i,$$

$$\Delta V = 4\pi e \sum_i \rho_i, \qquad (33.33)$$

$$\operatorname{div} \{\rho_i^0 v_i + \rho_i v_i^0\} + \frac{\partial \rho_i}{\partial t} = 0,$$

where v_i, ρ_i, V_i now denote deviations from the stationary values v_i^0, ρ_i^0, V_i^0, corresponding to the case of uniform motion. The linearized system has solutions in the wave form. Assuming for all the magnitudes, a dependence of the type $\exp\{i\omega t - i\kappa x\}$, we easily obtain the solvability conditions of the linearized system in the form

$$\sum_i \frac{\omega_i^2}{(\omega - kv_i^0)^2} = 1, \qquad (33.34)$$

where $\omega_i^2 = 4\pi \frac{e}{m} N_i$

Haeff goes over to the continuous velocities distribution and writes the dispersion equation down as follows:

$$\int_{v_1}^{v_2} \frac{\frac{d\omega_i^2}{dv} dv}{(\omega - kv_i)^2} = 1; \qquad (33.35)$$

here the velocity distribution function is implicitly introduced. But comparing this with our dispersion equation, which was given as early as in the 1938 work cited above:

$$\frac{4\pi e^2}{mk} \int \frac{\frac{\partial \Phi_0}{\partial \xi} d\xi}{\omega - k\xi} = 1 \quad \text{or} \quad \frac{4\pi e^2}{m} \int \frac{\Phi_0(\xi)\, d\xi}{(\omega - k\xi)^2} = 1, \qquad (33.35')$$

we find that relation (33.34) is correct. For, setting

$$\Phi_0(\xi) = \sum_i N_i \delta(\xi - v_i^0),$$

we obtain the correct dispersion equation for this case.

Nonetheless, expression (33.34) is partial. We have no guarantee that the initial velocity dispersion can be small in effect, since a small difference in the velocities may lead to great changes in the density of particles at sufficiently great distances, which is well known even from the approximation of noninteracting electrons.

Strictly speaking, the transition from the sum (33.34) to the integral (33.35) is not legitimate.

1. Initial equations (33.32) do not include the temperature dispersion in the velocities. Consequently, the final result as well should not contain parameters adequate to the temperature.

2. We shall arrive at the dispersion equation, if we introduce the condition ($\omega_i^2 = \frac{4\pi e^2}{m} N_i \rightarrow \frac{4\pi e^2}{m} \Phi_0(\xi)\, d\xi$) of an infinitely small concentration for each beam. But, it was postulated above that $\underline{v}_i^0$, ρ_i^0, V_i^0 are magnitudes of the same order (under linearization).

In precise calculations, therefore, the starting point must be our equation (33.35').

The dispersion equation for two beams has solutions in the form of nondamping waves (ω real; k real) only on the condition that the relationship is observed:

$$\frac{\delta\omega}{v\omega_1} < \sqrt{2},$$

where

$$\delta = \frac{v_1 - v_2}{3}, \quad v = \frac{v_1 + v_2}{2};$$

v_1 and v_2 being the initial velocities of the beams:

In the case of the contrary inequality, k is complex; and, consequently, the solutions increase (or decrease) with the distance. Haeff explains the existence of such solutions by the amplifier.

For the elimination of transverse perturbations in an electron gas, we have linearized equations of the form:

$$-\frac{\partial\varphi^{(t)}}{\partial t} = \operatorname{div}_r v\varphi^{(t)} + \frac{e}{m}\operatorname{div}_v \frac{e}{m}\left(e^{(t)} + \frac{1}{c}[vh^{(t)}]\right)_0,$$

$$\operatorname{rot} h^{(t)} - \frac{1}{c}\frac{\partial e^{(t)}}{\partial t} = \frac{4\pi e}{c}\int v\varphi^{(t)}\, dv,$$

$$\operatorname{rot} e^{(t)} - \frac{1}{c}\frac{\partial h^{(t)}}{\partial t} = 0,$$

$$\operatorname{div} e^{(t)} = 0, \quad \operatorname{div} h^{(t)} = 0. \tag{33.36}$$

Here, as in the case of longitudinal waves, we shall seek solutions for $\varphi^{(t)}(r, v, t), e^{(t)}(r, t), h^{(t)}(r, t)$ in the form

$$\left.\begin{aligned} e^{(t)} &= e_k b_k e^{i\omega t - ikr}, \\ h^{(t)} &= h_k c_k e^{i\omega t - ikr}, \\ \varphi^{(t)} &= g_k^{(t)}(v)\, e^{i\omega - ikr} \end{aligned}\right| \tag{33.37}$$

as a result of the condition

$$\operatorname{div} e^{(t)} = -i(ke_k) = 0; \quad \operatorname{div} h^{(t)} = -i(kh_k) = 0,$$

the waves are transverse. The interrelations vectors $\underline{e_k}$, $\underline{h_k}$, $\underline{k}$ is determined by the equation: $\operatorname{rot} e^{(t)} = -i\frac{\omega}{c}h^{(t)}$, which likewise links amplitudes $\underline{b_k}$ and c_k

$$[ke_k]\, b_k = h_k c_k \frac{\omega}{c},$$

from which

$$h_k = [ke_k]\frac{1}{|k|}; \quad b_k = \frac{\omega}{c|k|}c_k.$$

Making use of the third equation of the system for transverse waves, we have:

$$-i[kh_k]\, c_k = \frac{i\omega}{c} e_k b_k + \frac{4\pi e}{c}\int v g_k^{(t)}(v)\, dv.$$

Taking into account the preceding relationships for amplitude

c_k, b_k, we obtain:

$$-\frac{c}{\omega} i [k [k e_k]] b_k = i \frac{\omega}{c} e_k b_k + \frac{4\pi e}{c} \int v g_k^{(t)}(v)\, dv.$$

If we now take into account that

$$\frac{1}{|k|^2}[k [k e_k]] = \frac{k}{|k|^2}(k e_k) - e_k (kk) \frac{1}{|k|^2} = -e_k,$$

then the preceding relationship acquires the form:

$$i \frac{c^2 k^2 - \omega^2}{\omega} e_k b_k = \frac{4\pi e}{c} \int v g_k^{(t)}(v)\, dv, \tag{33.38}$$

or multiplying by $\underline{e_k}$ right and left, we have:

$$b_k = i\, 4\pi e \frac{\omega}{\omega^2 - c^2 k^2} \int (e_k v)\, g_k^{(t)}(v)\, dv. \tag{33.39}$$

Thus, the amplitudes of the fields: $\underline{e}^{(t)}$ and $\underline{h}^{(t)}$, by means of the relationships cited, express, by means of the amplitudes, the distribution function $g_k^{(t)}$. To determine the latter, we make use of the first of the initial system of equations for transverse waves. We obtain:

$$i [\omega - (kv)]\, g_k^{(t)}(v) = -\frac{e}{m} \left\{ (e_k \nabla_v) f_0 b_k + \frac{1}{c} [v h_k] \nabla_v f_0 \cdot c_k \right\}.$$

In view of $[v h_k] \nabla_v f_0 = 0$ the action of the force $\frac{e}{c}[vh]$ automatically drops from the calculations. Finally, the equation for the unknown amplitude $g_k^{(t)}(\underline{v})$ has the form

$$g_k^{(t)}(v) = \frac{4\pi e^2}{m} \frac{\omega}{\omega^2 - c^2 k^2} \frac{(e_k \nabla_v) f_0}{[(kv) - \omega]} \int (e_k v)\, g_k^{(t)}(v)\, dv, \tag{33.40}$$

$$\left(\nabla_v f_0 = m v \frac{\partial f_0}{\partial \varepsilon}, \quad \varepsilon = \frac{m v^2}{2} \right).$$

The dispersion equation for transverse waves is obtained by multiplying the preceding relationship by $\underline{e_k}$ and integrating for all values of velocities

$$4\pi e^2 \frac{\omega}{\omega^2 - c^2k^2} \int \frac{(e_k v)\frac{\partial\Phi_0}{\partial\varepsilon}}{(kv - \omega)} dv = 1, \tag{33.41}$$

where e_k is the unit vector directed to the strength of the electric field of the linearly polarized wave, Φ_0 is the unperturbed distribution function, which is considered to depend solely on the energy ε.

We set $e_k = e_k(y),\ k = k(x)$. The dispersion equation takes on the form

$$I(\omega, k) = 4\pi e^2 \frac{\omega}{\omega^2 - c^2k^2} \int \frac{\eta^2 \frac{\partial\Phi_0}{\partial\varepsilon}}{k\xi - \omega} d\xi d\eta d\zeta = 1. \tag{33.42}$$

For a Maxwell distribution

$$I(\omega, k) = \frac{4\frac{\pi e^2 N}{m}}{\omega^2 - c^2k^2} \frac{1}{\sqrt{2\pi}} \sqrt{\frac{m}{kT}} \int_{-\infty}^{+\infty} \frac{e^{-\frac{m\xi^2}{2kT}} d\xi}{1 - \frac{\xi k}{\omega}} = 1.$$

For the integral in $I(\omega, k)$, we thence have:

$$-\frac{1}{kT}\left(\frac{m}{2\pi kT}\right)^{\frac{3}{2}} N \int_{-\infty}^{+\infty} \frac{e^{-\frac{m\xi^2}{2kT}}}{k\xi - \omega} d\xi \int_{-\infty}^{+\infty} \eta^2 e^{-\frac{m\eta^2}{2kT}} d\eta \int_{-\infty}^{+\infty} e^{-\frac{m\zeta^2}{2kT}} d\zeta;$$

so that

$$\int_{-\infty}^{+\infty} \eta^2 e^{-\frac{m\eta^2}{2kT}} d\eta = \left(\frac{kT}{m}\right)^{\frac{3}{2}} \sqrt{2\pi}; \quad \int_{-\infty}^{+\infty} e^{-\frac{m\xi^2}{2kT}} d\xi = \left(\frac{kT}{m}\right)^{\frac{1}{2}} \cdot \sqrt{2\pi}.$$

We get:

$$\frac{\partial\Phi_0}{\partial\varepsilon} = -\frac{1}{kT}\Phi_0 = -\frac{1}{kT} N\left(\frac{m}{2\pi kT}\right)^{\frac{3}{2}} e^{-\frac{m(\xi^2+\eta^2+\zeta^2)}{2kT}} \tag{33.43}$$

Here, too, we introduce nondimensional quantities. We take as the unit frequency the characteristic frequency of the electron plasma; then the unit of the wave number will be

$$k_0 = \sqrt{4\frac{\pi N e^2}{mc^2}},$$

$$\omega^* = \frac{\omega}{\omega_0}, \quad k^* = \frac{k}{k_0}.$$

Introducing these quantities into the dispersion equation, we get:

$$\omega^{*2} = k^{*2} + \frac{1}{\sqrt{2\pi}} \int_{-\infty}^{+\infty} \frac{e^{-\frac{1}{2}x^2}\,dx}{1 - \frac{k^*}{\omega^*}\frac{v}{c}x}, \tag{33.44}$$

where v is the thermal velocity of the electrons $\frac{\sqrt{kT}}{m}$.

Assuming $\frac{v}{c} \ll 1$, we express the expression under the integral sign in the form of the series, as in the case of the longitudinal waves:

$$\frac{1}{1 - \frac{k^*}{\omega^*}\frac{v}{c}x} = 1 + \left(\frac{k^*}{\omega^*}\frac{v}{c}x\right) + \left(\frac{k^*}{\omega^*}\frac{v}{c}x\right)^2 + \ldots$$

In the case of longitudinal waves and the validity of such an expansion is limited by the condition $|k^*| < 1$. In the case under consideration, thanks to the presence of the factor $\frac{v}{c}$, the domain over which the expansion is valid is broadened. It is only required that

$$\frac{k^*}{\omega^*}\frac{v}{c} \ll 1,$$

or in ordinary units

$$\frac{\omega}{k} \gg v.$$

In the final formula

$$\left(\frac{\omega}{k}\right)^2 = c^2 + \left(\frac{\omega_0}{k}\right)^2.$$

Confining ourselves to the first two terms, we have:

$$\omega^{*2} = k^{*2} + \frac{1}{\sqrt{2\pi}} \int e^{-\frac{1}{2}x^2} dx + \left(\frac{k^*}{\omega^*}\frac{v}{c}\right)^2 \frac{1}{\sqrt{2\pi}} \int x^2 e^{-\frac{1}{2}x^2} dx.$$

The integration here must be performed only for $|x| < \frac{\omega^*}{k^*} \cdot \frac{c}{v}$,

but for the purpose of calculation, we shall, without serious error, because of the condition $\left(\frac{k^*}{\omega^*}\frac{v}{c} \ll 1\right)$ extend the limits of integration to infinity.

Under these conditions, if we likewise limit ourselves to the first member of the expansion, we obtain:

$$\omega^{*2} = 1 + k^{*2},$$

or, in ordinary units

$$k^2 = \frac{(\omega^2 - \omega_0^2)}{c^2},$$

that is, the well-known formula.

Section 34. Oscillations of Plasma With Fermi's Distribution Function in State of Degeneration

With Goldman[+] the vibrations were considered for the case in which an unperturbed electron gas is described by a Fermi distribution function (the stationary equation is satisfied by any function symmetrical in the velocities):

$$(\xi^2 + \eta^2 + \zeta^2) = \frac{2m^3}{h^3} \frac{1}{e^{\frac{m(\xi^2+\eta^2+\zeta^2)}{2\theta} - \frac{1}{\varepsilon}} + 1} \tag{34.1}$$

and is in the state of degeneration. The degree of degeneration is characterized by the smallness of the magnitude $\varepsilon \approx \frac{\theta}{E_0}$,

[+]1) See the author's monograph, cited above; 2) N. Goldman, Zh. E.T.F., 17, 681 (1947).

where E_0 is the maximum energy of the Fermi distribution, $\theta = kT$. We have, from the normalization condition:

$$\frac{1}{\varepsilon}=\frac{h}{2m\theta}\left(\frac{3N}{8\pi}\right)^{\frac{2}{3}}\left\{1-\frac{\pi}{12}\varepsilon^2+\dots\right\}. \qquad (34.2)$$

In metal, at room temperature, ε is of the order of several thousands, and the second term in parentheses, of the order of 10^{-5}, very small as compared to unity.

Dispersion equation (32.5) is written down as:

$$1+\frac{8\pi^2e^2}{k^2m}\cdot\frac{2m^3}{h^3}\int\limits_c \frac{1}{e^{\frac{m\xi^2}{2\theta}-\frac{1}{\varepsilon}}+1}\frac{ik\xi\,d\xi}{p+ik\xi}=0. \qquad (34.3)$$

Instead of ξ, we introduce the relative velocity $u = \frac{\xi}{b}$, where

$$b=\sqrt{\frac{2\theta}{m}\frac{1}{\varepsilon}}=\frac{h}{m}\left(\frac{3N}{8\pi}\right)^{\frac{1}{3}}\left\{1-\frac{\pi}{24}\varepsilon^2+\dots\right\}, \qquad (34.4)$$

and let

$$\omega_0^2=\frac{4\pi Ne^2}{m}, \quad v=-\frac{p}{ikb}. \qquad (34.5)$$

Dispersion equation (34.3) assumes the following form:

$$-\frac{1}{2}\int\limits_c \frac{1}{e^{\frac{u^2-1}{\varepsilon}}+1}\frac{u\,du}{u-v}=\frac{k^2b^2}{3\omega_0^2}\left(1+\frac{\pi}{8}\varepsilon^2+\dots\right), \qquad (34.6)$$

contour c enveloping the point $u = v$.

The function of a complex variable

$$\frac{1}{e^{\frac{u^2-1}{\varepsilon}}+1},$$

for $\varepsilon \longrightarrow 0$:

a) At the points $R(u^2) < 1$ tends to 1,

b) At the points $R(u^2) > 1$ tends to 0.

If $R(v^2) > 1$, then in the case of complete degeneration ($\varepsilon \to 0$)

$$-\frac{1}{2}\int \frac{1}{e^{\frac{u^2-1}{\varepsilon}}+1}\frac{udu}{u-v} = -\frac{1}{2}\int_{-1}^{+1}\frac{udu}{u-v} = \frac{v}{2}\ln\frac{v+1}{v-1} - 1. \tag{34.7}$$

Dispersion equation (34.7) will be satisfied if v is real, that is, $p = -i\omega$ is purely imaginary

$$\frac{\omega}{2kb}\ln\frac{\omega+kb}{\omega-kb} - 1 = \frac{k^2b^2}{3\omega_0^2}.$$

In the case of long waves ($k \to 0$) by expansion in a series, we obtain:

$$\omega^2 = \omega_0^2\left(1 + \frac{3k^2b^2}{5\omega_0^2} - \frac{114}{175}\frac{k^4b^4}{\omega_0^4} + \dots\right). \tag{34.8}$$

In the other extreme case, when k is large

$$\omega = kb\left(1 + 2e^{-2\left[\frac{k^2b^2}{3\omega_0^2}+1\right]}\right). \tag{34.9}$$

For the second type of solutions, the dispersion equation is written as:

$$1 + \frac{8\pi^2e^2}{k^2m}\int_{-\infty}^{\infty}\frac{k\xi\Phi(\xi^2)\,d\xi}{k\xi - \omega} = 0, \tag{34.10}$$

where the integral should be taken in the sense of the principal value. In the case of degeneration, calculation gives:

$$\frac{\omega}{2kb}\ln\left|\frac{\omega+kb}{\omega-kb}\right| - 1 = \frac{k^2b^2}{3\omega_0^2}, \tag{34.11}$$

so that for $\omega > kb$, the previous result is obtained. If $\omega < kb$, we obtain a new acoustic branch, passing through the point $k = 0$.

Section 35. Behavior of Plasma in External Electrical Field

The problem of the behavior of electron plasma in an external field may be set up in various ways depending on the conditions of the experiment. Thus, we may consider the static problem of the distribution of the potential in the plasma for given external charges, the problem as to the forced oscillations under given external fields variable with time. Finally, a special place should be given to the cases in which on the limiting surfaces it is possible in a definite way to specify the distribution function and the potential.

On the basis of the results obtained in Chapter III, we may write the equations for the potentials (for a given uniform distribution of the ions) in the form:

$$\Delta\varphi = -4\pi e\left\{\int f dv - N\right\} - 4\pi\rho(x, y, z)$$

$$\int_{-\infty}^{\infty} f dv = N e^{-\frac{e\varphi}{\theta}}$$

or

$$\Delta\varphi = -4\pi e N\{e^{-\frac{e\varphi}{\theta}} - 1\} - 4\pi\rho(x, y, z). \qquad (35.1)$$

If $e\varphi/\theta \ll 1$, then

$$\Delta\varphi = \frac{4\pi e^2 N}{\theta}\varphi + 4\pi\rho(x, y, z). \qquad (35.2)$$

Equations (35.1) and (35.2) coincide with the well-known equations of the theory of electrolytes.

At great distances from external charges, the solutions of equation (35.2) are expressed by exponentially decreasing or increasing functions. Both of these have a physical meaning, depending on the way in which the internal charges are located (to the left or to the right) with respect to the region under consideration. Equations (35.2) indicate the presence of a characteristic effect of screening of the electric field in the plasma. The depth of penetration of the field is defined by the formula

$$d^2 = \frac{mc^2}{4\pi N e^2}, \quad c^2 = \frac{\theta}{m}.$$

It is essential to note that in equation (35.1), the distribution function f is a simple symmetrical Maxwell function. Actually, only such a function, as was shown above (see Chapter II of the first part) can be a solution of the stationary equations in the case of spatially inhomogeneous distribution. Equation (35.1) cannot, therefore, answer the question as to the effect of the electric field on the distribution function when there is a drift velocity.

At the same time, it is immediately clear that the character of the screening must essentially depend on the presence of a velocity of translation of the electrons (or ions) relative to the external sources of the electric field, the position of which in space is taken as given.

The essential factor here is that the transition to the static case is not at all trivial in the theory being set forth. For example, solving any temporal equation and then going over to the static case in the solution obtained, we obtain a result that does not in general follow from the equation, in which previously we have set $\frac{\partial}{\partial t} = 0$.

We pose the problem as to the oscillations of plasma, assigning a frequency $\omega > 0$; after obtaining the solution, we shall set $\omega \to 0$. To solve the problem, we confine ourselves to a linear approximation

$$f = f_0 + \varphi, \quad \varphi \ll f_0,$$

and we seek a solution for the potential and for a small addition to the distribution function in the form of damping (or expanding) plane waves

$$\exp(i\omega t - px),$$

where p is to be determined, but ω is given.

In order for the solution to include the screening effect as well, we shall consider p as complex. The following dispersion equation linking and p, then is valid:

$$p^2 = -\frac{4\pi e^2}{m} \int_{-\infty}^{\infty} \frac{\frac{\partial f_0}{\partial \xi} d\xi}{\xi - i\frac{\omega}{p}}, \tag{35.3}$$

where f_0 is any unperturbed distribution function, depending only on the velocities.

We pass to nondimensional quantities

$$\omega^{+}=\frac{\omega}{\omega_0},\quad p^{+}=\frac{p}{p_0},\quad p_0=\frac{\omega_0}{c},\quad \xi^{+}=\frac{\xi}{c},$$

$$f_0^{+}=f_0/Nc,\quad \int_{-\infty}^{\infty} f_0^{+}\,d\xi^{+}=1,$$

where c is the characteristic velocity in the distribution f_0, N is the concentration of the particles, and ω_0 the characteristic frequency of the plasma. We then obtain:

$$-p^{+2}=\int_{-\infty}^{\infty}\frac{\dfrac{\partial f_0^{+}}{\partial \xi^{+}}\,d\xi^{+}}{\xi^{+}-i\dfrac{\omega^{+}}{p^{+}}}. \tag{35.4}$$

In what follows, we omit the + for all the quantities. In the transition $\omega \rightarrow 0$, three separate cases must be distinguished depending on the way in which the point

$$z=\frac{i\omega}{p}$$

is located relative to the real axis.

As was shown earlier (see Sec. 21), the limiting value of an integral of the Cauchy type, as z approaches the line of integration, is defined as:

$$\lim_{z\to x_0}\int_{-\infty}^{\infty}\frac{\Phi(x)\,dx}{x-z}=\int_{-\infty}^{\infty}\frac{\Phi(x)\,dx}{x-x_0}\pm i\pi\Phi(x_0), \tag{35.5}$$

where x_0 is any point on the real axis; the sign (+) is taken in the case that z tends to x_0 (by any path), remaining in the upper part of the plane of the complex variable, and correspondingly, the sign (-) if z remains in the lower part of the plane.

In formula (35.5), the integral on the right side is taken in the sense of the principal value

$$\int_{-\infty}^{\infty}\frac{\Phi(x)\,dx}{x-x_0}=\lim_{\varepsilon\to 0}\left(\int_{-\infty}^{x_0-\varepsilon}\frac{\Phi(x)\,dx}{x-x_0}+\int_{x_0+\varepsilon}^{\infty}\frac{\Phi(x)\,dx}{x-x_0}\right). \tag{35.6}$$

In the event that z is on the real axis, the relationship that has been given provides a method for determining the integral in the sense of the principal value

$$\left(\frac{1}{2}\lim_{\substack{z\to x_0\\ \mathrm{Im}\,z>0}}\int_{-\infty}^{\infty}\frac{\Phi(x)\,dx}{x-z}+\frac{1}{2}\lim_{\substack{z\to x_0\\ \mathrm{Im}\,z<0}}\int_{-\infty}^{\infty}\frac{\Phi(x)\,dx}{x-z}\right)=\int_{-\infty}^{\infty}\frac{\Phi(x)\,dx}{x-x_0}. \tag{35.7}$$

Setting for our case p = $\varkappa$ + ik, where $\varkappa$ and k are real numbers and, consequently,

$$z=\frac{i\omega}{p}=\frac{\omega}{k^2+\varkappa^2}(i\varkappa-k), \tag{35.8}$$

we shall make $\omega\to 0$ ($\omega > 0$). We have

$$-p^2=\lim_{\omega\to 0}\int_{-\infty}^{\infty}\frac{\frac{\partial f_0}{\partial\xi}\,d\xi}{\xi-\frac{i\omega}{p}}=\int_{-\infty}^{\infty}\frac{\frac{\partial f_0}{\partial\xi}\,d\xi}{\xi}\pm i\pi\left(\frac{\partial f_0}{\partial\xi}\right)_{\xi=0}, \tag{35.9}$$

where we assume the sign (+) in the case $\varkappa > 0$ and (-) for $\varkappa < 0$. If $\varkappa = 0$ and $\omega = 0$, then we have:

$$-p^2=\int_{-\infty}^{\infty}\frac{\frac{\partial f_0}{\partial\xi}}{\xi}\,d\xi. \tag{35.10}$$

The dependence of the functions being sought (the potential and the distribution function) on the coordinates and the time is determined by the expression

$$e^{i\omega t-px}=e^{i\omega t-\varkappa x-ikx}. \tag{35.11}$$

The sign of k must be determined from what follows. Substituting the expression p = $\varkappa$ + ik in formula (35.9), we have:

$$\begin{aligned}k^2-\varkappa^2&=J_1,\\ 2\varkappa k&=\mp J_2\end{aligned} \tag{35.12}$$

(where now the sign (-) is for the upper and (+) for the lower half plane), from which

$$\varkappa^2 = \frac{1}{2}\left(-J_1 + \sqrt{(J_1)^2 + (J_2)^2}\right);$$
$$k^2 = \frac{1}{2}\left(+J_2 + \sqrt{(J_1)^2 + (J_2)^2}\right) \tag{35.13}$$

(the sign (+) must be kept before the root, since by definition $\varkappa$ and k are real numbers), where

$$J_1 = \int_{-\infty}^{\infty} \frac{\frac{\partial f_0}{\partial \xi}}{\xi}\, d\xi; \qquad J_2 = \pi\left(\frac{\partial f_0}{\partial \xi}\right)_{\xi=0}. \tag{35.14}$$

In the case where $\varkappa = 0$ (with $\omega = 0$), p = ik and (35.10) gives

$$k^2 = \int_{-\infty}^{\infty} \frac{\frac{\partial f_0}{\partial \xi}}{\xi}\, d\xi. \tag{35.15}$$

In the first two cases, there are waves with decreasing and increasing amplitude

$$\left.\begin{aligned} &\exp(i\omega t - \varkappa x - ikx), \quad \omega > 0, \quad \varkappa > 0, \\ &\exp(i\omega t + \varkappa' x - ikx), \quad \omega > 0, \quad \varkappa' = -\varkappa > 0, \end{aligned}\right\} \tag{35.16}$$

and in the third case ($\varkappa = 0$), nondamping waves

$$\exp(i\omega t - ikx), \tag{35.17}$$

the equation of dispersion for which has been analyzed above.

The direction of propagation of the waves (the sign of k) is determined by the sign of $\varkappa$ k (35.12); thus for

$$\varkappa > 0 \quad 2\varkappa k = -\pi\left(\frac{\partial f_0}{\partial \xi}\right)_{\xi=0},$$

and for

$$\begin{matrix}\varkappa < 0 \\ (\varkappa' > 0)\end{matrix} \quad 2\varkappa k = +\pi\left(\frac{\partial f_0}{\partial \xi}\right)_{\xi=0}.$$

If we set $\left(\frac{\partial f_0}{\partial \xi}\right)_{\xi=0} > 0$, which corresponds to the direction of the velocity of translation to the right, we obtain $\varkappa k < 0$ for the first case and $\varkappa k > 0$ for the second. Thus, as in the case of the damping waves in the direction of the motion of the particles, so in the case of the expanding waves in the same direction, the sign of k is negative, and, hence, there are only waves being propagated against the motion of the particles.

We consider some particular cases, giving a definite form to function f:

1. Let $f_0 = f_0\left(\frac{\xi^2}{2}\right)$ — be an arbitrary symmetrical distribution function, decreasing as the velocity increases. For $\xi > 0$ and $\xi < 0$, respectively,

$$\frac{\partial f_0}{\partial \xi} \lesseqgtr 0 \text{ and } \left(\frac{\partial f_0}{\partial \xi}\right)_{\xi=0} = 0. \tag{35.18}$$

We have from (35.12): $k^2 = 0$

$$\left.\varkappa^2 = -J_1 = -\int_{-\infty}^{\infty} \frac{\frac{\partial f_0}{\partial \xi}\,d\xi}{\xi} = \int_{-\infty}^{\infty} f_0\,d\xi = 1 \right\} \tag{35.19}$$

or in dimensional units

$$\varkappa^2 = \frac{\omega_0^2}{c^2}.$$

The third case as well leads to the same result

$$k^2 = -1, \quad k'^2 = -k^2 = \frac{\omega_0^2}{c^2} = \varkappa^2. \tag{35.20}$$

Thus, each of the three cases leads to one and the same result, the Debye screening with the functions

$$\exp(\pm \varkappa x),$$

that is, in this case the limiting solution (for $\omega \rightarrow 0$) does not depend on the position of the pole on the complex plane.

2. Let $f_0 = \delta(\xi - \xi_0)$, in which case

$$\left(\frac{\partial f_0}{\partial \xi}\right)_{\xi=0} = 0,$$

$$J_1 = \int_{-\infty}^{\infty} \frac{\partial f_0}{\partial \xi}\frac{1}{\xi}\,d\xi = \frac{1}{\xi}f_0\Big|_{-\infty}^{\infty} + \int_{-\infty}^{\infty}\frac{f_0\,d\xi}{\xi^2} = \frac{1}{\xi_0^2} > 0, \tag{35.21}$$

consequently, (35.12) gives

$$\varkappa = 0,$$

$$k^2 = \frac{1}{\xi_0^2}.$$

Formula (35.15) leads to the same result in dimensional units

$$k^2 = \frac{\omega_0^2}{\xi_0^2}.$$

Accordingly, there appear nondamping spatially-periodic solutions with the period

$$d^2 = \frac{m\xi_0^2}{4\pi Ne^2}.$$

3. Let f_0 be the Maxwell distribution function

$$f_0 = \frac{1}{\sqrt{2\pi}}e^{-(\xi-\xi_0)^2},$$

in which case

$$J_1 = \int_{-\infty}^{\infty}\frac{\partial f_0}{\partial \xi}\frac{1}{\xi}\,d\xi = -\frac{1}{\sqrt{2\pi}}\int_{-\infty}^{\infty}\frac{x\,e^{-x^2/2}}{x+\xi_0}\,dx, \tag{35.22}$$

$$J_2 = \pi\left(\frac{\partial f_0}{\partial \xi}\right)_{\xi=0} = \frac{\pi}{\sqrt{2\pi}}\xi_0 e^{-\frac{\xi_0^2}{2}}. \tag{35.23}$$

The integral J_1 appearing here represents a special function introduced previously by us, which can be determined on the basis of formula (35.7)

$$J_1 = -\frac{1}{\sqrt{2\pi}} \int_{-\infty}^{\infty} \frac{xe^{-x^2/2}}{x+\xi_0}\,dx = -\int_0^{\infty} xe^{-x^2/2}\cos\xi_0 x\,dx =$$

$$= -\left(1-\sqrt{2}\,\xi_0 e^{-\xi_0^2/2}\int_0^{\xi_0/\sqrt{2}} e^{\eta^2}\,d\eta\right). \qquad (35.24)$$

For $0 \leqslant \xi_0 \leqslant \frac{1,85}{\sqrt{2}}$ this function varies from ($-\frac{\pi}{2}$) to zero; for $\xi_0 > \frac{1,85}{\sqrt{2}}$ it has a maximum and tends to zero; if $\xi_0 \to \infty$, as $\frac{1}{\xi_0^2}$. In dimensional units

$$\left.\begin{aligned} \varkappa^2 &= \frac{\omega_0^2}{2c^2}\left(-J_1+\sqrt{J_1^2+J_2^2}\right), \\ k^2 &= \frac{\omega_0^2}{2c^2}\left(+J_1+\sqrt{J_1^2+J_2^2}\right), \end{aligned}\right\} \qquad (35.25)$$

$$J_1 = -\int_0^{\infty} xe^{-x^2/2}\cos\frac{\xi_0}{c}x\,dx, \qquad (35.26)$$

$$J_2 = \frac{\pi}{\sqrt{2\pi}}\frac{\xi_0}{c}e^{-\xi_0^2/2c^2}, \qquad c^2 = \theta/m. \qquad (35.27)$$

for $\frac{\xi_0}{c} \ll 1$

$$k^2 \to \frac{\omega_0^2}{4c^2}\frac{\xi_0^2}{c^2}\sqrt{\frac{\pi}{2}}, \quad \varkappa^2 \to \frac{\omega_0^2}{c^2},$$

that is, т. е. $\varkappa^2 \gg k^2$. For nondamping waves in this case, there are no periodic solutions. For $\frac{\sqrt{2}\,\xi_0}{c} = 1,85 \quad (J_1 = 0)$

$$k^2 = \varkappa^2 = \frac{\omega_0^2}{2c^2}J_2 = 1,85\sqrt{\pi}\frac{\omega_0^2}{4c^2}e^{-\frac{1}{4}(1,85)^2};$$

for $\frac{\xi_0}{c} \gg 1$

$$k^2 = \frac{\omega_0^2}{\xi_0^2}, \quad \varkappa^2 \to 0.$$

We now show that solutions decreasing in amplitude in the direction of translation of the particles correspond to what are called "leading solutions." Our starting point will not be the dispersion equation but a particular representation of the initial linearized equation, in the following form:

$$\left.\begin{aligned} \Delta\varphi &= -4\pi e \int_{-\infty}^{\infty} (f - f_0)\, d\xi, \\ f - f_0 &= \frac{e}{m}\frac{\partial f_0}{\partial \xi} \int_{+\infty}^{t} \frac{\partial \varphi}{\partial x}(x - \xi(t-\tau), \tau)\, d\tau, \end{aligned}\right\} \tag{35.28}$$

when t $< + \infty$ (see Sec. 9), then

$$\Delta\varphi = -\frac{4\pi e^2}{m} \int_{-\infty}^{\infty} \frac{\partial f_0}{\partial \xi}\, d\xi \int_{+\infty}^{t} \frac{\partial \varphi}{\partial x}(x - \xi(t-\tau), \tau)\, d\tau. \tag{35.29}$$

In conformity with what has preceded, we shall seek a solution in the form

$$\varphi = \varphi_0 e^{i\omega t - px}, \quad p = \varkappa + ik, \quad \omega > 0, \quad \varkappa > 0. \tag{35.30}$$

This solution, as we have seen, corresponds in the dispersion equation to a position of the pole in the upper portion of the plane of the complex variable

$$z = \frac{\omega}{k^2 + \varkappa^2}(i\varkappa - k).$$

Upon substitution of the proposed solution (35.30) in (35.29), the integral appears

$$(-p) \int_{+\infty}^{t} e^{i\omega(\tau - t) + p\xi(t - \tau)}\, dt$$

or, carrying out the substitution $\tau - t = \eta > 0$,

$$(-p) \int_{0}^{+\infty} e^{(i\omega - p\xi)\eta}\, d\eta = (-p)\left\{ \frac{1}{i\omega - p\xi} - \frac{e^{(i\omega - p\xi)\eta}}{i\omega - p\xi} \Big|_{\eta = \infty} \right\}$$

that is, on condition that p = $\varkappa$ + ik, $\varkappa > 0$, then the second integral disappears, and we get the following condition for the solvability of equation (35.29):

$$p^2 = -\frac{4\pi e^2}{m} \int_{-\infty}^{\infty} \frac{\frac{\partial f_0}{\partial \xi}\, d\xi}{\xi - \frac{i\omega}{p}}. \qquad (35.31)$$

This condition coincides exactly with the dispersion equation, with the pole situated in the upper portion of the plane of the complex variable. Hence, the "leading solution" has the form

$$\exp(i\omega t - \varkappa x + ikx), \quad \omega > 0, \quad \varkappa > 0.$$

The direction of the velocity of drift propagation, as we have seen, is determined only by the sign of $\left(\frac{\partial f_0}{\partial \xi}\right)_{\xi=0}$. In the case where the translation takes place to the right, then $k' = -k > 0$, and, consequently, the waves are propagated in the direction opposite to the motion of the particles.

In a similar way, it is easy to see that increasing solutions, for which the pole is situated on the real axis, correspond to "lagging solutions." Finally, the half sum of the "lagging" and "leading" solutions correspond to a pole situated on the real axis, and leads to nondamping waves.

CHAPTER III

THEORY OF STRIATIONS[+]

Section 36. Introduction

It is known that under the conditions existing in the positive column of an electric discharge at sufficiently low pressures, the constant flow of electrons in this column in certain cases ceases to be spatially homogeneous. Periodically alternating regions of maximum luminosity are formed, which can be visually observed. There is also a periodic structure along the tube for the basic quantitites characterizing the discharge (electric potential, density of electrons, average kinetic energy of electrons). This periodicity arises for quite definite values of the gas pressure, current density and difference in potentials at the ends of the tube. A discharge having such a structure in the positive column has been called a striated discharge, and the individual layers in it are called striations.

The fundamental properties of this phenomenon are:

1. The striations are always formed abruptly, as soon as the parameters of the discharge attain strictly defined values.

2. The phenomenon of striated discharge has so sharply marked a periodic character that diffusion schemes are seen at once to be inadequate to describe it.

3. Striations are observed even under experimental conditions where the ordinary kinetic gas theory picture of "collisions" of electrons with atoms obviously does not correspond to reality, for example, in cases where the length of the free path of the electrons is considerably greater than the dimensions of the bulb. This last fact indicates definitely that the striation phenomenon cannot be understood and explained on the basis of the ordinary pattern of kinetic theory. Rozhansky, in his 1937 monograph "Fizika gazovovo razriada" /Physics of gaseous discharges/, describes the theoretical problem as follows: "At the present time, there is no satisfactory theory of the stratification of the positive column. Obviously, it can only be explained by a theory that shows why a uniform distribution of ions and electrons becomes unstable under certain conditions." In our theory, the explanation of striations is based on the following fundamental physical ideas:

1. It is assumed that the decisive factor is the collective interaction between electrons and their motion of translation, while all the numerous other effects (ionization, recombination, radiation,

[+] See also the monograph cited above, and Zh. MGU, No. 3-4, 63 (1946).

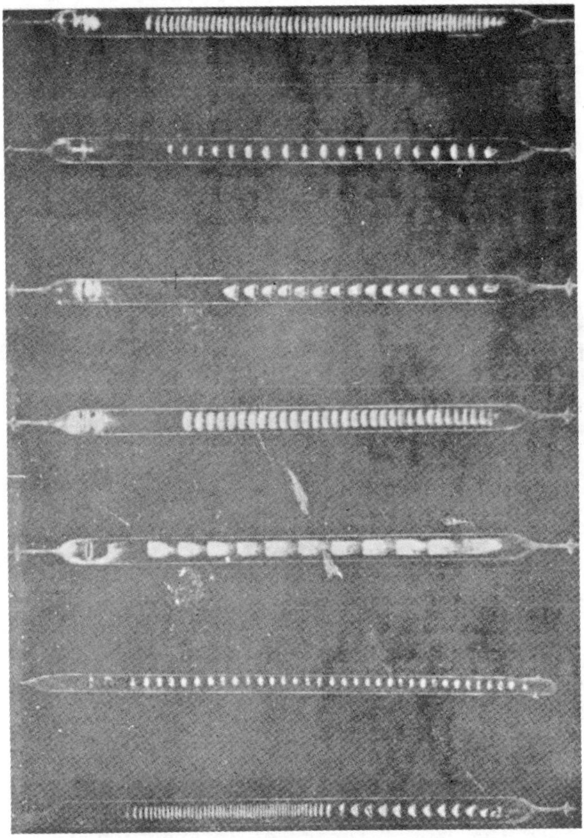

Fig. 8. General form of striations in positive column of electric discharge.

etc.) are only ancillary, secondary phenomena. Therefore, the theory is based on the preceding equations of motion, as applied to electrons and ions.

2. When the electrons have a drift velocity, the spatially homogeneous distribution of electrons among immobile ions is no longer the only stationary solution of the problem. As the drift velocity increases, at a certain value, a spatial irregularity appears abruptly, to a certain extent analogous to what takes place in crystallization processes. Consequently, from the mathematical point of view, the problem reduces to the problem of finding branching solutions for the initial nonlinear equations. It is necessary to find out whether or not spatially periodic solutions branch off from the uniform distribution. This problem is solved positively in what follows.

The theory of striations given below was published in a monograph of the author in 1945, and perfected in the work of his students: Bazarov and others (1947 -1949).

Section 37. Formation of Striations

At first, we shall confine ourselves to the framework of linear theory. Since the stationary solutions belong to the class of solutions specified above, for which no initial moment of time is specified, we shall take a solution in the form (21.1):

$$\left.\begin{aligned}\varphi(t, r, v) &= \frac{1}{2}\int_{-\infty}^{t} F(\tau, r - v(t-\tau), v)\, d\tau + \\ &\quad + \frac{1}{2}\int_{+\infty}^{t} F(\tau, r - v(t-\tau), v)\, d\tau, \\ F(t, r, v) &= \frac{1}{m}\nabla_v f_0 \nabla_r \int K(|r - r'|)\rho(r', t)\, dr',\end{aligned}\right\} \tag{37.1}$$

where φ is the perturbation of the spatially homogeneous distribution function f_0. For the time being, we shall not specify its form, nor that of the law of forces of interaction.

For the perturbation of the density $\rho(r, t) = \int_{-\infty}^{\infty} \varphi(r, v, t)\, dv$ we obtain an analogous equation, if we integrate (37.1) with respect to the velocities:

$$\begin{aligned}\rho(r, t) &= \\ &= \frac{1}{2}\int_{-\infty}^{\infty} av\, \nabla_v f_0 dv \left\{\frac{1}{2}\int_{-\infty}^{t} d\tau\, \nabla_r \int_{-\infty}^{\infty} K(|r - v(t-\tau) - r'|)\rho(r', \tau)\, dr' - \right. \\ &\quad \left. + \frac{1}{2}\int_{\infty}^{t} d\tau\, \nabla_r \int_{-\infty}^{\infty} K(|r - v(t-\tau) - r'|)\rho(r', \tau)\, dr' \right\}.\end{aligned} \tag{37.2}$$

We set a solution in the form

$$\rho(\mathbf{r}') = \sum_{\mathbf{k}} q_{\mathbf{k}} e^{i\mathbf{k}\mathbf{r}}, \tag{37.3}$$

where the q_k are constant. Then

$$\operatorname{grad}_r \int K(|\mathbf{r} - \mathbf{v}(t-\tau) - \mathbf{r}'|)\, e^{i\mathbf{k}\mathbf{r}'} d\mathbf{r}' = \operatorname{grad}_r \left(e^{i\mathbf{k}\mathbf{r} - i\mathbf{k}\mathbf{v}(t-\tau)} \sigma(\mathbf{k})\right) = \\ = i\mathbf{k} e^{i\mathbf{k}\mathbf{r} - i\mathbf{k}\mathbf{v}(t-\tau)} \sigma(\mathbf{k}), \tag{37.4}$$

where, as in the preceding chapters,

$$\sigma(k) = 4\pi \int_0^\infty K(\rho) \frac{\sin k\rho}{k\rho} \rho^2 \, d\rho, \tag{37.5}$$

and we obtain from equation (37.2) the condition for its solvability in the given class of function

$$1 = \frac{1}{2} \int_{-\infty}^{t} G_k(t-\tau)\, d\tau + \frac{1}{2} \int_{+\infty}^{t} G_k(t-\tau)\, d\tau, \tag{37.6}$$

in which we have introduced the notation

$$G_k(t-\tau) = \frac{\sigma(\mathbf{k})}{m} \int_{-\infty}^{\infty} e^{-i\mathbf{k}\mathbf{v}(t-\tau)} (i\mathbf{k} \operatorname{grad}_v f_0)\, d\mathbf{v}. \tag{37.7}$$

By the substitution $t - \tau = x$, we may bring equation (37.6) into the form

$$\int_0^\infty G_k(t)\, dt = 1, \tag{37.8}$$

taking into account the fact that $G_k(x) = -G_k(-x)$.

Into conditions (37.6, 37.8), there enter the forces of interaction between the particles (expressed in $\sigma(k)$), the period of the striations (expressed by k), the temperature, density and drift velocity (all included in f_0). Satisfaction of condition

(37.6) guarantees the appearance of periodic solutions.

We now make the expression that has been obtained concrete, introducing into the initial function the distribution of the velocities of regular motion $\underline{v_0}$. Then $f_0 = f_0(\underline{v'^2})$, where $\underline{v'} = \underline{v} - \underline{v_0}$. Noting that f_0 enters only into $G_k(t)$, and therefore denoting this function by the expression given above for f_0 in terms of $G'_k(t)$, we obtain:

$$G'_k(t) = \frac{\sigma(k)}{m}\int_{-\infty}^{\infty} e^{-ikvt}\, ik\, \mathrm{grad}_v f_0(v'^2)\, dv, \tag{37.9}$$

but

$$\begin{gathered}\mathrm{grad}_v f_0(v'^2) = \mathrm{grad}_{v'} f_0(v'^2)\\ dv = dv'.\end{gathered} \tag{37.10}$$

Hence,

$$G'_k(t) = e^{-ikv_0t}\frac{\sigma(k)}{m}\int_{-\infty}^{\infty} e^{-ikv't}\, ik\, \mathrm{grad}_{v'} f_0(v'^2)\, dv' = e^{-ikv_0t} G_k(t). \tag{37.11}$$

In this way, the drift velocity of the electrons enters into the factor e^{-ikv_0t}, for $G_k(t)$.

Substituting $G'_k(t)$ in the fundamental formula (37.8), we have:

$$\int_0^{\infty} G_k(t)\cos kv_0 t\, dt = 1, \tag{37.12}$$

and we see that the velocity $\underline{v_0}$ enters directly into the condition for the formation of periodic solutions. Without any limitation of generality, we direct the x axis along $\underline{v_0}$, and then

$$\int_0^{\infty} G_k(t)\cos k\xi_0 t\, dt = 1. \tag{37.13}$$

We apply this criterion to an electron gas. We take, for the function f_0, the Maxwell distribution

$$f_0(v) = Ae^{-\frac{v^2}{c^2}}, \tag{37.14}$$

where

$$A = \frac{N}{c^3}\pi^{-\frac{3}{2}}, \quad c = \sqrt{\frac{2kT}{m}}.$$

Here, as we likewise saw in Sec. 32,

$$\sigma(k) = 4\pi\int_0^\infty K(\rho)\frac{\sin k\rho}{k\rho}\rho^2 d\rho =$$
$$= \lim_{x \to 0}\frac{4\pi e^2}{k}\int_0^\infty e^{-x\rho}\sin k\rho = \frac{4\pi e^2}{k^2}. \tag{37.15}$$

We now calculate $G_k(t)$; in our case

$$\operatorname{grad}_v f_0 = -\frac{2N}{\pi^{3/2}c^5}e^{-\frac{v^2}{c^2}}v$$

and, consequently,

$$G_k(t) = -\frac{2\sigma(k)}{mc^2}A\int_{-\infty}^{\infty} e^{-ikvt}(ikv)e^{-\frac{v^2}{c^2}}dv. \tag{37.16}$$

Here there is the tabular integral

$$\int_0^{+\infty} e^{-\frac{v^2}{c^2}} v\sin kvt\, dv = \frac{kc^3}{4}\sqrt{\pi}\, te^{-\frac{k^2c^2t^2}{4}} \tag{37.17}$$

and, consequently, we have at last

$$G_k(t) = -\frac{\sigma(k)N}{m}tk^2e^{-\frac{k^2c^2t^2}{4}} \tag{37.18}$$

Formula (37.13) assumes the form

$$\frac{1}{\frac{N\sigma(k)}{mc^2}} = -4\int_0^\infty xe^{-x^2}\cos\Omega x\,dx, \tag{37.19}$$

where $\Omega = \frac{2\xi_0}{c}$, if $\underline{v}$ ($\xi-\xi_0, \eta, \zeta$) or, taking into account (37.15) and going over to nondimensional magnitudes:

$$k^{*2} = 4I(\Omega) = -4\int_0^\infty xe^{-x^2}\cos\Omega x\,dx, \tag{37.20}$$

where

$$k^{*2} = \frac{k^2}{k_0^2},\quad k_0^2 = \frac{4\pi Ne^2}{mc^2},\ \text{a}\ \Omega = \frac{2\xi_0}{c}.$$

We take note of certain properties of the special function I (Ω):

1. $I(\Omega) = I(-\Omega)$. (37.21)

2. For $\Omega \gg 1$

$$I(\Omega) = -\frac{1}{\Omega^2}\int_0^\infty ze^{-\frac{z^2}{\Omega^2}}\cos z\,dz = = -\frac{1}{\Omega^2}\int_0^\infty z\left(1-\frac{z^2}{\Omega^2}+\ldots\right)\cos z\,dz. \tag{27.22}$$

Defining

$$\int_0^\infty z\cos z\,dz = \lim_{\alpha\to 0}\int_0^\infty e^{-\alpha z} z\cos z\,dz = -1,$$

we have for the term decreasing least rapidly

$$\underset{\Omega\to\infty}{I(\Omega)} \sim \frac{1}{\Omega^2}. \tag{27.23}$$

3. For $\Omega \ll 1$, assuming that $\frac{0}{c}$ is greater than the critical value 0.925, we have:

$$I(\Omega) = -\int_0^\infty x e^{-x^2}\left(1 - \frac{\Omega^2 x^2}{2} + \ldots\right) dx, \tag{37.24}$$

from which

$$I(\Omega) = \frac{1}{2}\left(\frac{\Omega^2}{2} - 1\right)$$
$$\Omega \to 0. \tag{37.25}$$

4. At the point $\Omega = \frac{2\xi_0}{c}$ = 1.85, the function I (Ω) changes sign, going from negative values to positive values, as Ω increases.

It is immediately seen from equation (37.20) that a spatially periodic distribution of the density arises only on the condition that the velocity ξ_0 exceeds a certain limit, defined by the relation

$$\xi_0 = \frac{1.85}{2} c.$$

The criterion gives not only the condition for the formation of periodic solutions, but also the numerical value of the length of the periods. Their dependence on the velocity ξ_0, shown in Fig. 9, was obtained by numerical integration.

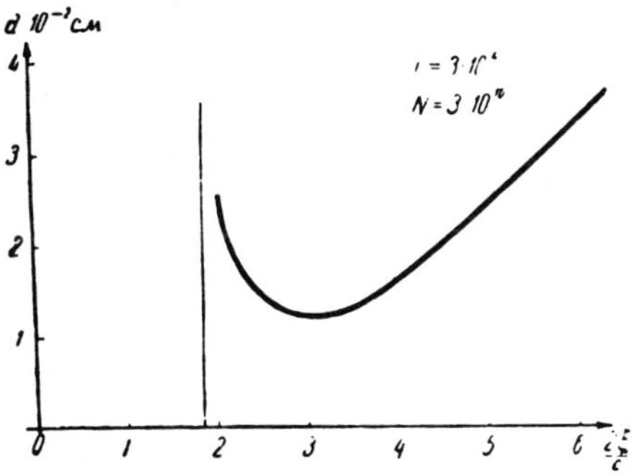

Fig. 9.

The size of the period is defined by the formula

$$d = \left(\frac{2\pi\theta}{Ne^2}\right)^{\frac{1}{2}}\left[I\left(\frac{2\xi_0}{c}\right)\right]^{-\frac{1}{2}}. \tag{37.26}$$

Usually, for the striations in a gaseous discharge,

$N \sim 10^{10}$ cм$^{-3}$, $T = 3 \cdot 10^4$ K;

$\frac{\xi_0}{c} > \frac{1{,}85}{2}$

(there are no precise experimental data for the magnitude $\frac{\xi_0}{c}$), and the length of the periods calculated from these data, is on the order of magnitude of millimeters and centimeters (depending on the magnitude of $\frac{\xi_0}{c}$), which corresponds to the visually observed magnitudes.

We compare the theoretical size of the periods with the experimental data of Merrill and Webb:

$$T = 3 \cdot 10^{4} \, ^\circ K, \quad c = \frac{2\theta}{m} \sim 0{,}9 \cdot 10^{8} \; \frac{\text{cm}}{\text{sec}} \; .$$

As will be seen from Table 3, the agreement may be considered as satisfactory.

TABLE 3

Concentration N(cm $^{-3}$)	Velocity of beam ξ_0 $\frac{\text{cm}}{\text{sec}}$	Experimental period (mm)	Calculated from the formula $d = \frac{2\pi}{k}$ (mm)
$1.93 \cdot 10^{t0}$	$2.45 \cdot 10^{8}$	0.75	1.5
$3.85 \cdot 10^{t0}$	$2.45 \cdot 10^{8}$	0.4	1.0
$2.56 \cdot 10^{t0}$	$2.72 \cdot 10^{8}$	0.5	1.3

The periodicity of the expression for the density guarantees the periodicity of all the mean quantitites characterizing the electrons. This follows from the fact that the distribution function φ is directly expressed by ρ on the basis of (37.1). In this way, we have obtained a periodicity of distribution formed abruptly under certain conditions. We therefore are justified in identifying the theoretical result obtained with the experimentively observed phenomenon of striations.

Section 38. Striations in Metals

We have considered the spontaneous formation of periodical solutions in a flow of electron gas, if its "unperturbed" distribution function is Maxwell's distribution function. We now assume that the "unperturbed" distribution function is a Fermi-Dirac function. The method of solution of the initial equation remains the same. The specialization of the distribution function is expressed only in calculating $G_k(t)$, and it is natural that the dispersion equation $\int\limits_0 G_k(t)\cos k\xi_0 t\,dt = 1$ as well is changed. As we shall see below, the results can only be obtained in a convenient form for low temperatures.

We must calculate (37.7):

$$G_k(t) = \frac{\sigma(k)}{m}\int\limits_{-\infty}^{\infty} e^{-ikvt} ik\,\mathrm{grad}_v f_0 d\mathbf{v}. \tag{38.1}$$

We note first that

$$f_0 = f_0(\varepsilon),$$

where $\varepsilon = \frac{mv^2}{2}$; hence

$$\mathrm{grad}_v f_0 = \frac{\partial f_0}{\partial \varepsilon}\,\mathrm{grad}_v \varepsilon = \frac{\partial f_0}{\partial \varepsilon} m\mathbf{v}.$$

We may therefore write (38.1) again as follows:

$$G_k(t) = \frac{\sigma(k)}{m}\int\limits_{-\infty}^{\infty} e^{-ikvt} ikm\mathbf{v}\frac{\partial f_0}{\partial \varepsilon}\,d\mathbf{v} = i\sigma(k)\int\limits_{-\infty}^{\infty} e^{-ikvt}(k\mathbf{v})\frac{\partial f_0}{\partial \varepsilon}\,d\mathbf{v}. \tag{38.2}$$

We calculate this integral:

$$G_k(t) = i\sigma(k)\int\limits_0^{\infty} v^3 dv\,\frac{\partial f_0}{\partial \varepsilon}\int\limits_0^{2\pi} d\varphi\int\limits_0^{\pi} e^{-ikvt\cos\theta} k\sin\theta\cdot d\theta =$$

$$= 2\pi i\sigma(k)\,k\int\limits_0^{\infty} v^3 dv\,\frac{\partial f_0}{\partial \varepsilon}\,\frac{d}{dt}\left(\frac{i}{kv}\int\limits_0^{\pi} e^{-ikvt\cos\theta}\sin\theta\,d\theta\right) =$$

$$= 2\pi i\sigma(k)\,k\int\limits_0^{\infty}\frac{\partial f_0}{\partial \varepsilon}\,v^3 dv\left(\frac{2i}{k^2v^2}\,\frac{d}{dt}\,\frac{\sin kvt}{t}\right) =$$

$$= -\frac{4\pi\sigma(k)}{k}\,\frac{d}{dt}\left(\frac{1}{t}\int\limits_0^{\infty}\frac{\partial f_0}{\partial \varepsilon}\sin kvt\cdot v\cdot dv\right), \tag{38.3}$$

since
$$\int_0^\pi e^{-ikvt\cos\theta}\sin\theta\, d\theta = \frac{2\sin kvt}{kvt}.$$

We make one more transformation. Since $\varepsilon = \frac{mv^2}{2}$, then

$$\frac{d\varepsilon}{m} = v\,dv, \qquad \text{and} \qquad v = \sqrt{\frac{2\varepsilon}{m}},$$

hence equation $G_k(t)$ may be written as:

$$G_k(t) = -\frac{4\pi\sigma(k)}{mk}\frac{d}{dt}\left(\frac{\int_0^\infty \frac{\partial f_0}{d\varepsilon}\sin\sqrt{\frac{2\varepsilon}{m}}\,kt\,d\varepsilon}{t}\right). \tag{38.4}$$

The integral

$$\int_0^\infty \frac{\partial f_0}{\partial\varepsilon}\sin\sqrt{\frac{2\varepsilon}{m}}\,kt\,d\varepsilon$$

cannot be taken exactly if f_0 is a Fermi-Dirac distribution function

$$f_0 = 2\left(\frac{m}{h}\right)^3 \frac{1}{e^{\frac{\varepsilon-\varepsilon_0}{\theta}}+1}. \tag{38.5}$$

However, for low temperatures ($\theta \to 0$) a formula given by Sommerfeld may be employed. Its derivation is elementary and not long; for completeness, we give it here:

$$-\frac{1}{2\left(\frac{m}{h}\right)^3}\int_0^\infty \frac{\partial f_0}{\partial\varepsilon}\varphi(\varepsilon)\,d\varepsilon = \int_0^\infty \frac{e^{\frac{\varepsilon-\varepsilon_0}{\theta}}\frac{1}{\theta}}{\left[e^{\frac{\varepsilon-\varepsilon_0}{\theta}}+1\right]^2}\varphi(\varepsilon)\,d\varepsilon =$$
$$= \frac{1}{\theta}\int_0^\infty \frac{\varphi(\varepsilon)\,d\varepsilon}{\left(e^{\frac{\varepsilon-\varepsilon_0}{\theta}}+1\right)} = \frac{1}{\theta}\int_0^\infty \frac{\varphi(\varepsilon)\,d\varepsilon}{\left(e^{\frac{\varepsilon-\varepsilon_0}{\theta}}+1\right)\left(e^{\frac{-\varepsilon+\varepsilon_0}{\theta}}+1\right)}.$$

Introducing a new variable $x = \frac{\varepsilon - \varepsilon_0}{\theta}$ and expanding

$$\varphi(\varepsilon) = \varphi(x) = \sum_{n=0}^{\infty} \frac{x^n}{n!} \left\{ \frac{d^n}{dx^n} \varphi(x) \right\}_{x=0},$$

we substitute this expression for $\varphi(x)$ in the integral, replacing the lower limit $-\frac{\varepsilon_0}{\theta}$ by $-\infty$, keeping the low temperatures in mind.

We calculate the integral

$$I = \int_{-\infty}^{\infty} \frac{\sum_{n=0}^{\infty} \frac{x^n}{n!} \left\{ \frac{d^n \varphi(x)}{dx^n} \right\}_{x=0}}{(e^x + 1)(e^{-x} + 1)} dx. \tag{38.6}$$

It is easy to see that after integration only members with an even index for x remain.

Since the integrals

$$I_{2n} = \int_0^{\infty} \frac{x^{2n} e^{-x}}{(1+e^{-x})^2} dx = \sum_{k=1}^{\infty} (-1)^{k+1} k \int_0^{\infty} x^{2n} e^{-kx} dx = 2 \sum_{k=1}^{\infty} (-1)^{k+1} \frac{\Gamma(2n+1)}{k^{2n}};$$

$$\frac{x}{(1+x)^2} = -x \frac{d}{dx} \left(\frac{1}{1+x} \right) = -x \frac{d}{dx} \sum_{k=0}^{\infty} (-1) x^k =$$

$$= \sum_{k=1}^{\infty} (-1)^{k+1} k x^k; \quad |x| < 1,$$

$$\int_0^{\infty} e^{-kx} x^{2n} dx = \frac{1}{k^{2n-1}} \int_0^{\infty} e^{-z} z^{2n} dz = \frac{\Gamma(2n+1)}{k^{2n-1}}$$

are taken in a finite form, as is seen from the result obtained,

$$I_0 = 2 \int_0^{\infty} \frac{e^{-x} dx}{(1+e^{-x})^2} dx = \int_0^1 \frac{dz}{(1+z)^2} = 1,$$

then we obtain:

$$I = \varphi(0) + 2\sum_{n=1}^{\infty} C_{2n}\left\{\frac{d^{2n}}{dx^{2n}}\varphi(x)\right\}_{x=0} =$$

$$= \varphi(\varepsilon_0) + 2\sum_{n=1}^{\infty} C_{2n}\,\theta^{2n}\left\{\frac{d^{2n}}{dx^{2n}}\varphi(\varepsilon)\right\}_{\varepsilon=\varepsilon_0}.$$

Here $C_{2n} = \sum_{k=1}^{\infty} \frac{(-1)^{k+1}}{k^{2n}}$. We at once introduce the values for C_2, C_4:

$$C_2 = \sum_{k=1}^{\infty} \frac{(-1)^{k+1}}{k^2} = \frac{\pi^2}{12}; \qquad C_4 = \sum_{k=1}^{\infty} \frac{(-1)^{k+1}}{k^4} = \frac{7\pi^4}{6!}.$$

In this way, we obtain the formula:

$$-\int_0^{\infty} \varphi(\varepsilon)\frac{\partial f_0}{\partial \varepsilon}\,d\varepsilon = \varphi(\varepsilon_0) + 2\sum_{n=1}^{\infty} C_{2n}\,\theta^{2n}\left\{\frac{d^{2n}}{d\varepsilon^{2n}}\varphi(\varepsilon)\right\}_{\varepsilon=\varepsilon_0}. \tag{38.7}$$

This is the formula that we employ; in it ε_0 denotes the zero energy of a Fermi gas:

$$\varepsilon_0 = \frac{h^2}{8m}\left(\frac{3N}{\pi}\right)^{\frac{2}{3}}. \tag{38.8}$$

We can now return to the calculation of $G_k(t)$. From formulas (38.4) and (38.7), we obtain in a zero approximation with respect to θ, corresponding to $\theta = 0$:

$$G_k(t) = \frac{8\pi\sigma(k)}{mk}\left(\frac{m}{h}\right)^3 \frac{d}{dt}\left(\frac{\sin kt\sqrt{\frac{2\varepsilon_0}{m}}}{t}\right).$$

We introduce the designation:

$$\xi = \sqrt{\frac{2\varepsilon_0}{m}}\,k.$$

In the zero approximation (in the same sense) (37.12) has the form

$$2\left(\frac{m}{h}\right)^{3}\frac{4\pi\sigma(k)}{mk}\int_{0}^{\infty}\frac{d}{dt}\left(\frac{\sin\beta t}{t}\right)\cos k\xi t\,dt = 1. \tag{38.9}$$

This integral is taken by parts

$$\int_{0}^{\infty}\frac{d}{dt}\left(\frac{\sin\beta t}{t}\right)\cos k\xi t\,dt = \Big|_{0}^{\infty}\frac{\sin\beta t}{t}\cos k\xi t + k\xi\int_{0}^{\infty}\frac{\sin\beta t\cos k\xi t}{t}\,dt =$$

$$= -\beta + \frac{k\xi}{2}\int_{0}^{\infty}\frac{\cos(\beta-k\xi)t-\cos(\beta+k\xi)t}{t}\,dt. \tag{38.10}$$

The last integral is the limit of the integral

$$\int_{0}^{\infty} e^{-\alpha x}\frac{\cos\beta_1 x-\cos\beta_2 x}{x}\,dx = \frac{1}{2}\ln\frac{\alpha^2+\beta_2^2}{\alpha^2+\beta_1^2} \tag{38.11}$$

for $\alpha \to 0$.

In this way (38.10) equals

$$-\beta+\frac{k\xi}{4}\ln\frac{(\beta+k\xi)^2}{(\beta-k\xi)^2} = -\beta+\frac{k\xi}{2}\ln\frac{\beta+k\xi}{|\beta-k\xi|}.$$

We transform this expression:

$$\beta\left(-1+\frac{k\xi}{2\beta}\ln\frac{\frac{k\xi}{\beta}+1}{\left|\frac{k\xi}{\beta}-1\right|}\right) = \beta\left[-1+\frac{\xi}{2\sqrt{\frac{2\varepsilon_0}{m}}}\ln\frac{\frac{\xi}{\sqrt{\frac{2\varepsilon_0}{m}}}+1}{\left|\frac{\xi}{\sqrt{\frac{2\varepsilon_0}{m}}}-1\right|}\right].$$

Then equation (38.9) assumes the form

$$2\left(\frac{m}{h}\right)^3 \frac{4\pi\sigma(k)}{mk}\beta\left[-1+\frac{\xi}{2\sqrt{\frac{2\varepsilon_0}{m}}}\ln\frac{\frac{\xi}{\sqrt{\frac{2\varepsilon_0}{m}}}+1}{\left|\frac{\xi}{\sqrt{\frac{2\varepsilon_0}{m}}}-1\right|}\right]=1. \tag{38.12}$$

Setting the value $\sigma(k)=\frac{4\pi e^2}{k^2}$ and introducing the designation $k^{*2}=\frac{k^4}{k_0^2}$, where $k_0^3=\frac{2\pi Ne^2}{\varepsilon_0}$,

we reduce equation (38.12) to the form

$$k^{*2}=3\left(-1+\frac{\xi^*}{2}\ln\frac{\xi^*+1}{|\xi^*-1|}\right), \tag{38.13}$$

where

$$\xi^*=\frac{\xi}{\sqrt{\frac{2\varepsilon_0}{m}}}.$$

If $x \to \infty$, then we have

$$\lim_{x\to\infty}\frac{x}{2}\ln\frac{x+1}{x-1}=\lim_{x\to\infty}\frac{x}{2}\ln\frac{1+\frac{1}{x}}{1-\frac{1}{x}}=\lim_{x\to\infty}\frac{x}{2}\left[\sum_{k=1}^{\infty}\frac{2}{2k-1}\left(\frac{1}{x}\right)^{2k-1}\right]=$$
$$=\lim_{x\to\infty}\left(1+\frac{1}{3x^2}+\dots\right), \tag{38.14}$$

that is,

$$k^{*2}\sim\frac{1}{\xi^{*2}} \quad \text{when} \quad \xi^*\to\infty. \tag{38.15}$$

If x $\longrightarrow$ 0, then

$$\lim_{x \to 0} 3\left(-1+\frac{x}{2} \ln \frac{x+1}{x-1}\right) = -3. \qquad (38.16)$$

We give below a table of function (38.13):

$$k^{+2} = 3\left(-1+\frac{x}{2} \ln \frac{x+1}{|x-1|}\right)$$

x	0.33	0.5	0.66	0.75	0.8	0.83	0.85	0.9	1.5	2	4
k^{+2}	-2.7	-2.1	-1.5	-0.48	-0.39	-0.15	-0.15	0.96	0.6	0.27	0.06

It will be seen from this table that a spatially periodic distribution of the density arises after the velocity of translation has exceeded a given limit: k^{+2} changes its sign, becoming positive at $\xi^+ = 0{,}84$.

Function (38.13) has a discontinuity at $\xi^+ = 1$. Up to the present, we have considered the situation for absolute zero. We now consider temperatures close to absolute zero. In contrast to the previous discussion, which assumed that all the electrons are carried along in a stream, we now consider a different case, in which only a part of the electrons take part in the movement of translation, namely, that part consisting of electrons on the boundary of the Fermi distribution.

To this end, we take as the initial distribution function the difference between the distribution functions for a finite temperature and absolute zero.

$$f' = f_\theta - f_{\theta=0}.$$

It will easily be seen in that case that we shall obtain a result for $G_k(t)$ that coincides with the first approximation in Sommerfeld's formula. We therefore proceed to calculate the first approximation for $G_k(t)$.

Since

$$G_k(t) = \frac{4\pi a(k)}{mk} \frac{d}{dt}\left(\frac{\int_0^\infty \frac{\partial f}{\partial \varepsilon} \sin kt \sqrt{\frac{2\varepsilon}{m}}\, d\varepsilon}{t}\right), \qquad (38.17)$$

we obtain as a first approximation:

$$G_k^{(1)}(t) = -\frac{2\pi\sigma(k)}{m\varepsilon_0}\sqrt{\frac{2}{m}}\frac{d}{dt}\left(2\sqrt{\frac{2}{m}}kt\sin\beta t + \frac{\cos\beta t}{\sqrt{\varepsilon_0}}\right). \qquad (38.18)$$

We now consider the equation:

$$\int_0^\infty G_k^{(1)}(t)\cos k\xi t\,dt = 1, \qquad (38.19)$$

which, after performing elementary calculations, gives:

$$k^{*2} = 2\sqrt{2}\pi^3\left(\frac{\theta}{\varepsilon_0}\right)^2 \frac{\xi^{*2}(\xi^{*2}-5)}{(\xi^{*2}-1)^2}, \qquad (38.20)$$

where

$$\xi^* = \frac{\xi}{\sqrt{\frac{2\varepsilon_0}{m}}}, \quad k_0^2 = \frac{2\pi Ne^2}{\varepsilon_0} \text{ и } k^* = \frac{k}{k_0}.$$

It will be seen from the last formula that as the velocity of translatory motion increases, a spatially periodic structure arises. For this, we must have $\xi^{+2} > 5$, or $\xi > 2{,}4\sqrt{\frac{2\varepsilon_0}{m}}$. We write the formula for the period of the striation

$$d^2 = d^2\frac{1}{2\sqrt{2}\pi^2}\left(\frac{\varepsilon_0}{\theta}\right)^2 \frac{(\xi^{*2}-1)^2}{\xi^{*2}(\xi^{*2}-5)}, \qquad (38.21)$$

where

$$d_0 = \frac{1}{k_0^2}.$$

We evaluate the periods obtained for silver:

$$\varepsilon_0 = 8 \cdot 10^{-12} \text{ ergs}$$

$$d_0^2 = \frac{\varepsilon_0}{2\pi N e^2} = \frac{8 \cdot 10^{-12}}{2\pi \cdot 5{,}19 \cdot 10^{22} \cdot 20 \cdot 10^{-20}} \sim 10^{-16}\, \text{cm}^2.$$

For temperature T = 300° K, we have:

$$\frac{\varepsilon_0}{\theta} = \frac{6 \cdot 10^4}{T} = 2 \cdot 10^2,$$

$$\frac{1}{2\sqrt{2}\pi^2} \sim \frac{1}{30}.$$

As a result, we obtain:

$$d^2 = 10^{-16} \cdot \frac{1}{30} \cdot 4 \cdot 10^4 \frac{(\xi^{*2} - 1)^2}{\xi^{*2}(\xi^{*2} - 5)},$$

$$d \sim 4 \cdot 10^{-6} \frac{\xi^{*2} - 1}{\xi^* \sqrt{\xi^{*2} - 5}}.$$

The second factor is of the order of unity for ξ^*, for example, equal to 3. In this way, we obtain an order of periodicity of the magnitude of 10^{-6} cm for a metal.

Section 39. Striations and Oscillations of Electron Plasma

We shall show that there is a connection between the oscillatory properties of plasma and the striations formed in it, and this connection is of the very closest nature.

Solutions of the type $\exp\{i\omega t - ikx\}$ change their character in the presence of a current. The curve of dependency of ω on k is deformed in such a way that for $\omega = 0$, it gives a root with $k \neq 0$, which also indicates the formation of stationary periodic solutions.

In its most general form, the dispersion relationship that is valid for any form of central forces and any distribution function has the form (33.5)

$$\int_0^\infty G_k(t) \cos \omega t \, dt = 1, \qquad (39.1)$$

where

$$G_k(t) = \frac{\sigma(k)}{m} \int_{-\infty}^{\infty} e^{-ikvt} i(k\nabla_v) f_0 \, dv. \tag{39.2}$$

If we introduce into the distribution function $f_0(\underline{v})$ the velocity of regular motion in the form $f_0 = f_0(\underline{v} - \underline{v}^0)$, then as we have seen for this case in Sec. 37,

$$G_k'(t) = e^{-ik\xi_0 t} G_k(t); \quad v_0 = (\xi_0, 0, 0).$$

Then, equation (39.1) is reduced to the form (ibid.)

$$\int_0^{\infty} G_k(t) \cos(\omega - k\xi_0) \, dt = 1, \tag{39.3}$$

which connects ω and k, ξ_0 entering as a parameter. This equation is a generalization of the dispersion equation for the case when velocities of translation are present.

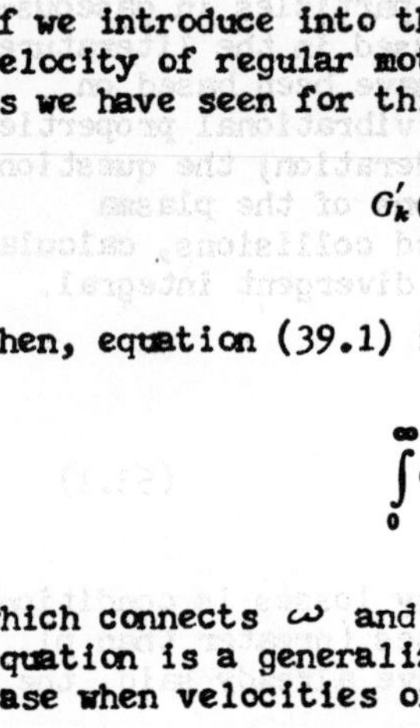

Fig. 10. Dispersion curves for the presence of a drift.

For a Maxwell distribution function, as we see in (37.18),

$$G_k(t) = -\frac{\sigma(k)}{m} N k^2 t e^{-\frac{k^2 c^2 t^2}{4}}. \tag{39.4}$$

For a Coulomb law of forces

$$\sigma(k) = 4\pi \frac{e^2}{k^2}, \tag{39.5}$$

and then the dispersion equation takes on the form

$$k^2 = k_0^2 \int_0^{\infty} x e^{-x^2} \cos 2\left(\frac{\omega}{kc} - \frac{\xi_0}{c}\right) x dx \tag{39.6}$$

$$\left(k_0^2 = \frac{4\pi N e^2}{m c^2}, \quad c^2 = \frac{2\theta}{m}\right)$$

We have already seen what this gives for the case $_0 = 0$. Figure 10 shows the dependence of on k for the various values of $_0$. These graphs illustrate the proposition stated at the beginning of this section with respect to the connection between the oscillatory properties of plasma and the phenomena of striations.

Section 40. Shifting Striations

It is known from experimental observations that the striations are not always immobile with respect to the tube; stroboscopic investigation of the positive column, which seems at first sight to be homogeneous, sometimes shows a motion of periodic distribution of luminosity from the anode to the cathode. These striations are often called "shifting striations," and sometimes "wandering layers." The literature does not show any explanations whatever of this phenomenon. The "shifting striations" have the following notable properties:

1. Shifting striations and immobile striations are not observed simultaneously.

2. Shifting striations are formed abruptly (that is, as soon as the microscopic conditions for their existence have arisen).

3. Shifting striations have a stable velocity of translation on the order of magnitude of the velocity of sound, or their velocity is comprised within a narrow interval of velocities, this interval being determined by the macroscopic conditions of the discharge.

4. The conditions of formation, the size of the periods and the velocity of translation of the striations are highly sensitive to the size of the space occupied by the gas.

From the point of view of our treatment, it is natural to hold that shifting striations are perturbations, being propagated in space, of the equilibrium density of the particles. In other words, shifting striations, like stationary striations, belong to the class of solutions of the type $\text{Exp}\ i(\omega t \pm \kappa x)$, $\frac{\omega}{k}$ being different from zero for shifting striations, and constituting the velocity of their propagation.

It is clear, in connection with what has been said, what kind of solutions must be sought for shifting striations. In the first place, these solutions must be valid for those values of the macroscopic characteristics for which there are no immobile striations. This means that this class of solutions should not, for $\omega = 0$, give real values for k. In the second place, the solutions characterizing the striations should not occur either for very small values of k, since in this case (for finite ω), a small change in k would lead to a great change in the phase velocity $v = \frac{\omega}{k}$, and, as we have shown, the interval of velocities of propagation of the shifting striations should be a narrow one in every case.

Accordingly, even qualitative analysis leads to the conclusion that the approach to the consideration of this phenomenon should comprise the search for solutions of the type $\exp i(\omega t \pm kx)$ such that they will be represented in the (ω, k) plane either by separate points or by curves comprised within an interval of variation of ω and k and not intersecting the axes of coordinates.

Our previous solutions do not meet these requirements. For, as we saw, the dispersion curves obtained for an electron gas intersect the ω axis and the k axis and are continuous functions of the parameter ξ_0.

Although practically immobile striations are not affected by the length of the tube, there are precise experimental indications that there is such an influence in the case of shifting striations: shifting striations are more "sensitive" to the finiteness of the space. We therefore endeavor to calculate (in a primitive manner), the finiteness of the volume of electron gas. We do this as follows: in our previous calculations, all the integrations with respect to spatial coordinates were made over an infinite space. We shall now extend these integrals only over a finite interval of the order of the dimensions of the vessel. More exactly, for the integrals appearing in the theory, of the type

$$\int K(|r-r'|)\, f(r')\, dr',$$

we shall consider that the function $K(|\underline{r} - \underline{r}'|)$ is different from zero only in a finite interval around the point $\underline{r}$. Thereby a linear quantity will be explicity introduced into the theory, that characterizes the volume occupied by the gas.

The calculations made hitherto are unchanged in any way; it is only the form of function (37.5) that is changed

$$\sigma(k) = \int K(|r-r'|)e^{ik(r-r')}dr' = 4\pi \int_0^\infty K(\rho)\, \frac{\sin k\rho}{k\rho}\, \rho^2 d\rho.$$

That is, if before we had for Coulomb's law $K(\rho) = \frac{e^2}{\rho}$ (37.15)

$$\sigma(k) = e^2 \int_0^\infty \frac{1}{\rho} \frac{\sin k\rho}{k\rho} \rho^3 d\rho = \frac{4\pi e^2}{k^2},$$

we shall now have:

$$\sigma(k) = e^2 \int_0^a \frac{1}{\rho} \frac{\sin k\rho}{k\rho} \rho^2 d\rho = \frac{4\pi e^2}{k^2}(1 - \cos ka) = \frac{8\pi e^2}{k^2} \sin^2 \frac{ka}{2}, \qquad (40.1)$$

where a is a length equal in order of magnitude to the dimension of the space. From this there follows, not the previous equation (37.20)

$$k^2 = 4k_0^2 I(\Omega),$$

but the equation

$$\frac{k^2}{2\sin^2 \frac{ka}{2}} = 4k_0^2 I(\Omega). \qquad (40.2)$$

Here $\Omega = \frac{2\omega}{kc}$; for the left side of this equation, we introduce the notation $\frac{k^2}{2\sin^2 \frac{ka}{2}} = f(k)$. Equation (40.2), although very similar to (37.20), nonetheless gives essentially new results. The function f(k) breaks up into separate branches at the points of discontinuity $k = \frac{\pi}{a} 2n \ (n \neq 0)$

For n = 0, we have $\lim_{k \to 0} f(k) = \frac{2}{a^2}$. At the points $k = \frac{\pi}{a}(2n+1), \quad n = 0, 1, 2, \ldots$, the function $f(k)$ coincides with the corresponding points of the parabola $\frac{k^2}{2}$. Comparing the graph of the left and the right sides of (40.2), we can easily obtain the conditions for the formation of the solutions that interest us.

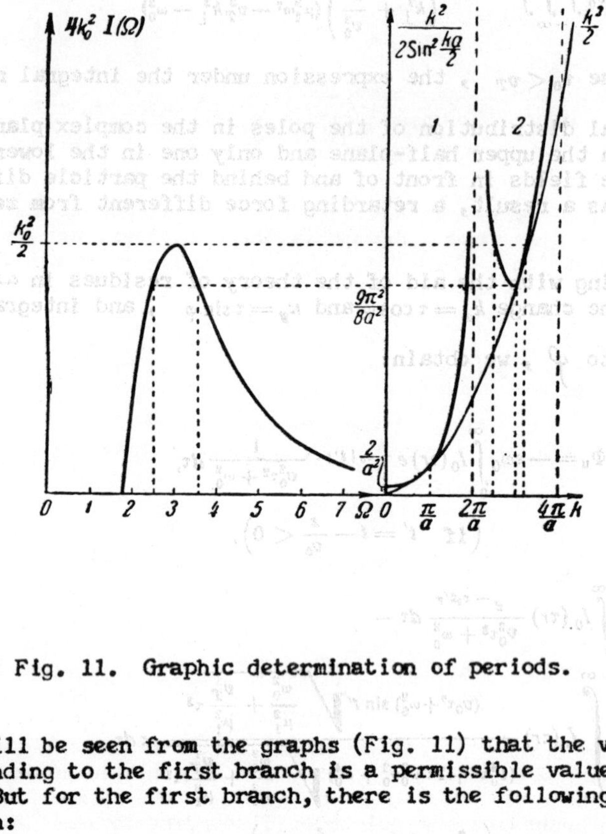

Fig. 11. Graphic determination of periods.

It will be seen from the graphs (Fig. 11) that the value corresponding to the first branch is a permissible value $k = 0$. But for the first branch, there is the following condition:

$$k < \frac{2\pi}{a},$$

or the period of the striations $d > a$, that is, greater than the dimensions of the vessel. It is impossible to observe a periodicity of greater dimensions than those of the vessel, and we therefore completely exclude the first branch from consideration. Thereby the point $k = 0$ is excluded as a possibility.

It will be seen from the graph that the equation (40.3)

holds true only for Ω for which $I(\Omega) > 0$, so that $f(k) > 0$, from which it is clear, since in this case $\Omega > 1.85$, $\omega = \frac{\Omega}{2}kc$ and $k \neq 0$, that the dispersion curves formed upon consideration of the remaining branches (excluding the first) will have no solutions for $\omega = 0$ and $k = 0$, that is, we shall always have $\frac{\omega}{k} \neq 0$. Furthermore, for every branch k lies in a finite interval of its own, as does ω. This is precisely what we were seeking.

Figure 11 illustrates the method of finding the regions of simultaneous solutions for the right and the left sides sides of the equation: $\frac{k^2}{2\sin^2\frac{ka}{2}} = 4k_0^2 I(\Omega)$, on the left the function $4k_0^2 I(\Omega)$, on the right $\frac{k^2}{2\sin^2\frac{ka}{2}}$. On the right, the first and second branches of the curve are represented. The regions k and Ω are shown for which there are solutions.

With respect to the periodicity corresponding to the subsequent branches, we have for the n-th branch, the inequality

$$k > \frac{2\pi}{a}(n-1), \text{ that is, } d < \frac{a}{2(n-1)} \quad (n \geqslant 2).$$

It is likewise to be seen from the curve that in order for equation (40.2) to apply, it is necessary that

$k_0^2 \max I(\Omega) > \min_a f(k)$. Hence, we obtain condition

$$k_0^2 > \left[\frac{\pi}{a}(2n-1)\right]^2$$

for the n-th branch. In other words, $a > \pi\frac{2n-1}{k_0}$, which signifies that the dimensions of the vessel must exceed the Debye radius $\left(\frac{1}{d^2} = k_0^2 = \frac{4\pi N e^2}{mc^2}\right)$. This condition can be satisfied.

Up to the present, we have considered the case in which there was no velocity of translation. In order to take it into account, it is sufficient to consider the formula

$$\frac{\omega}{k} = \left(\frac{\Omega}{2} \pm \frac{\xi_0}{c}\right)c.$$

It follows from this formula that the influence of the velocity of translation depends on its sign, which is natural, inasmuch as there is a specified direction(the direction of the motion of a stratified structure). But it will be seen from equation (40.2) that there are regions where there are both a movement of translation and shifting striations. At the same time, some value for the velocity of translation makes possible the existence of real k even for $\omega = 0$. Consequently, we arrive at the conclusion that shifting striations may for some values of ξ_0 go over into stationary striations (we have considered the minus sign in (40.2), and the plus sign does not give any essentially new results).

As for the magnitude $\frac{\omega}{k}$, and defining the velocity of propagation of the stratified structure, it is always of the order of the velocity of heat, as is likewise visible from the graph.

Let us draw some conclusions. The attempt was made here to take into account the finiteness of the volume of gas, and thereby go over to an explanation of the formation of shifting striations. This was done in an extremely primitive manner. However, obviously, the direction in which the attempt was made is correct.

We obtain an equation of the type that should correspond to shifting striations. Nonetheless, it is clear both from the results and from the calculations made that we need here a more correct and more precise consideration of the finiteness of the volume.

We have obtained the following new results as the consequence of our discussion:

1. The necessary condition for the formation of solutions of the type $i(\omega t \pm kx)$ with $c\,\frac{\omega}{k} \neq 0$ consists in:

$$a > \frac{(2n-1)\pi}{k_0},$$

where $\frac{1}{k_0}$ is the Debye radius, a the dimensions of the vessel, n a whole number ($n \gg 2$).

2. The periods of the shifting striations and the velocity of their propagation are defined by the formula

$$\frac{k^2}{\sin\frac{ka}{2}} = 4k_0^3 / (\Omega), \quad \text{where} \quad \Omega = 2\left(\frac{\omega}{kc} \pm \frac{\xi_0}{c}\right),$$

where ξ_0 is the velocity of translation. The velocity of the motion of the striations $\frac{\omega}{k}$.

To conclude this chapter, we should bring out three points:

1. The theory expounded above, since it is based on the linearization of the initial equation, does not furnish the possibility of calculating the amplitude of the periodic structure. The next stage should be treatment of the nonlinear temporal equation.

2. The theory does not take into account the role of many factors that are characteristic for the positive column (ionization, recombination and other phenomena). Taking these phenomena into account requires "localization" of the description of the interactions either in the sense of classic, statistical, or quantum mechanics. In the theory set forth, we have rejected the localized description of particles. As we have stressed several times above, this denial is of a general nature. Accordingly, the next step in the theory is elucidation of the question of the kind of macroscopic conditions for an atomic aggregate of many particles in a positive column for which a partial introduction of localization is necessary, what kind of mechanism there is for the interaction of these point particles with the "extended" particles. The details of these facts must be brought out consistently and leads to introduction into the theory of the phenomena in question. Naturally, it would be possible to apply some kind of "hybrid" means of describing the motion of electrons, by various methods from various theories and putting them together. We have not taken this path since, when working in this direction, it is never possible to be sure of the internal consistency of the methods of investigation.

3. The boundary conditions in the theory have been discussed only in part. It should be noted that the problem of calculating the boundary conditions still remains to be solved. In this book, we have succeeded in solving it only in particular cases (see Secs. 21, 45, etc.).

CHAPTER IV

WAVE PROPERTIES OF MEDIA CONSISTING OF NEUTRAL PARTICLES

Section 41. Some Peculiarities of Wave Propagation

The dispersion equation defining the dependence of ω on $|k|$ in the most general form for any forces of interaction and any initial distribution function has the form

$$\frac{1}{\frac{4\pi}{m}\int\limits_0^\infty K(\rho)\rho^2\frac{\sin k\rho}{k\rho}\,d\rho}=\int\limits_{-\infty}^{\infty}\frac{(k\nabla)f_0}{(kv-\omega)}\,dv. \qquad (41.1)$$

for f_0, the Maxwell distribution function we have:

$$\frac{1}{\frac{4\pi N}{\theta}\int\limits_0^\infty K(\rho)\rho^2\frac{\sin k\rho}{k\rho}\,d\rho}=-\int\limits_0^\infty xe^{-x^2}\cos vx\,dx=I(v), \qquad (41.2)$$

where

$$v=\frac{2\omega}{kc},\quad c=\sqrt{\frac{2\theta}{m}}.$$

Formula (41.2) implicitly gives the dependence of ω on k. In Fig. 12, the right graph refers to function I(v), and the left curve the dependence of the left part of (41.2) on k, defined by the character of the force of interaction. Assigning some value k, we can graphically determine v, and hence by the formula

$v=\frac{2\omega}{kc}$, we can find ω. Here v is the phase velocity of propagation. As has already been pointed out, the course for the relationship I(v) that has been given remains valid in its general outline not only for a Maxwell distribution function f_0, but also for a very broad class of functions. Accordingly, what will be obtained from equation (41.2) will in general outline retain its significance for a broad class of distribution functions. We indicate a number of general properties and peculiarities of the formation of wave formations.

1. <u>The abrupt disappearance of waves at a definite degree of rarefaction of the medium and sufficiently high temperature</u>. This result is usually obtained for the simplest law of short-

acting forces of repulsion between particles

$$K(\rho)=\frac{g}{\rho}e^{-\varkappa\rho}, \qquad (41.3)$$

where g and $\varkappa$ are constant parameters of the molecular interactions. In this case

$$\int_0^\infty \frac{g}{\rho}e^{-\varkappa\rho}\frac{\sin k\rho}{k\rho}\rho^2 d\rho=\frac{g}{k^2+\varkappa^2} \qquad (41.4)$$

and, consequently,

$$\frac{\theta}{4\pi Ng}(k^2+\varkappa^2)=I(v). \qquad (41.5)$$

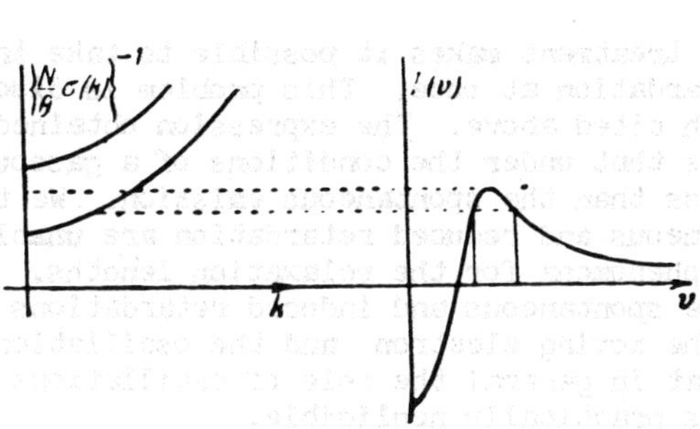

Fig. 12. Graphic determination of conditions for formation of waves.

We see in Fig. 12 that for a sufficiently large value of $\frac{\theta}{4\pi Ng}$, the curve for the dependence on k lies entirely above the maximum of the curve for I(v). In this case, the curves will have no points with the same ordinate, and, consequently, there will be no real solutions for ω and k. In this way, the values for the concentration of N, temperature θ and wave numbers k, at which no wave solutions can exist, are defined by the inequality

$$\frac{\theta}{4\pi Ng}(k^2+\varkappa^2)>\max I(v), \qquad (41.6)$$

or in the more general case:

$$\frac{\theta}{4\pi N}\left\{\int_0^\infty K(\rho)\frac{\sin k\rho}{k\rho}\rho^2 d\rho\right\}^{-1} > \max I(v). \tag{41.7}$$

The critical values of concentrations and temperatures at which the waves disappear, are defined by the obvious relationship

$$\frac{\theta}{4\pi Ng}\varkappa^2 = \max I(v) \tag{41.8}$$

and similarly for the general case.

2. <u>The existence of two velocities of sound for molecular forces, with forces of repulsion prevailing.</u> The dispersion curve for dependence of on k, obtained by the graphic path mentioned above, is given in Fig. 13.

This figure shows that for small k there exist two angles of inclination for the dispersion curve to the k axis, which correspond to the presence of two velocities of sound (density waves). More general laws for the forces may, of course, show the presence not of two, but of a large number of wave lengths, corresponding to the same frequency. This result likewise depends on the initial distribution function involved in each case.

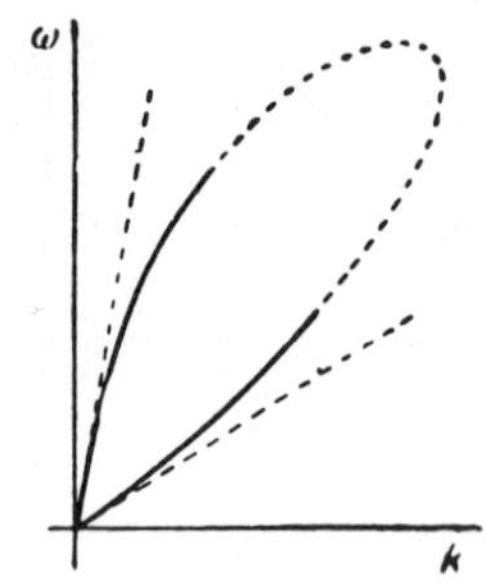

Fig. 13. General form of dispersion curve for case of short-acting forces of repulsion.

3. <u>The existence of maximum frequencies at wave lengths for the general case of several forces and any initial distribution function.</u> This important property follows from the fact that the right side of the dispersion equation, the function I(v), is bounded over the entire interval v from 0 to ∞ . Consequently, the left side as well is bounded, which is possible only when k cannot take on infinitely large values, and this in turn lead to the boundedness of the values of ω^2.

The maximum value for k^2 occurs when I(v) is the maximum, but the maximum of this function lies at a finite distance; thus, v^2, corresponding to the maximum value of the wave number is finite, and therefore ω^2 as well is finite.

For any forces, calculation of the "distant" interactions leads automatically to the impossibility of the propagation of wave formations with frequencies as high as desired and with wave numbers as high as desired. This limitation is connected with the integral character of the interactions in the theory given, and holds true for any forces and for a broad class of initial distribution functions.

It is in this that there lies the radical difference of the method used for describing the media from those formerly used, in which the construction of the equations is of such a nature that they permit wave solutions for frequencies as high as desired and with as short a wave length as desired.

From what has been said, the value for the maximum wave number should be defined from the equation

$$\frac{1}{\frac{4\pi}{m}\int\limits_0^\infty K(\rho)\,\rho\,\frac{\sin k\rho}{k}\,d\rho} \leqslant \max I(v), \tag{41.9}$$

or

$$\frac{4\pi}{m}\int\limits_0^\infty K(\rho)\,\rho\,\frac{\sin k\rho}{k}\,d\rho \geqslant \frac{1}{\max I(v)}, \tag{41.10}$$

and the value of the maximum frequency from the equation

$$\omega^2 \leqslant k^2 \max v_m^2, \tag{41.11}$$

where v_m is a root of the equation

$$\frac{dI(v)}{dv} = 0. \tag{41.12}$$

4. The existence of a critical velocity of translation of the medium, beginning with which the possibility of propagation of waves vanishes. As has already been noted, the dispersion

equation for a stream of particles has the form (39.3-39.6).

$$I(v)=\int_0^\infty G_k(x)\cos vx\,dx=$$

$$=\frac{1}{\frac{4\pi N}{\theta}\int_0^\infty K(\rho)\frac{\sin k\rho}{k\rho}\rho^2\,d\rho}\geqslant\frac{1}{\frac{4\pi N}{\theta}\int_0^\infty K(\rho)\rho^2\,d\rho} \quad (41.13)$$

$$v=2\left(\frac{\omega}{kc}-\frac{\zeta_0}{c}\right),\quad c=\sqrt{\frac{2\theta}{m}},$$

where ζ_0 is the velocity of the stream. We see that for constant values of the parameters of the medium N and θ, as the velocity of the stream ζ_0 changes and at a fixed value for $\frac{\omega}{K}$, it is always possible to achieve doing away with the inequality sign, since the function on the left, I(v) may change its sign. Consequently, the class of wave functions indicated will not exist. The critical velocity is defined from the relationship

$$\int_0^\infty G_k(x)\cos 2\left(\frac{\omega}{kc}-\frac{\zeta_0}{c}\right)x\,dx=\frac{1}{\frac{4\pi N}{\theta}\int_0^\infty K(\rho)\rho^2 d\rho}. \quad (41.14)$$

Section 42. Striations in a Stream of Neutral Particles

The formation of a spatially periodic structure is also possible in a system consisting of uncharged particles that interact not according to Coulomb's law, wheh there is a motion of translation. For this, it is necessary that the forces of repulsion prevail in the interactions among the particles.

We consider an aggregate of (neutral) particles, interacting according to a central law. We have seen that the equation that gives the connection between k and ξ_0 in this case, if the Maxwell function is taken for the distribution function $f_0(v)$, is

$$\frac{1}{\int_0^\infty K(\rho)\frac{\sin k\rho}{k\rho}\rho^2\,d\rho}=\frac{4\pi N}{mc^2}I(\Omega), \quad (42.1)$$

where $K(\rho)$ is the potential function for the interaction of the particles, and

$$I(\Omega) = -\int_0^\infty xe^{-x^2}\cos\Omega x\,dx, \quad \Omega = \frac{2\xi_0}{c}. \qquad (42.2)$$

We consider the integral depending on $K(\rho)$:

$$\sigma(k) = 4\pi\int_0^\infty K(\rho)\frac{\sin k\rho}{k\rho}\rho^2 d\rho.$$

Some general conclusions can be drawn as to the sign of $\sigma(k)$.

If function $K(\rho) > 0$ decreases monotonously (monotonously decreasing forces of repulsion), while integral $\sigma(k)$ converges, then $\sigma(k) > 0$. This is immediately obvious if we break up the integrals of integration into intervals in which the sign is positive and negative. The first interval will then give a positive value, and the second a negative one, but known to be smaller. Each successive "positive" interval will have a greater value than the succeeding "negative" interval. In this way, the entire interval will be positive.

We shall now assume that this condition $\sigma(k) > 0$ is satisfied for any k. Then, taking into consideration

$$\frac{\sin x}{x} \leqslant 1,$$

we may write the inequality:

$$\int_0^\infty K(\rho)\frac{\sin k\rho}{k\rho}\rho^2 d\rho \leqslant \int_0^\infty K(\rho)\rho^2 d\rho. \qquad (42.3)$$

It follows from equation (41.1) that if there is a real root k, then the condition is satisfied:

$$\frac{4\pi N}{mc^2} I(\Omega) \geqslant \frac{1}{\int_0^\infty K(\rho)\rho^2 d\rho}. \qquad (42.4)$$

But if $\sigma(k) > 0$, then the quantity on the right is known to be positive. Consequently, a necessary condition for the formation of a spatially periodic solution is

$$I(\Omega) \geqslant 0.$$

But this also means that spatially periodic solutions are possible only in the presence of a velocity of regular motion, the criterion for the appearance of periodic solutions being the condition

$$\frac{2\xi_0}{c} \geqslant 1{,}85.$$

The periods for the striations that are formed may be evaluated from equation (42.1)

$$\frac{4\pi N}{mc^2} I(\Omega) = \frac{1}{\int\limits_0^\infty K(\rho) \frac{\sin k\rho}{k\rho} \rho^2 d\rho},$$

if there is specified some definite law of the forces of interaction between the particles.

We have thus come to the conclusion that the motion of particles repelling each other according to some central law can lead to the formation of a spatial periodicity in the direction of the motion. This spatial periodicity is described as striations in a neutral gas.

CHAPTER V

TOWARDS THE THEORY OF ELECTRON-RAY HIGH-FREQUENCY GENERATORS

Section 43. Propagation of Modulation Along Beam of Electrons for the Case of Low Concentration

In new types of high-frequency, electron-ray generators, an essential role is played by the propagation of modulations of density, current and other characteristics along the beam of electrons, the propagation being evoked by the imposed perturbation at a given point on the beam. A typical case of such a perturbation is a variable potential difference on a pair of grids set perpendicularly to the beam.

We consider this question from the point of view of the theory being presented.

The initial equation of the problem should be the kinetic equation in one dimension or the sought-for distribution function of the electrons $f(x, \xi, t)$:

$$\frac{\partial f}{\partial t} + \xi \frac{\partial f}{\partial x} + F(x, t) \frac{\partial f}{\partial \xi} = \left[\frac{\partial f}{\partial t}\right]^{st} *), \tag{43.1}$$

where $F(x, t) = \frac{e}{m} E(x, t)$ is defined by the given field intensity on the pair of grids. In the case where the distance between the grids is small, so that the transit time of the electrons is enormously less than the period imposed on the field grids, there may be a passage to a double electric layer.

The potential can then be expressed by Dirac's delta-function in the following way:

$$\varphi(x,t) = V(t) \int_{-\infty}^{x} \delta(x)\, dx; \tag{43.2}$$

from which the field strength will be:

---

* This chapter was written at a time when the author was not entirely clear as to the relationship between "collisions" and integral interactions. The two are introduced on the same basis in equation (43.1). However, it will be seen from what follows that the results obtained in this chapter do not depend on the presence of the term $\left(\frac{\partial f}{\partial t}\right)^{st}$.

$$E(x, t) = -\frac{\partial\varphi}{\partial x} = V(t)\,\delta(x), \qquad (43.3)$$

and, consequently, the expression for the force F is

$$F(x, t) = \frac{e}{m} E(x, t) = \frac{e}{m} V(t)\,\delta(x). \qquad (43.4)$$

Calculation of the variation of the distribution function

$\left[\frac{\partial f}{\partial t}\right]^{st}$ due to collisions with atoms of gas is not fundamental, if the tube operates at a high vacuum, but even taking into account the "impact member" does not complicate the calculations

$$\left[\frac{\partial f}{\partial t}\right]^{st} = -\pi\sigma^2 N_A \xi f = -\alpha\xi f, \qquad (43.5)$$

where σ is the effective atomic cross-section for the dispersion of electrons, which for a short interval of velocities can be considered as independent of the velocity, N_A is the number of atoms in the unit volume and ξ the component of the velocity.

Without the presence of the perturbing field of the grids, solution (43.1), taking into consideration (43.3) for the case of stationary motion, is

$$f = f_0 = \psi_0(\xi)\, e^{-ax}, \qquad (43.6)$$

where $\psi_0(\xi)$ — is the distribution function in the unperturbed beam, which is supposed to be given.

We shall seek solution for (43.1) in the form of an expansion of the parameter $\frac{eV}{m\xi^2} = \varepsilon$, taking $\varepsilon \ll 1$, which is possible on the condition that the minimum velocities in ψ_0 are still sufficiently large to eliminate the potential barrier on the grids; otherwise ψ_0 is arbitrary.

The equation for the first approximation is

$$\frac{\partial f_1}{\partial t}+\xi\frac{\partial f_1}{\partial x}+\alpha\xi f_1=-F(x,\ t)\ \frac{\partial f_0}{\partial \xi}=\Phi_1(x,\ \xi,\ t). \tag{43.7}$$

We seek a solution for equation (43.7) in the form of a double Fourier integral

$$f_1(x,\ \xi,\ t)=\int\int_{-\infty}^{+\infty} f_{1\omega}(k,\ \xi)\, e^{i\omega t+ikx}\, d\omega\, dk. \tag{43.8}$$

The right side in (43.7) will be

$$\Phi_1(x,\ \xi,\ t)=\int\int_{-\infty}^{+\infty}\Phi_{1\omega}(k,\ \xi)\, e^{i\omega t+ikx}\, d\omega\, dk, \tag{43.9}$$

where the coefficients of the Fourier expansion will be, taking into consideration (43.4) and (43.6),

$$\Phi_{1\omega}(k,\ \xi)=-\frac{1}{2\pi}\frac{e}{m}V_\omega\frac{\partial\psi_0}{\partial\xi}; \tag{43.10}$$

where the V_ω are the Fourier coefficients for the expansion of function V(t) (they do not depend on k).

Substituting in (43.7) and equating corresponding Fourier coefficients on the right and on the left, we obtain a solution in the form:

$$f_1=-\frac{e}{m}\frac{\partial\psi_0}{\partial\xi}\frac{1}{2\pi}\int\int_{-\infty}^{+\infty} V_\omega e^{i\omega t}\frac{e^{ikx}}{i\omega+(ik+\alpha)\xi}\, d\omega\, dk. \quad (43.11) \tag{43.11}$$

The integral is easily taken from the theory of residues

$$\frac{1}{2\pi i\xi}\int_{-\infty}^{+\infty}\frac{e^{ikx}dk}{\left(\frac{\omega}{\xi}-i\alpha\right)+k}=\frac{1}{\xi}e^{-i\frac{\omega}{\xi}x-\alpha x} \tag{43.12}$$

for x > 0 and equal to zero for x < 0, so that

$$f_1(x, \xi, t) = -\frac{e}{m} \frac{\partial \psi_0}{\partial \xi} \frac{1}{\xi} e^{-\alpha x} \int_{-\infty}^{+\infty} V_\omega e^{i\omega\left(t - \frac{x}{\xi}\right)} d\omega =$$

$$= -\frac{e}{m} \frac{1}{\xi} \frac{\partial f_0}{\partial \xi} V\left(t - \frac{x}{\xi}\right). \tag{43.13}$$

The solution of the subsequent approximations is obtained by the same method. The law of recurrence is easily seen from comparison of the first three approximations. Designating operation $\frac{1}{\xi} \frac{\partial}{\partial \xi} = L$ for the approximation of order n, we shall have for $x > 0$:

$$f_n = \left(-\frac{e}{m}\right)^n V^n \left(t - \frac{x}{\xi}\right) L^{(n)} f_0, \tag{43.14}$$

where $L^{(n)}$ denotes the application of operation L n times.

In this way, the solution of initial equation (43.1) will be in the region $x > 0$

$$f(x, \xi, t) = \sum_{n=0}^{\infty} \left(-\frac{e}{m}\right)^n V_n \left(t - \frac{x}{\xi}\right) L^{(n)} f_0 \tag{43.15}$$

and equal to f_0 for $x < 0$. Formula (43.15) gives the solution to the problem.

For the particular case of high velocities for the electrons in the beam, when it is possible to ignore the heat distribution of the electrons according to velocities, the relationships are simpler. In this case, we may assume

$$f_0 = N_0 \delta(\xi - \xi_0) e^{-\alpha x}, \tag{43.16}$$

where N_0 is the concentration of electrons in the unperturbed beam for $x = 0$.

The expression for the value of density ρ_ψ of any quantity $\psi(x, \xi)$, connected with the electrons in the beam is

$$\rho_\psi = N(x, t)\overline{\psi(x, \xi)} = \int_{-\infty}^{+\infty} \psi(x, \xi) f(x, \xi, t)\, d\xi, \qquad (43.17)$$

from which, making use of (43.15) and (43.16) and performing successive partial integration, we have:

$$\rho_\psi = N(x, t)\overline{\psi(x, \xi)} = $$
$$= N_0 e^{-ax} \sum_{n=0}^{\infty} \left(\frac{e}{m}\right)^n L^{*(n)} \left\{\psi(\xi_0)\, V^{(n)}\left(t - \frac{x}{\xi_0}\right)\right\}, \qquad (43.18)$$

where L^+ denotes the operation $\frac{\partial}{\partial \xi_0} \frac{1}{\xi_0}$.

As for the fundamental characteristics of the beam, we obtain the densities of the number of particles $N(x, t)$, the current densities $j(x, t)$ and the densities of the kinetic energy $w(x, t)$ from (43.18), setting $\psi(x, \xi) = 1,\ \xi, \frac{m\xi^2}{2}$, respectively, and confining ourselves to the first two terms:

$$N(x, t) = N_0 e^{-ax}\left\{1 + \frac{ex}{m\xi_0^3} \frac{\partial V\left(t - \frac{x}{\xi_0}\right)}{\partial t} - \frac{eV\left(t - \frac{x}{\xi_0}\right)}{m\xi_0^2} + \ldots\right\},$$

$$j(x, t) = N_0 \xi_0 e^{-ax}\left\{1 + \frac{ex}{m\xi_0^3} \frac{\partial V\left(t - \frac{x}{\xi_0}\right)}{\partial t} + \ldots\right\}, \qquad (43.19)$$

$$w(x, t) = N_0 \frac{m\xi_0^2}{2} e^{-ax}\left\{1 + \frac{ex}{m\xi_0^3} \frac{\partial V\left(t - \frac{x}{\xi_0}\right)}{\partial t} + \frac{eV\left(t - \frac{x}{\xi_0}\right)}{m\xi_0^2} + \ldots\right\}.$$

The solutions presented clearly comprise a retardation effect and an increase in the depth of modulation as x increases. For the harmonically changing field at the grids, the limit of applicability of the first approximation is limited by the condition

$$\frac{eV}{m\xi_0^2} \frac{\omega x}{\xi_0} \gg 1.$$

Taking into account the distance 2a between the grids, we shall have for the strength of the perturbing field, instead of (43.4)

$$F(x,t) = \frac{e}{m}E(t)\,D(x) = \frac{e}{m}E(t)\,\frac{1}{\pi}\int_{-\infty}^{+\infty}\frac{\sin ak}{k}\,e^{ikx}\,dk, \qquad (43.20)$$

where D(x) is the Dirichlet function. The solution of equation (43.7) is conducted as previously. For the distribution function, we obtain in the first approximation, setting $E = E_0 \cos\omega x$ for x $>$ 0,

$$f_1(x,\xi,t) = -\frac{2eE_0}{m}\,\frac{\sin\frac{\omega}{\xi}a}{\omega}\,\sin\omega\left(t-\frac{x}{\xi}\right)\frac{\partial f_0}{\partial\xi}, \qquad (43.21)$$

from which we get, for the fundamental characteristics of the beam:

$$\Delta N = N_0\frac{eV_0}{m\xi_0^2}e^{-\alpha x}\left\{\frac{x}{a}\sin\frac{\omega}{\xi_0}a\cos\omega\left(t-\frac{x}{\xi_0}\right) - \cos\frac{\omega}{\xi_0}a\cdot\sin\omega\left(t-\frac{x}{\xi_0}\right)\right\};$$

$$\Delta j = N_0\xi_0\frac{eV_0}{m\xi_0^2}e^{-\alpha x}\left\{\frac{x}{a}\sin\frac{\omega}{\xi_0}a\cos\omega\left(t-\frac{x}{\xi_0}\right) + \left[\frac{\sin\frac{\omega}{\xi_0}a}{\frac{\omega}{\xi_0}a} - \cos\frac{\omega}{\xi_0}a\right]\sin\omega\left(t-\frac{x}{\xi_0}\right)\right\},$$

$$\Delta w = N_0\frac{m\xi_0^2}{2}\frac{eV_0}{m\xi_0^2}e^{-\alpha x}\left\{\frac{x}{a}\sin\frac{\omega}{\xi_0}a\cos\omega\left(t-\frac{x}{\xi_0}\right) + \left[2\frac{\sin\frac{\omega}{\xi_0}a}{\frac{\omega}{\xi_0}a} - \cos\frac{\omega}{\xi_0}a\right]\sin\omega\left(t-\frac{x}{\xi_0}\right)\right\}, \qquad (43.22)$$

where Δ N, Δ j, Δ w denote the deviations of the corresponding quantities from their stationary value (at $V_0 = 0$). Where $\frac{\omega}{\xi_0}a \ll 1$ the formulas obtain go over into (43.19) for the case $V = V_0\sin\omega t$. . The reduced formulas clearly show the influence of the finiteness of the difference on the modulation properties of the electron beam. For example, we have for the current density:

$$\frac{\Delta j}{(\Delta j)_{a=0}} = \frac{\sin \frac{\omega}{\xi_0} a}{\frac{\omega}{\xi_0} a}. \tag{43.23}$$

The considerations introduced here should be taken as applying not only to the operation of high-frequency generators, but also to amplifying tubes operating on an analogous principle.

Section 44. Study of Propagation of Modulation Along Concentrated Beams, Taking Into Consideration Interaction Among Electrons

In this section, we should like to consider the role played by interaction among electrons in the problem of the propagation of modulation in an electron ray, which we discussed in Sec. 43.

The initial system of equations, taking into account only Coulomb forces, should be the following fundamental equation of the theory:

$$\left.\begin{aligned} &\frac{\partial f}{\partial t} + \operatorname{div}_r \boldsymbol{v} f + \frac{e}{m} \boldsymbol{E} \operatorname{grad}_v f = -F(x, t) \frac{\partial f}{\partial \xi} + \left[\frac{\partial f}{\partial t}\right]^{st}, \\ &\operatorname{div} \boldsymbol{E} = 4\pi e \int f d\omega + 4\pi \rho_0 (x, y, z) \\ &(d\omega = d\xi d\eta d\zeta), \end{aligned}\right\} \tag{44.1}$$

where E comprises the fields generated by the electrons of the beam, and likewise by any external charges $\rho_0 (x, y, z)$. F is the field on the grids (separated from E only for convenience) and is defined, as in Sec. 43, by the representation of a pair of grids in the form of a double electrical layer [formula (43.4)], and $\left[\frac{\partial f}{\partial t}\right]^{st}$ takes into account the effect of the dispersion of electrons on the atoms, and is defined by formula (43.5).

The ordinary method of approximate solutions of such problems consists in assuming that the interactions among electrons are small. For in the zero approximation, the distribution function is taken for noninteracting electrons; by means of the Maxwell equations, this function is used to calculate the fields; then substitution of them in the equation of motion makes it possible to find a correction to the distribution function due to the interaction among the electrons.

This mode of approximation, which formally corresponds to development according to the size of the charge, cannot be applied, however, in the case of considerable densities, where the interaction among the electrons plays an essential part. A different approximation method is possible in the problem under consideration, a method that does not require that the interactions among the electrons be small. What is required is that the changes in the energy of the electrons as they pass through the pair of grids be small as compared with the kinetic energy of the electrons in the unperturbed beam (which corresponds to an expansion in terms of the parameter $\frac{eV}{m\xi^2}$).

In this way, we shall solve system (44.1), setting $f = f_0 + f_1$ and $\underline{E} = \underline{E}_0 + \underline{E}_1$, where the zero approximation includes the influence on the beam both of the characteristic field and of the external field (for example, from the focusers), excluding only the grid fields.

As a first approximation for f_1 and $\underline{E}_1$, we get:

$$\left.\begin{aligned} &\frac{\partial f_1}{\partial t} + \operatorname{div}_r \boldsymbol{v} f + \frac{e}{m} \boldsymbol{E}_1 \operatorname{grad}_v f_0 + \frac{e}{m} \boldsymbol{E}_0 \operatorname{grad}_v f_1 = \\ &\qquad = -F \frac{\partial f_0}{\partial \xi} + \left[\frac{\partial f}{\partial t}\right]^{st}, \\ &\operatorname{div} \boldsymbol{E}_1 = 4\pi e \int f_1 \, d\boldsymbol{v}. \end{aligned}\right\} \tag{44.2}$$

An important complication is introduced into the problem by the dispersion of the beam, due to the presence of space charges (complication of $\underline{E}_0$ and f_0), but in this problem the dispersion of the beam is not a factor of primary importance, and can be compensated by requiring a suitable distance for the external fields (which is usually done in various kinds of focusers). In any case, at distances where the scattering of the beam can still be ignored, we set:

$$f_0 = \varphi_0(\xi) e^{-ax} = N_0 e^{-ax} \delta(\xi - \xi_0), \tag{44.3}$$

and at the same approximation $E_0 = 0$.

The system of equations here obtained corresponds to the system that applies for electron plasma, which is not in a field of external forces. The physical cause of this fact is that neglecting the scattering is equivalent to introducing fictitious positive forces, exactly compensating the space charge of the electron beam in stationary motion. This fact indicates the existence of qualities common to a beam of electrons and electron plasma, and raises the question as to the appearance of the oscillatory properties of electron plasma in the problem here under consideration.

For the plane case, to which we confine ourselves here, the system of equations (44.2), under the given condition of a zero approximation, allows of an exact solution for the case $\alpha \longrightarrow 0$.

We shall seek the solution in the form of a double Fourier integral

$$\left.\begin{aligned} f_1(x, \xi, t) &= \iint\limits_{-\infty}^{+\infty} f_1(\omega, k, \xi)\, e^{i\omega t + ikx}\, d\omega\, dk, \\ E_1(x, t) &= \iint\limits_{-\infty}^{+\infty} E_1(\omega, k)\, e^{i\omega t + ikx}\, d\omega\, dk. \end{aligned}\right\} \quad (44.4)$$

We obtain the following system of equations for the Fourier components:

$$\left.\begin{aligned} [i\omega + (ik + \alpha)\xi] f_1(\omega, k, \xi) + \frac{e}{m} b_k \frac{\partial \psi_0}{\partial \xi} &= -\frac{e}{2\pi m} V(\omega) \frac{\partial \psi_0}{\partial \xi}, \\ ikE(\omega, k) &= 4\pi e \int\limits_{-\infty}^{+\infty} f_1(\omega, k, \xi)\, d\xi, \end{aligned}\right\} \quad (44.5)$$

where the coefficients b_k and $E(\omega, k)$ are linked by the relation

$$b(\omega, k) = \frac{1}{(2\pi)^2} \iint\limits_{-\infty}^{+\infty} E_1(x, t)\, e^{-i\omega t - ikx - \alpha x} dx\, dt = E_1(\omega, k - i\alpha). \quad (44.6)$$

Excluding the b_k gives the following equation to define the Fourier coefficients of the distribution function:

$$f_1(\omega, k, \xi) + \frac{4\pi \frac{e^2}{m}}{[i\omega + (ik+\alpha)\xi](ik+\alpha)} \frac{\partial \psi_0}{\partial \xi} \int_{-\infty}^{+\infty} f_1(\omega, k - i\alpha, \xi)\, d\xi =$$

$$= -\frac{eV(\omega)}{2\pi m} \frac{\frac{\partial \psi_0}{\partial \xi}}{[i\omega + (ik+\alpha)]}, \qquad (44.7)$$

which is solved in an elementary way, and in a limiting case $\alpha \rightarrow 0$, we obtain for $f_1(x, \xi, t)$:

$$f_1(x, \xi, t) = \lim_{\alpha \to 0} \frac{-e}{2\pi m} \frac{\partial \psi_0}{\partial \xi} \left\{ \int_{-\infty}^{+\infty}\int \frac{V(\omega)\, e^{i\omega t + ikx} d\omega dk}{i\omega + (ik+\alpha)\xi} + \right.$$

$$\left. + \omega_0^2 \int_{-\infty}^{+\infty}\int \frac{V(\omega)\, e^{i\omega t + ikx} d\omega dk}{[i\omega + (ik+\alpha)\xi]\, \{[(i\omega + (ik+\alpha)\xi_0]^2 + \omega_0^2\}}, \right. \qquad (44.8)$$

where $\omega_0^2 = 4\pi N_0 \frac{e^2}{m}$ — is the square of the characteristic frequency of the electron plasma.

All the mean quantities obtained from the distribution function (44.8) are expressed by elementary functions.

The expression for the value of the density ρ_ψ of any quantity ψ connected with the electrons in the beam is:

$$\rho_\psi = N(x, t)\overline{\psi(x, \xi)} = \int_{-\infty}^{+\infty} \psi(x, \xi) f(x, \xi, t)\, d\xi, \qquad (44.9)$$

from which, by means of (44.8), we obtain, for $x > 0$:

$$\rho_\varphi = N_0 \psi(x, \xi_0) \left\{ 1 + \frac{e}{m\xi_0^3} \frac{1}{\omega_0} \sin \frac{\omega_0}{\xi_0} x \frac{\partial V\left(t - \frac{x}{\xi_0}\right)}{\partial t} + \right.$$

$$\left. + \frac{e}{m_0}\left(\frac{1}{\psi} \frac{\partial \psi}{\partial \xi_0} - \frac{1}{\xi_0} \right) \cos \frac{\omega_0}{\xi_0} x V\left(t - \frac{x}{\xi_0}\right) \right\}. \qquad (44.10)$$

The fundamental characteristics of the beam: the density of the particles, current density and density of kinetic energy are obtained from (44.10), if we take $\psi(x,\xi_0) = 1,\ \xi_0,\ \frac{m\xi_0^2}{2}$, respectively.

For noninteracting electrons, we have from Sec. 43, the first two terms of the formula (43.18):

$$\rho_\psi = N_0\psi(x,\ \xi_0)\left\{1+\frac{e}{m\xi_0^2\xi_0}\,x\,\frac{\partial V\left(t-\frac{x}{\xi_0}\right)}{\partial t}+\right.$$
$$\left.+\frac{e}{m\xi_0}\left(\frac{1}{\psi}\frac{\partial\psi}{\partial\xi_0}-\frac{1}{\xi_0}\right)V\left(t-\frac{x}{\xi_0}\right)\right\}. \qquad (44.11)$$

Making use of (44.10) in the second relation of the system (44.5), we have as $\alpha \to 0$, an expression for E_1:

$$E_1 = -\frac{\omega_0}{\xi_0}\sin\frac{\omega_0}{\xi_0}x\,V\left(t-\frac{x}{\xi_0}\right). \qquad (44.12)$$

Comparison of (44.10) and (44.11) shows:

1. The limit of applicability of the theory of noninteracting electrons. Formula (44.10) goes over into (44.11) in the case $\frac{\omega_0}{\xi_0}x \ll 1$. In this way, only in the case that the length of the beam is considerably less than the characteristic length $D=\frac{\xi_0}{\omega_0}$, will the calculations based on the hypothesis of noninteracting electrons be correct. In practical units

$$D \cong 3{,}2\cdot 10^{-8}\frac{V^{\frac{3}{4}}}{j_0^{\frac{1}{2}}} \quad (V \text{ in volts, } j_0 \text{ in amperes}).$$

If the interval from V = 10^3 volts and current density $j_0 = 10^{-4}\ \frac{\text{amp.}}{\text{cm}^2}$ to V = 100 v and $j_0 = 10\ \frac{\text{amp.}}{\text{cm}^2}$, D changes from about 40 cm to 0.3 mm.

2. A characteristic property of the solutions obtained (44.10) and (44.11) is the presence of spatial periodicity of the properties in a beam of electrons, the period being equal to $2\pi D$. The existence of this periodicity is a direct manifestation of the oscillatory properties of electron plasma in the case of electron flow that has been discussed. Solution (44.12) represents a superposition of longitudinal waves with the dispersion law $\omega = \xi_0 k + \omega_0$, while the law of dispersion of longitudinal waves in electron plasma (with no flow) is $\omega \sim \omega_0$.

3. The variables with time going to make up current density and charge density in the beam are connected with the value of the field from (44.10) and (44.11) by the simple relationships:

$$\frac{\partial E_1}{\partial t} = -4\pi j_1(x, t) \quad \text{и} \quad \frac{\partial E_1}{\partial x} = 4\pi \rho_1(x, t). \tag{44.13}$$

4. In the region in which interaction among electrons plays an essential party (small D), the magnitude of the electromotive force with which the modulated beam acts on the resonator:
a) has a greater value than the potential difference on the first pair of grids; b) selectively depends on the relationship of the modulation frequency ω to the "plasma frequency" ω_0. We prove this.

Knowing the electric field in the beam, we are able to define the electromotive force induced by the beam. Let us conceive that a second pair of grids are placed in the beam, connected with an oscillatory system. We calculate the magnitude of the electromotive force induced in it by the varying space charge of the beam. For simplicity's sake, let the oscillatory system represent a Thompson circuit of the usual type. Then in the quasi-stationary region, by Ohm's law

$$\int_1^2 E_e dl = \int_1^2 \left(-\frac{1}{c}\frac{\partial A_e}{\partial t} - \text{grad}_e \Phi\right) dl =$$
$$= -\frac{1}{c}\frac{\partial}{\partial t}\int_1^2 A_e dl - (\Phi_2 - \Phi_1) = IR, \tag{44.14}$$

where A is the vector potential and Φ_2 minus Φ_1 the potential difference between the grids. In the usual approximation

$$\frac{1}{c}\int_1^2 A_e dl \cong \frac{1}{c}\oint A_e dl = LI, \tag{44.15}$$

where L is the coefficient of self-induction of the oscillatory circuit.

Thus,

$$IR = -L\frac{dI}{dt} - (\Phi_2 - \Phi_1); \qquad (44.16)$$

where $\Phi_2 - \Phi_1$ is the complete potential difference on the grids, conditioned in the case concerning us both by the superficial charges on the place of the condensor (grids) and by the space charges on the beam:

$$\Phi_2 - \Phi_1 = -\int_1^2 E_e^s\,dl - \int_1^2 E_e^\rho\,dl = \frac{e}{C} - \int_1^2 E_e^\rho\,dl, \qquad (44.17)$$

where C is the capacitance of the plane capacitor (grids) in vacuo: E_e the strength of the electric field in the beam, previously calculated (44.12). Consequently, the equation for the current of the circuit will be:

$$L\frac{dI}{dt} + RI - \frac{e}{C} = \int_1^2 E_e^\rho\,dl, \qquad (44.18)$$

from which it is seen that the role of the electromotive force is played by the expression

$$\varepsilon^{cm} = \int_1^2 E_e^\rho\,dl. \qquad (44.19)$$

Substituting the expression for the strength of the electric field in the beam (44.12) and assuming that on the first pair of grids $V(t) = V_0 \cos \omega t$, we obtain for the electromotive force:

$$E^{cm} = \int_{x_2}^{x_1} E(x, t)\,dx =$$

$$= V_0\frac{\omega_0}{\xi_0}(x_2 - x_1)\left\{\frac{\sin\frac{\omega-\omega_0}{\xi_0}\frac{x_2-x_1}{2}}{\frac{\omega-\omega_0}{\xi_0}\frac{x_2-x_1}{2}} = \sin\left(\omega t - \frac{\omega-\omega_0}{\xi_0}\frac{x_1+x_2}{2}\right) - \right.$$

$$\left. - \frac{\sin\frac{\omega+\omega_0}{\xi_0}\frac{x_2-x_1}{2}}{\frac{\omega+\omega_0}{\xi_0}\frac{x_2-x_1}{2}} - \sin\left(\omega t - \frac{\omega+\omega_0}{\xi_0}\frac{x_2+x_1}{2}\right)\right\}; \qquad (44.20)$$

for the case $\frac{\omega_0}{\xi_0}(x_2-x_1)\gg 1$, the amplitude of the electromotive force exceeds the potential difference on the first pair of grids by $\frac{\omega_0}{\xi_0}(x_2-x_1)$ times. In the opposite case $\frac{\omega_0}{\xi_0}(x_2-x_1)\ll 1$, it is that much less. This effect and the presence of a resonance factor can obviously be used to increase the power of the new type of transmitting and amplifying tubes.[+]

Section 45. Diode Theory

Vainshtein, in his dissertation, used the initial equations of the theory being presented to define the static characteristics of the diode[++].

His results are of importance for the theory being set forth not only as one of the possible applications of the theory, but also as an example in which discontinuous solutions of the original equations appear.

For a plane diode, the problem reduces to the integration of the equation:

$$\left.\begin{aligned} &\frac{\partial f}{\partial t}+u\frac{\partial f}{\partial x}+\frac{\varepsilon}{m}\frac{\partial \varphi}{\partial x}\frac{\partial f}{\partial u}=0,\\ &\frac{\partial^2\varphi}{\partial x^2}=4\pi\varepsilon\int_{-\infty}^{\infty} f\,du\\ &(\varepsilon<0), \end{aligned}\right\} \tag{45.1}$$

under the following boundary conditions:

$$\varphi(0, t)=0, \quad \varphi(e, t)=V$$

$$f=g_1(u, t) \quad \text{for} \quad x=0, \quad u>0, \tag{45.2}$$

$$f=g_2(u, t) \quad \text{for} \quad x=l, \quad u<0 \tag{45.3}$$

$\varphi(x, t)$ is the electric potential, $g_1(u, t)$ and $g_2(u, t)$ are arbitrary functions to be given.

[+] The questions discussed here have been further developed by G. Myakishev. See Zh. T. F. 18, 1063 (1948).

[++] L. Vainshtein Sbornik nauchnykh trudov, izd. Sov. Radio, Vol. X Moscow, 1948.

Here $g_1(u, t)$ is the distribution function of the electrons entering the diode from the cathode side ($u > 0$ for $x = 0$), and $g_2(u, t)$ of those coming from the anode side ($u < 0$, for $x = 1$), although in most practical problems $g_2 = 0$, but the general case as well can be effected experimentally.

For the static case ($\partial / \partial t = 0$), as has been stated several times above, the first equation of (45.1) has the integral

$$f(x, u) = \Omega\left(\frac{mu^2}{2} - \varepsilon\varphi(x)\right),$$

where Ω is any function. It is obvious that it is impossible to satisfy under the conditions with two arbitrary functions by means of a solution with one arbitrary function. However, it is known from the general theory of linear differential equations of the first order that the general solution of equations of this type is not always necessarily a single-valued continuous function of one integral. It therefore seems possible to seek the solution of the diode problem in the class of discontinuous functions.

To solve the problem that has been posed, we pass first to nondimensional quantities. The characteristic units of length and time for an electron gas are:

$$[x] = \sqrt{\frac{kT}{4\pi\varepsilon^2[n]}}, \quad [t] = \sqrt{\frac{m}{4\pi\varepsilon^2[n]}}, \qquad (45.4)$$

where $[n]$ is a unit of electron concentration, still to be established and T the temperature of the cathode.

The role of the units of velocity of potentials must be played by the following quantities:

$$[u] = \frac{[x]}{[t]} = \sqrt{\frac{kT}{m}}, \qquad (45.5)$$

$$[\varphi] = \frac{m}{\varepsilon}[u]^2 = \frac{kT}{\varepsilon}. \qquad (45.6)$$

We now express the unit of electron concentration by an arbitrarily chosen unit of current density. We require that

$$[j] = \varepsilon[u][f] = \sqrt{2\pi}j_0, \qquad (45.7)$$

where j_0 is a current density characteristic for the problem. For example, the density of the anode current may be taken for j_0.

On the other hand, because of the dimensionality of distribution function f, and the inner electron concentration should be expressed as:

$$[n] = [f]\,[u]. \tag{45.8}$$

Eliminating $[f]\,[u]$ from the two preceding relationships, we find:

$$[n] = \sqrt{2\pi}\sqrt{\frac{m}{kT}}\,\frac{j_0}{\epsilon}, \tag{45.9}$$

and likewise

$$[f] = \frac{[n]}{\sqrt{\frac{kT}{m}}}. \tag{45.10}$$

In the units of measurement chosen, the initial equation has the form:

$$\frac{\partial f^*}{\partial t^*} + u^*\frac{\partial f^*}{\partial x^*} + \frac{\partial \varphi^*}{\partial x^*}\frac{\partial f^*}{\partial u^*} = 0; \quad \frac{\partial^2 \varphi^*}{\partial x^{*2}} = \int_{-\infty}^{\infty} f^* du^*, \tag{45.11}$$

in what follows we omit the * for all the quantities.

Assuming that the potential has the form of the curve given in Fig. 14 (this must be supported by further calculation), by virtue of the condition $\frac{\partial^2 \varphi}{\partial x^2} > 0$, the potential may also be represented by curves of the type Figs. 15 and 16.

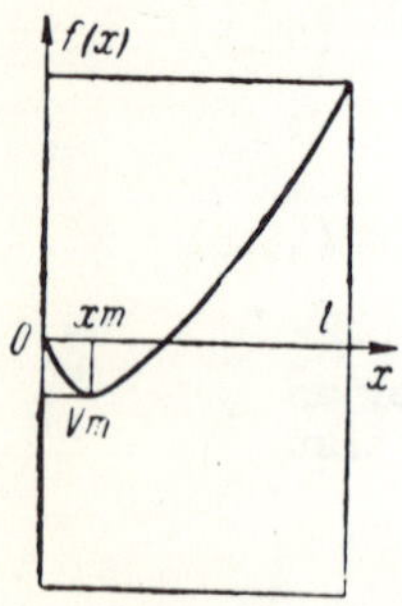

Fig. 14

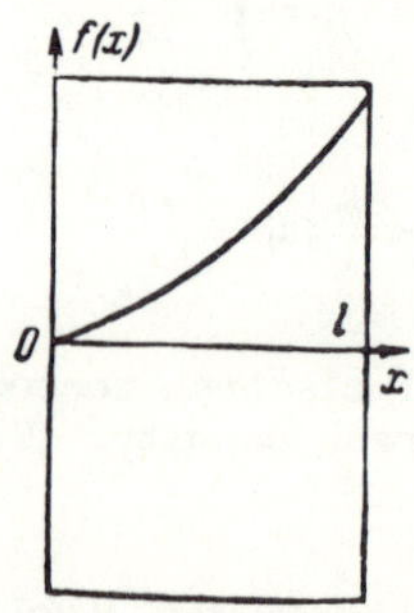

Fig. 15

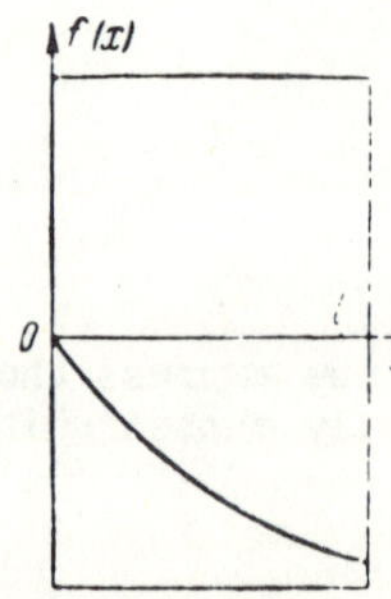

Fig. 16

At a distance x_m from the cathode, there is a potential minimum of depth V_m. We introduce the quantities

$$\xi = x^* - x^*_m, \qquad (45.12)$$

$$\eta = \varphi^* - V^*_m \qquad (45.13)$$

and we shall seek a stationary solution, integrating the equations

$$u \frac{\partial f}{\partial \xi} + \frac{\partial \eta}{\partial \xi} \frac{\partial f}{\partial u} = 0, \qquad (45.14)$$

$$\frac{d^2 \eta}{d\xi^2} = \int_{-\infty}^{\infty} f \, du. \qquad (45.15)$$

We write down the boundary conditions as:

$$f = g_1\left(\frac{u^2}{2}\right) \quad \text{for } u > 0, \ \xi = \xi_1,$$

$$(45.16)$$

$$f = g_2\left(\frac{u^2}{2}\right) \quad \text{for } u < 0, \ \xi = \xi_2.$$

By the indices $_1$ and $_2$, we designate quantities referring to the cathode and the anode, respectively, thus, for instance, η_1 is the magnitude of the cathode potential with reference to the minimum.

Equation (45.14) has the obvious solution

$$\lambda = \frac{u^2}{2} - \eta(\xi). \qquad (45.17)$$

We set this solution at the base of the construction of an integral satisfying boundary conditions (45.16).

We consider the electron trajectories in the phase plane $[\xi, u]$ (see Fig. 17). The electrons emitted from the cathode

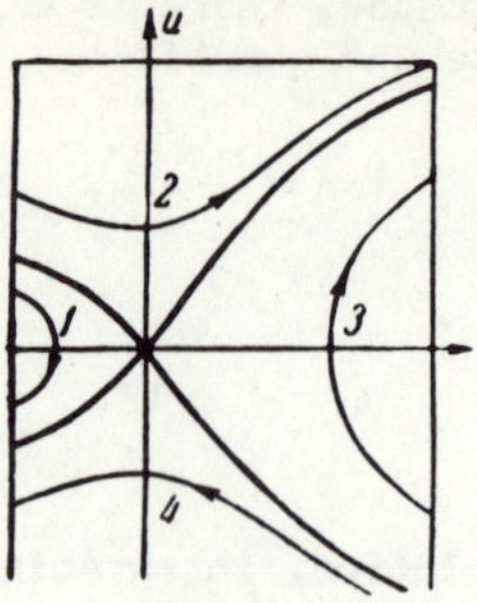

Fig. 17.

either return to the cathode (trajectory type 1) or go to the anode (trajectory type 2), depending on whether α is smaller or larger than η_1 (that is, $\lambda < 0$ or $\lambda > 0$); similar phase trajectories for the anode electrons will either be curves of type 3 (for $\alpha < \eta_2$, that is, $\lambda < 0$), or curves of type 4 (for $\alpha > \eta_2$, that is $\lambda > 0$). The trajectories of all four types

are separated by two intersecting curve $= 0$, that is, $u = \pm\sqrt{2\eta}$.

Of particular importance is the curve

$$\Sigma = 0$$

where Σ designates the function

$$\Sigma = u - \operatorname{sgn}\xi\sqrt{2\eta}. \tag{45.18}$$

Designating $\rho_0 = \left(\frac{d^2\eta}{d\xi^2}\right)_{\xi=0}$ for small ξ, we shall take

$$\Sigma = u - \sqrt{\rho_0}\,\xi \quad (\xi \to 0), \tag{45.19}$$

so that Σ is a continuous and continuously differentiable function.

By sgn ξ we mean a function

$$\operatorname{sgn}\xi = \begin{cases} +1 & \text{for } \xi > 0, \\ -1 & \text{for } \xi < 0, \\ 0 & \text{for } \xi = 0. \end{cases} \tag{45.20}$$

The curve $\Sigma = 0$ separates the phase trajectories departing from the cathode from the trajectories coming in from the anode. Σ satisfies the relationship

$$\left\{ u\frac{\partial}{\partial\xi} + \frac{\partial\eta}{\partial\xi}\frac{\partial}{\partial u} \right\}\Sigma = \left\{ 1 - \frac{u \operatorname{sgn}\xi}{\sqrt{2\eta}} \right\}\frac{d\eta}{d\xi} = -\Sigma\frac{\operatorname{sgn}\xi}{\sqrt{2\eta}}\frac{d\eta}{d\xi}, \tag{45.21}$$

from which, it will be seen, that if we introduce the function $\epsilon(\Sigma)$

$$\epsilon(\Sigma) = \begin{cases} 1 & \text{for } \Sigma > 0, \\ 0 & \text{for } \Sigma < 0, \\ 1/2 & \text{for } \Sigma = 0, \end{cases} \tag{45.22}$$

$$\frac{d}{d\Sigma}\epsilon(\Sigma) = \delta(\Sigma),$$

then, along with the continuous solution λ for equation (45.14), we may point out discontinuous solutions of the type

$$\epsilon(\Sigma) \text{ and } \epsilon(-\Sigma),$$

which formally satisfy this equation. Thus, for instance,

$$\left\{ u\frac{\partial}{\partial\xi} + \frac{\partial\eta}{\partial\xi}\frac{\partial}{\partial u} \right\}\epsilon(\Sigma) = -\Sigma\,\delta(\Sigma)\frac{\operatorname{sgn}\xi}{\sqrt{2\eta}}\frac{d\eta}{d\xi} = 0,$$

so that

$$\Sigma\,\delta(\Sigma) \equiv 0.$$

By means of these discontinuous integrals, we may construct solution (45.14), satisfying conditions (45.16), namely,

$$f=\in(\Sigma)g_1\left(\frac{u^2}{2}-\eta+\eta_1\right)+\in(-\Sigma)g_2\left(\frac{u^2}{2}-\eta+\eta_2\right). \qquad (45.23)$$

Each of the summands satisfies (45.14), just as each factor separately satisfies it; on the other hand, conditions (45.16) are likewise satisfied since, for example, for $\xi=\xi_1$, $u>0$

we have $\in(\Sigma)=1,\quad \in(-\Sigma)=0$ and $f=g_1\left(\frac{u^2}{2}\right)$.

If the course of potential $\eta(\xi)$ were known, then (45.23) would solve the equation, but $\eta(\xi)$ is not known, so that it is necessary to substitute (45.23) in the second equation (45.15) and solve the equation obtained for $\eta(\xi)$. This equation has the form

$$\frac{d^2\eta}{d\xi^2}=\int\limits_{\operatorname{sgn}\xi\sqrt{2\eta}}^{\infty}g_1\left(\frac{u^2}{2}-\eta+\eta_1\right)du+\int\limits_{-\infty}^{\operatorname{sgn}\xi\sqrt{2\eta}}g_2\left(\frac{u^2}{2}-\eta+\eta_2\right)du. \qquad (45.24)$$

Introducing the variable of integration $\lambda=\frac{u^2}{2}-\eta$, we transform it into the form

$$\frac{d^2\eta}{d\xi^2}=2\int\limits_{-\eta}^{0}\{\in(-\xi)g_1(\lambda+\eta_1)+\in(\xi)g_2(\lambda+\eta_2)\}\frac{d\lambda}{\sqrt{2(\lambda+\eta)}}+$$
$$+\int\limits_{0}^{\infty}\{g_1(\lambda+\eta_1)+g_2(\lambda+\eta_2)\}\frac{d\lambda}{\sqrt{2(\lambda+\eta)}}. \qquad (45.25)$$

This is an equation of the type

$$\frac{d^2\eta}{d\xi^2}=F(\eta), \qquad (45.26)$$

so that, performing the integration

$$\frac{1}{2}\left(\frac{d\eta}{d\xi}\right)^2=\int\limits_0^{\eta}F(\eta)d\eta, \qquad (45.27)$$

we obtain (the first summand in (45.25) is transformed into Dirichlet's formula for multiple integrals):

$$\left(\frac{d\eta}{d\xi}\right)^2 = \begin{cases} \Phi_\alpha(\eta) & \text{for } \xi < 0, \\ \Phi_\beta(\eta) & \text{for } \xi > 0, \end{cases} \tag{45.28}$$

where

$$\Phi_\alpha(\eta) = 2^{\frac{3}{2}}\{2\int_{-\eta}^{\infty} g_1(\lambda+\eta_1)\sqrt{\lambda+\eta}\,d\lambda + \int_0^{\infty}[g_1(\lambda+\eta_1)+g_2(\lambda+\eta_2)][\sqrt{\lambda+\eta}-\sqrt{\lambda}]\,d\lambda\},$$

$$\Phi_\beta(\eta) = 2^{\frac{3}{2}}\{2\int_{-\eta}^{0} g_2(\lambda+\eta_2)\sqrt{\lambda+\eta}\,d\lambda + \int_0^{\infty}[g_1(\lambda+\eta_1)+g_2(\lambda+\eta_2)][\sqrt{\lambda+\eta}-\sqrt{\lambda}]\,d\lambda\}. \tag{45.29}$$

The course of the potential is given by the formulas

$$\xi = \xi_\alpha(\eta) = -\int_0^{\eta}\Phi_\alpha^{-\frac{1}{2}}(\eta)\,d\eta \quad \text{for } \xi < 0,$$

$$\xi = \xi_\beta(\eta) = \int_0^{\eta}\Phi_\beta^{-\frac{1}{2}}(\eta)\,d\eta \quad \text{for } \xi > 0. \tag{45.30}$$

Since $\xi_2 - \xi_1$ is equal to the distance between the cathode and the anode (in the new unit), we obtain the following relation between η_1 and η_2:

$$\xi_\beta(\eta_2) - \xi_\alpha(\eta_1) = l^*. \tag{45.31}$$

Further, the anode voltage equals

$$\eta_2 - \eta_1 = V^*. \tag{45.32}$$

The expression for the density of the complete current in the diode is as follows:

$$j = \varepsilon \int_{-\infty}^{\infty} fu\, du - \frac{1}{4\pi} \frac{\partial^2 \varphi}{\partial t\, \partial x}$$

or, in nondimensional units

$$j^* = \int_{-\infty}^{\infty} f^* u^*\, du^* - \frac{\partial^2 \varphi^*}{\partial t^*\, \partial x^*}.$$

This expression includes both the convection current and the displacement current. For the anode current, we have:

$$j^* = \int_{\eta_1}^{\infty} g_1(\alpha)\, d\alpha - \int_{\eta_2}^{\infty} g_2(\beta)\, d\beta. \tag{45.33}$$

In the stationary case, the anode current thus does not depend on the course of the potential, but only on η_1 and η_2, as well as on the emission functions of the velocities distribution.

Equations (45.29) - (45.33) give, in principle, the static characteristic of the diode, that is, the dependence of the anode current on the anode voltage (and emission).

We consider the following particular case:

$$g_1(\alpha) = j_{s1} e^{-\alpha}; \quad g_2(\beta) = j_{s2} e^{-\beta}. \tag{45.34}$$

This means that both electrodes have the same temperature, and current density j^+ equal to

$$j = j_{s1} e^{-\eta_1} - j_{s2} e^{-\eta_2}. \tag{45.35}$$

If $j = 0$, that is,

$$j_{s1} e^{-\eta_1} = j_{s2} e^{-\eta_2} = \frac{1}{\sqrt{2\pi}}, \tag{45.36}$$

then the distribution function f obtained by formula (45.23) equals

$$f = \frac{1}{\sqrt{2\pi}} e^{-\frac{u^2}{2} + \eta} \tag{45.37}$$

and

$$\Phi_\alpha(\eta) = \Phi_\beta(\eta) = 2(e^\eta - 1),$$

$$|\xi| = \frac{1}{\sqrt{2}}\int_0^\eta \frac{d\eta}{\sqrt{e^\eta - 1}} = \sqrt{2}\arccos e^{-\frac{\eta}{2}},$$

so that the distribution potential is given by the formula

$$\eta = -2\ln\cos\frac{|\xi|}{\sqrt{2}}. \qquad (45.38)$$

The graph of this function is shown in Fig. 18. Here, we have a statistically balanced case; the temperatures of cathode and anode are equal, the mean anode current equals zero, so that the anode voltage is virtually equal to the contact potential difference between the anode and the cathode, and the potential distribution (45.38) satisfies Laue's equation+, which in our notation takes on the form

$$\frac{d^2\eta}{d\xi^2} = e^\eta. \qquad (45.39)$$

The most interesting from the practical point of view is a diode in which only the cathode emits; it may be considered as a particular case of (45.34), when $j_{2s} = 0$. Then, the density of the anode current is given by the expression

$$j = j_{s1}e^{-\eta_1}. \qquad (45.40)$$

This case is worked out in detail in works cited by Vainshtein, the results obtained being equivalent to (45.29) - (44.34), but by a different method. We adduce the results of study of these formulas.

In usual units, we have:

$$g_1(\alpha) = \frac{1}{\varepsilon} j_s e^{-\eta}, \quad g_2 = 0, \quad j = j_s e^{-\eta_1} \qquad (45.41)$$

+ Laue, M. Jahrbuch d. Radiokt. u. Elektronik 15, 205 (1918).

and we take for j_0 the mean value of the density of the anode current, so that in the new unit

$$j_0^* e^{-\eta_1} = \frac{1}{\sqrt{2\pi}} \tag{45.42}$$

and

$$\Phi_\beta(\eta) = \frac{2}{\sqrt{\pi}} \int_0^\infty e^{-\lambda} [\sqrt{\lambda+\eta} - \sqrt{\lambda}]\, d\lambda,$$

or

$$\Phi_\alpha(\eta) = \frac{1}{\sqrt{\pi}} \left\{ \int_0^\infty e^{-\lambda} [\sqrt{\lambda+\eta} - \sqrt{\lambda}]\, d\lambda + 2 \int_{-\eta}^0 e^{-\lambda} \sqrt{\lambda+\eta}\, d\lambda \right\}$$

$$\Phi_{\alpha,\beta} = e^\eta - 1 \pm \left\{ \frac{2}{\sqrt{\pi}} \eta^{\frac{1}{2}} - e^\eta \operatorname{Erf}(\eta^{\frac{1}{2}}) \right\}, \tag{45.43}$$

where

$$\operatorname{Erf} y = \frac{2}{\sqrt{\pi}} \int_0^y e^{-x^2} dx. \tag{45.44}$$

From this, there follow the approximate formulas for small and large

$$\Phi_{\alpha,\beta}(\eta) = \eta \quad \text{for} \quad \eta \ll 1 \tag{45.45}$$

$$\Phi_\alpha(\eta) = 2e^\eta - \frac{2}{\sqrt{\pi}} \eta^{\frac{1}{2}} - 1 + O(\eta^{-\frac{1}{2}}) \quad \text{for } \eta \gg 1,$$

$$\Phi_\beta(\eta) = \frac{2}{\sqrt{\pi}} \eta^{\frac{1}{2}} \left\{ 1 - \frac{\sqrt{\pi}}{2} \eta^{-\frac{1}{2}} + \frac{1}{2} \eta^{-1} + O(\eta^{-2}) \right\} \text{ for } \eta \gg 1.$$

The course of the potential, as obtained as a result of numerical integration by formula (45.43), is represented in Fig. 19.

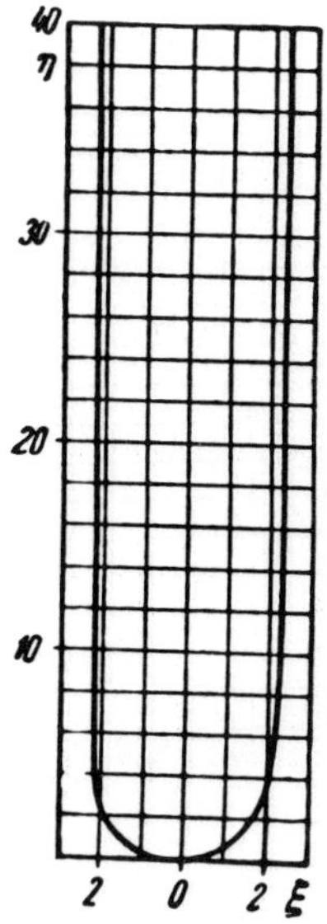

Fig. 18.

For small and large values of η, the following approximate expressions are obtained for functions $\xi_\alpha(\eta)$ and $\xi_\beta(\eta)$:

$$-\xi_\alpha(\eta)=\xi_\beta(\eta)=2\sqrt{\eta} \quad \text{for} \quad \eta\ll 1,$$

$$-\xi_\alpha(\eta)=2{,}554-\sqrt{2}\,e^{-\frac{\eta}{2}}+O(e^{-\eta}) \quad \text{for} \quad \eta\gg 1, \tag{45.46}$$

$$\xi_\beta(\eta)=\frac{2^{\frac{3}{2}}\pi^{\frac{1}{4}}}{3}\,\eta^{\frac{3}{4}}\left\{1+\frac{3\sqrt{\pi}}{4}\,\eta^{-\frac{1}{2}}+O\left(\eta^{-\frac{3}{4}}\right)\right\} \quad \text{for } \eta\gg 1.$$

We see that the course of the potential for $\xi<0$ coincides more or less with the course of the potential of the "equilibrium" diode (Fig. 18), and if in the equilibrium case, the minimum value of ξ equals

$$\xi_{\min}=-\frac{\pi}{\sqrt{2}}=-2{,}22,$$

then here

$$\xi_{\min}=-2{,}55.$$

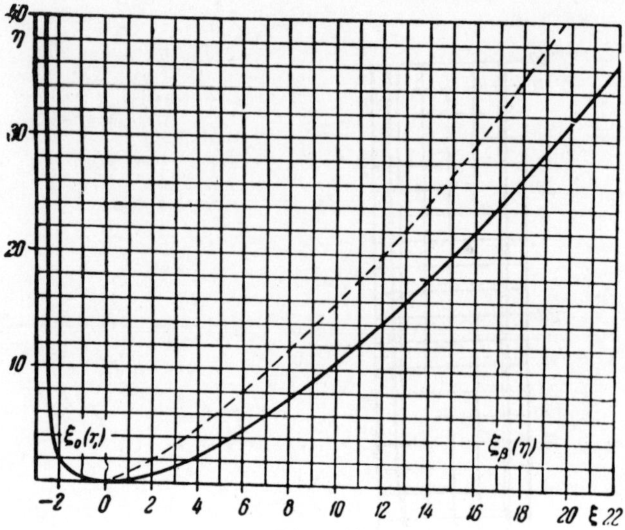

Fig. 19.

Usually $\xi_1 \sim \xi_{min}$, in which case the new unit of length [x] is equal in order of magnitude to the distance between the cathode and the potential minimum, and [t] is the transit time of this distance at velocity $\sqrt{\frac{kT}{m}}$, that is, roughly, the mean time in which the electrons are in the space between the cathode and the minimum.

For $\xi > 0$, that is, between the potential minimum and the anode, the course of the potential is approximated by another function, that is, taking the first term in the expansion (45.46), for $\eta \gg 1$, we have:

$$\eta = \frac{3^{\frac{4}{3}}}{4\pi^{\frac{1}{3}}} \xi^{\frac{4}{3}}. \qquad (45.47)$$

This formula is asymptotic for the large ξ, but it satisfies the correct boundary condition $\eta = \frac{d\eta}{d\xi} = 0$ for $\xi = 0$, so that it correctly represents the course of the potential for any $\xi > 0$. The potential distribution, according to (45.38) is shown by the broken line on Fig. 19.

If we assume in (45.47) that $\xi = \xi_2$ and $\eta = \eta_2$ and take into consideration that as a rule $\eta_2 \gg 1$ and $\xi_2 \gg 1$, while $\eta_1 \sim 1$ and $\xi_1 \sim 1$, then we obtain approximately, passing to ordinary unit, the "three halves" formula

$$j = \frac{1}{9\pi}\sqrt{\frac{2e}{m}}\frac{V^{\frac{3}{2}}}{l^2}.$$

In general, knowing functions ξ_α and ξ_β, it is possible to obtain the exact connection between j and V.

The transit time of the electron from the minimum to the anode

$$t_2^* = \int_0^{\xi_2} \frac{d\xi}{\sqrt{2[\lambda + \eta_1(\xi)]}} \sim \frac{2^{\frac{1}{2}}\pi^{\frac{1}{6}}}{3^{\frac{2}{3}}}\int_0^{\xi_2} \xi^{-\frac{2}{3}}\,d\xi = 2^{\frac{1}{2}}\pi^{\frac{1}{6}}3^{\frac{1}{3}}\xi_2^{\frac{1}{3}} = 2\pi^{\frac{1}{4}}\eta_2^{\frac{1}{4}} \qquad (45.48)$$

in a first approximation for $\eta_2 \gg 1$ does not depend on λ and gives the principal part of the transit time of the electrons in the tube.

This is the nature of the mean course of the potential in a plane diode, whose electrode emission is on the average distributed according to Maxwell.

It should be noted that the calculations adduced above are not based on the idea of strict spatial localization of particles, since our starting point was solely equation (45.1).

Fundamentally, Vainshtein's work cited above is devoted, in addition to the results given, to the theory of the auto-electronic effect, taking into consideration the space charge. In this part of his work, Vainshtein clearly takes the point of view of strict point particles, and therefore holds that the initial equations of the theory should be completed by terms that take into account "collisions" between electrons. But, it is difficult to say to what degree the analysis he makes actually makes use of the "localization" of particles and to what degree his results in the theory of the auto-electronic effect are obligated to the theory being set forth.

CHAPTER VI

ELECTRON PLASMA IN HYDRODYNAMIC APPROXIMATION

It was pointed out previously (Sec. 14) that the initial equations of the theory make it possible, under certain conditions, to describe the motion of a system of particles not in phase space of six dimensions, but in ordinary space (x, y, z). It is important to remark once again at this point that we do not with this obtain a complete apparatus for calculations, since the elements of the tensor obtained themselves depend on an unknown distribution function. It is only in two special cases that we can reduce the description in f ($\underline{r}$, $\underline{v}$, t) to mean spatial quantities.

1. The case of what is called the one-velocity approximation, in which we ignore the effect of thermal fluctuations in the velocities, which is equivalent to completely ignoring all the elements of the tensor in question.

2. The case in which the velocities distribution function is close to the Maxwellian. In this case, only the diagonal elements of the tensor differ from zero, its divergence reduces only to the gradient of the pressure, while under this condition, the pressure will be a function of the density, whose form is defined from the equation of state for an ideal gas.

These two postulates cannot be established strictly, and they must therefore serve only as a starting point for orienting investigations. It was shown in the monograph cited above that the second condition is <u>approximately</u> realized for an electron plasma in the case where the gradient of the spatial inhomogeneity satisfies the relationship

$$D\frac{1}{\varphi}\frac{d\varphi}{dx}(x, y, z, t) \ll 1,$$

where φ is any function describing the spatial inhomogeneity, and D is the Debye length.

In this chapter, we shall try systematically to discuss the role of the oscillatory properties of electron plasma under various conditions, on the basis of the postulates mentioned above.

It is of importance to stress the following matter of principle. A new aspect of the theory being set forth is, as has been pointed out several times above, the assertion that there exist physical conditions such that under them the particles act not as material points, but as formations extended over the entire region under consideration. Consequently, obtaining the hydrodynamic form of the equations is not connected with averaging over any physically infinitely small space, containing a sufficient number of particles; this form of equations is valid even for two particles.

Section 46. Fundamental Equations. Expression for Energy and Energy Flow

We write the equations for the hydrodynamic form, obtained in Sec. 14, as follows:

$$\frac{Du}{Dt} = \frac{1}{\rho} \operatorname{div} T + \frac{1}{\rho} F, \qquad (46.1)$$

where T_{ik} are the elements of the tensor of stresses, which are defined there, as well.

These equations apply independently of the degree of rarefaction of the medium.

The forces of interaction between the particles enter into the equation in two ways: implicitly they are contained in the components T_{ik}, since these components depend on the initial function f, whose form is determined by the initial equation of the theory, containing these forces; and explicitly in the expression

$$F = -\nabla_r \int_{-\infty}^{+\infty} K(|r - r'|) \rho(r') \, dr', \qquad (46.2)$$

where K is the kernel of interaction among the particles and $\rho(\underline{r}) = \int f(\underline{r}, \underline{v}) \, dv$.

First of all, we consider the potential motions taking into account the Coulomb forces of interaction between the various elements of the charged fluid. Such interactions occur in the theory being set forth for the case of as large a number as desired of particles, when the absence of interactions between the elements of a single particle no longer play any role.

In analyzing the motions in electron plasma, thanks to the high characteristic frequencies of the electrons, the positive fluid, consisting of ions, since it is heavy, may be regarded as filling the space with a constant density. Taking the motion of the positive fluid into account will be done subsequently.

It should be stressed that for what follows the presence of the positive fluid is not strictly necessary, since its role reduces to that of compensating the forces of repulsion between electrons.

This compensation may be experimentally obtained by the use of various focusers without the presence of positive charges.

The initial system of equations for potential movement of the negative fluid will have the form:

$$\Delta\varphi + \frac{1}{\rho}(\nabla\rho\nabla\varphi) = -\frac{1}{\rho}\frac{\partial\rho}{\partial t} \quad \text{(equation of continuity)}, \qquad (46.3)$$

$$\Delta\Phi = -4\pi\frac{e}{m}(\rho - \rho_0) \quad \text{(Poisson's equation)}, \qquad (46.4)$$

$$\frac{\partial\varphi}{\partial t} + \frac{1}{2}\,\text{grad}^2\,\varphi + \frac{e}{m}\Phi + \int\frac{dP}{\rho} = 0 \quad \text{(Bernoulli-Euler equation)} \qquad (46.5)$$

where φ is the potential of velocities, $u = \text{grad}\ \varphi$, and Φ is the electric potential. We shall consider that the pressure is a function of the density alone. For the time being, we shall not specify the form of the equation of state.

The system of equations (46.3 - 46.5) is nonlinear. We simplify it by the assumption that the state of the electron fluid differs only slightly from a state of equilibrium with density ρ_0, that is, we postulate

$$\frac{\rho - \rho_0}{\rho_0} = \frac{\sigma}{\rho_0} \ll 1. \qquad (46.6)$$

In the future, quantity σ will be called the <u>polarization</u>.

In that case, we shall have for the pressure, limiting ourselves to the first order,

$$P(\rho) = P(\rho_0 + \sigma) = P(\rho_0) + \left(\frac{\partial P}{\partial\rho}\right)_{\rho=\rho_0}\cdot\sigma = P(\rho_0) + v_T^2\,\sigma, \quad (46.7) \qquad (46.7)$$

where $v_T^2 = \left(\frac{\partial P}{\partial\rho}\right)_{\rho=\rho_0}$ plays the role of the velocity of sound in the negative fluid.

To the same approximation

$$\int\frac{dP}{\rho} = v_T^2\int\frac{d\rho}{\rho} = v_T^2\ln\frac{\rho}{\rho_0} = v_T^2\ln\left(1 + \frac{\sigma}{\rho_0}\right) = \frac{v_T^2}{\rho_0}\sigma. \qquad (46.8)$$

Consequently, the Bernoulli-Euler equation will have the form

$$\frac{\partial \varphi}{\partial z}+\frac{1}{2}\operatorname{grad}^2\varphi+\frac{e}{m}\Phi+\frac{v_T^2}{\rho_0}\sigma=0. \qquad (46.9)$$

In addition to conditions (46.7), (46.8), we may assume that velocities **u** are small, in order to linearize the equations. In this case, the terms of the second order will be

$$\frac{1}{\rho_0}(\nabla\varphi\nabla\rho)=\frac{1}{\rho_0}(u\operatorname{grad}\sigma) \text{ and } \operatorname{grad}^2\varphi=u^2.$$

After the indicated simplifications, the system of initial linearized equations has the form

$$\left.\begin{aligned}\Delta\varphi&=-\frac{1}{\rho_0}\frac{\partial\sigma}{\partial t},\\ \Delta\Phi&=-4\pi\frac{e}{m}\sigma,\\ \frac{\partial\varphi}{\partial t}+\frac{e}{m}\Phi+\frac{v_T^2}{\rho_0}\sigma&=0.\end{aligned}\right\} \qquad (46.10)$$

The medium described by system (46.10) is characterized by only three properties: inertia, compressability, and the presence of electric forces, which appear upon change in the electron density at some point.

In this connection, three characteristics of the medium likewise only play a part: the material density $\rho_0=mN_0$, the density of of electric charge $\frac{e}{m}\rho_0=eN_0$ (where N_0 is the concentration of electrons in the plasma) and the square of the velocity of sound $v_T^2=\frac{\partial P}{\partial\rho}$; we note that even these three characteristics suffice to describe a significant complex of phenomena in electron plasma. When external charges are present, a term $\rho_e(x, y, z, t)$ must be added to the second equation of system (46.10).

The total energy of the electron plasma is made up of the kinetic, elastic and electrical energies:

$$
\left.
\begin{aligned}
W_{\kappa} &= \frac{1}{2} \rho_0 \int \left[\left(\frac{\partial \varphi}{\partial x} \right)^2 + \left(\frac{\partial \varphi}{\partial y} \right)^2 + \left(\frac{\partial \varphi}{\partial z} \right)^2 \right] d\tau, \\
W_{y} &= \frac{1}{2\rho_0} v_T^2 \int (\rho - \rho_0)^2 \, d\tau, \\
W_{\text{эл}} &= \frac{1}{8\pi} \int \left[\left(\frac{\partial \Phi}{\partial x} \right)^2 + \left(\frac{\partial \Phi}{\partial y} \right)^2 + \left(\frac{\partial \Phi}{\partial z} \right)^2 \right] d\tau,
\end{aligned}
\right\} \qquad (46.11)
$$

where $d\tau$ is an element of tridimensional space. We find an expression for the flow of the complete energy. The magnitude of the energy flow should be defined, as usual, representing the space integral for the energy by the surface integral

$$
-\frac{\partial W}{\partial t} = \oint S \, df, \qquad (46.12)
$$

which serves to define $\underline{S}$.

Employing the expression for the kinetic and electric energies, we have:

$$
\frac{\partial (W_{\kappa} + W_{\text{эл}})}{\partial t} = \rho_0 \int \operatorname{grad} \dot{\varphi} \cdot \operatorname{grad} \varphi \cdot d\tau + \frac{1}{4\pi} \int \operatorname{grad} \Phi \cdot \operatorname{grad} \dot{\Phi} \, d\tau, \qquad (46.13)
$$

where the dots indicate derivatives with respect to the time.

The change with time of the elastic energy is given by the formula

$$
\frac{\partial W_y}{\partial t} = \frac{v_T^2}{\rho_0} \int \sigma \dot{\sigma} \, d\tau. \qquad (46.14)
$$

On the basis of the third and first equations of motion of the system (46.10)

$$
\sigma = -\frac{\rho_0}{v_T^2} \dot{\varphi} - \frac{e\rho_0}{m v_T^2} \Phi,
$$

$$
\dot{\sigma} = -\rho_0 \Delta \varphi.
$$

Substitution in (46.14) gives

$$
\frac{\partial W_y}{\partial t} = \rho_0 \int \dot{\varphi} \Delta \varphi \, d\tau + \frac{\rho_0 e}{m} \int \Phi \Delta \varphi \, d\tau. \qquad (46.15)
$$

making use once again of the first equation of system (46.15), as well as the second equation, we have:

$$\Delta\dot{\varphi} = -\frac{1}{\rho}\dot{\sigma} = \frac{1}{4\pi\frac{e}{m}\rho_0}\Delta\Phi \tag{46.16}$$

from which, after substitution in the second term of (46.15), we have:

$$\frac{\partial W_y}{\partial t} = \rho_0 \int \dot{\varphi}\,\Delta\varphi\, d\tau + \frac{1}{4\pi}\int \Phi\,\Delta\dot{\Phi}\, d\tau. \tag{46.17}$$

Combining with (46.13) and making use of Green's theorem

$$\int\limits_V (\operatorname{grad}\varphi \operatorname{grad}\psi + \varphi\Delta\psi)\, d\tau = \oint\limits_S \varphi \frac{\partial\psi}{\partial n}\, df,$$

we have, assuming

$$\varphi = \dot{\varphi}, \quad \varphi = \Phi,$$

$$\psi = \varphi, \quad \psi = \dot{\Phi},$$

$$\frac{\partial (W_k + W_y + W_{эл})}{\partial t} = \oint \left(\rho_0\dot{\varphi}\frac{\partial\varphi}{\partial n} + \frac{1}{4\pi}\Phi\frac{\partial\dot{\Phi}}{\partial n}\right) df, \tag{46.18}$$

from which the density of flow on the total energy is defined

$$S = \rho_0\dot{\varphi}\operatorname{grad}\varphi + \frac{1}{4\pi}\Phi\operatorname{grad}\dot{\Phi}. \tag{46.19}$$

It is easy to generalize this result to the case of the presence of friction in the motion of an electron fluid. We shall consider the force of friction as being proportional to the velocity with the phenomenologically introduced coefficient of proportionality α. Hence, the third equation of system (46.10), taking into account the friction, has the form

$$\frac{\partial\varphi}{\partial t} + \alpha\varphi + \frac{e}{m}\Phi + \frac{v_T^2}{\rho_0}\sigma = 0. \tag{46.20}$$

This involes a change as well in the expression for the flow. In the first equation of (46.16), the term $-\frac{\rho_0}{v_T^2}\alpha\varphi$, is added, and in (45.15) $\alpha\rho_0\int\varphi\Delta\dot{\varphi}\,d\tau$; using Green's theorem

$$\alpha\rho_0\int\dot{\varphi}\Delta\varphi\,d\tau = -\alpha\rho_0\int(\mathrm{grad}\,\varphi)^2\,d\tau + \alpha\rho_0\oint\varphi\frac{\partial\dot{\varphi}}{\partial n}\,df. \tag{46.21}$$

The first term must obviously be interpreted as the spatial loss of energy due to friction, and the second term is added in the original expression (46.19) for the flow.

Hence, when friction is present, the expression for the energy flow is:

$$S = \rho_0\dot{\varphi}\,\mathrm{grad}\,\varphi + \frac{1}{4\pi}\Phi\,\mathrm{grad}\,\dot{\Phi} + \alpha\rho_0\varphi\,\mathrm{grad}\,\varphi. \tag{46.22}$$

Section 47. Surface and Volume Oscillations

1. Equations for longitudinal oscillations. The general solution for a longitudinal field will be sought in the form of Fourier integrals:

$$\left.\begin{aligned}\sigma &= \int_{-\infty}^{+\infty}\sigma_\omega e^{i\omega t + ik_\omega r}\,d\omega,\\ \varphi &= \int_{-\infty}^{+\infty}\varphi_\omega e^{i\omega t + ik_\omega r}\,d\omega,\\ \Phi &= \int_{-\infty}^{+\infty}\Phi_\omega e^{i\omega t + ik_\omega r}\,d\omega.\end{aligned}\right\} \tag{47.1}$$

The character of the solutions is defined by the dependence of ω on k_ω, that is, the law of dispersion of the longitudinal waves.

Substituting (47.1) in system (46.10) gives the following equations to determine the amplitudes:

$$\left.\begin{aligned} -k_\omega^2 \varphi_\omega &= -\frac{i\omega}{\rho_0}\sigma_\omega, \\ -k_\omega^2 \Phi_\omega &= -4\pi\frac{e}{m}\sigma_\omega, \\ i\omega\varphi_\omega + \frac{e}{m}\Phi_\omega + \frac{v_T^2}{\rho_0}\sigma_\omega &= 0. \end{aligned}\right\} \tag{47.2}$$

The condition for the solvability of this system is

$$\begin{vmatrix} \frac{i\omega}{\rho_0} & -k_\omega^2 & 0 \\ 4\pi\frac{e}{m} & 0 & -k_\omega^2 \\ \frac{v_T^2}{\rho_0} & i\omega & \frac{e}{m} \end{vmatrix} = 0. \tag{47.3}$$

The roots of this equation give dispersion laws of the form:

$$1)\ k_\omega^2 = 0. \tag{47.4}$$

$$2)\ k_\omega^2 = \frac{\omega^2 - \omega_0^2}{v_T^2}, \tag{47.5}$$

where

$$\omega_0^2 = \frac{4\pi N e^2}{m}.$$

The first root, after insertion in (47.2), gives the following relationship between the amplitudes:

$$\sigma_\omega = 0, \quad \varphi_\omega = -\frac{e}{mi\omega}\Phi_\omega. \tag{47.6}$$

Physically, this solution corresponds to a trivial case, the translation of the entire plasma as a whole without any changes in density. The relationships among the amplitudes of (47.6) express nothing but the equation of motion for this case.

The second root gives the dispersion law for longitudinal

oscillations. For $\omega < \omega_0$, $k_\omega^2 < 0$, that is, k_ω — is purely imaginary. In this way, we do not obtain solutions of the type of shifting waves for frequencies from zero to ω_0. The solution is similar to the case of static polarization, when all the quantities characterizing the polarization state of the medium depend exponentially on the coordinates.

For $\omega = 0$, that is, in the static case, the distance in which the field decreases e times, is equal to

$$|k_\omega| = \frac{\omega_0}{v_T} = \sqrt{\frac{4\pi Ne^2}{mv_T^2}}, \tag{47.7}$$

which coincides with the familiar Debye polarization formula, if we take as the equation of state, which has been left unspecified, the equation for an ideal gas. In this case

$$v_T^2 = \frac{kT}{m} \quad \text{и} \quad |k_\omega| = \sqrt{\frac{4\pi Ne^2}{kT}}. \tag{47.8}$$

The physical meaning of the exponential solutions is obvious; they correspond to polarization induced by the varying field (see examples below) in the neighborhood of external charges or boundaries.

For $\omega > \omega_0$, the polarized state of the medium is propagated in space from the source of polarization in the form of longitudinal waves with a definite dispersion law.

For the case of the second root, the relationships among the amplitudes will be

$$\left.\begin{aligned} \sigma_\omega &= \frac{k_\omega^2}{4\pi \frac{e}{m}} \Phi_\omega, \\ \varphi_\omega &= \frac{i\omega}{4\pi \frac{e}{m} \rho_0} \Phi_\omega. \end{aligned}\right\} \tag{47.9}$$

Hence, the general solutions is composed of an exponential and a wave solution. For example, we have for Φ in one dimension:

$$\Phi(x, t) = \int_0^{+\omega_0} \Phi_\omega e^{i\omega t \pm \frac{x\sqrt{\omega_0^2 - \omega^2}}{v_T}} d\omega + \int_{\omega > \omega_0} \Phi_\omega e^{i\omega t \pm \frac{ix\sqrt{\omega^2 - \omega_0^2}}{v_T}} d\omega. \tag{47.10}$$

Exponential solutions arise where there are external charges, or on the bounding surfaces. In the latter case, the bounding surfaces may or may not be charged. For the existence of solutions, it is only necessary that

$$-\frac{\partial \Phi}{\partial n} = E_n \neq 0,$$

which last may take place either because of the presence of external charges on the surface of the region or because of the charges of the plasma itself, as they separate out from the volume under consideration and form an uncompensated charge in the neighborhood of the surface. In the latter case, we are justified in interpreting the exponential solutions as singular surface oscillations of the plasma.

The wave solutions likewise may be linked either with the presence of charges outside of the region under consideration, or with characteristic charges of the plasma, formed under the influence of fluctuations or by other causes. Since they represent a polarized state of the medium, which is propagated within the volume, we are justified in interpreting them as spatial characteristic oscillations of the plasma.

Thus, the spectrum of characteristic oscillations of plasma is divided into 2 parts: from zero to ω_0 (surface oscillations) and from ω_0 up (volume oscillations).

It is quite characteristic that over a large interval of wave lengths, the frequencies of the space oscillations are only slightly dependent on the wave length. For wave lengths from $\lambda = \infty$ to $\lambda = \lambda_0 = 2\pi \frac{v_T}{\omega_0}$, ω varies from zero to $\sqrt{2}\omega_0$.

We consider a concrete example: the excitation of plasma by a charged ball whose charges varies periodically with the time.

Because of the spherical symmetry of the problem, we shall seek a solution in the form $\sigma = \sigma_\omega \frac{1}{R} e^{i\omega t - i\kappa R}$ (similarly for Φ and φ). Substitution in the system (46.10) gives the following equations to determine the amplitudes:

$$\left.\begin{aligned} \varkappa_\omega^2 \varphi_\omega &= -\frac{i\omega}{\rho_0}\sigma_\omega, \\ \varkappa_\omega^2 \Phi_\omega &= -4\pi\frac{e}{m}\sigma_\omega, \\ i\omega\varphi_\omega + \frac{e}{m}\Phi_\omega &+ \frac{v_T^2}{\rho_0}\sigma_\omega = 0. \end{aligned}\right\} \tag{47.11}$$

The solvability condition for this system gives the expression

$$\varkappa_\omega^2 = \frac{\omega_0^2 - \omega^2}{v_T^2}.$$

The relationship between the amplitudes will be:

$$\left.\begin{aligned} \sigma_\omega &= -\frac{\varkappa_\omega^2}{4\pi\frac{e}{m}}\Phi_\omega, \\ \varphi_\omega &= \frac{i\omega}{4\pi\frac{e}{m}}\Phi_\omega. \end{aligned}\right\} \tag{47.12}$$

Φ_ω is determined from the boundary conditions. Assuming that as $R \to a \;\; \frac{\partial \Phi}{\partial r} \to \frac{\varepsilon}{a^2}$ (where a is the radius and ε the charge of the sphere), we have:

$$\Phi_\omega = \frac{\varepsilon}{1 + \varkappa_\omega \cdot a} e^{-\varkappa_\omega a}. \tag{47.13}$$

Accordingly, the formula for the potential of the field of the sphere with a pulsating charge will be for $\omega < \omega_0$

$$\Phi = \frac{\varepsilon}{1 + \frac{a}{v_T}\sqrt{\omega_0^2 - \omega^2}} \cdot \frac{1}{R} e^{-\frac{\sqrt{\omega_0^2 - \omega^2}}{v_T}(R-a) + i\omega t}, \tag{47.14}$$

and for $\omega > \omega_0$

$$\Phi = \frac{\varepsilon}{1+\frac{a^2}{v_T}(\omega^2-\omega_0^2)}\frac{1}{R}\left\{\cos\left(\omega t-\frac{\sqrt{\omega^2-\omega_0^2}}{v_T}\cdot(R-a)\right)+\right.$$

$$\left.+\frac{a}{v_T}\sqrt{\omega^2-\omega_0^2}\sin\left(\omega t-\frac{\sqrt{\omega^2-\omega_0^2}}{v_T}\cdot(R-a)\right)\right\}. \tag{47.15}$$

Hence, for frequencies above ω_0, the Debye polarization becomes unstable. The charge changing periodically with time becomes the source of longitudinal spherical waves of polarization.

Introducing the wave number of these waves $ik_\omega = \varkappa_\omega$, we have the dispersion law for longitudinal waves:

$$k_\omega^2 = \frac{\omega^2-\omega_0^2}{v_T^2}. \tag{47.16}$$

Using the expression for an energy

$$S_R = \rho_0 \cdot \dot{\varphi}\frac{\partial\varphi}{\partial R} + \frac{1}{4\pi}\Phi\frac{\partial\dot{\Phi}}{\partial R}, \tag{47.17}$$

we define the energy lost by the charge per second.

Substituting in (47.17) the expression for Φ from (47.15), and correspondingly for φ, we find:

$$-\frac{dW}{dt} = 4\pi R^2\overline{S_R} = \frac{\varepsilon_2}{1+\frac{a^2(\omega^2-\omega_0^2)}{v_T^2}}\cdot\frac{\omega\sqrt{\omega^2-\omega_0^2}}{v_T}\left(\frac{\omega^2}{\omega_0^2}-1\right), \tag{47.18}$$

where the bar indicates the mean value for the period.

Expression (47.18) shows that for frequencies $\omega^2-\omega_0^2 \sim \omega_0^2$ and for

$$a \ll \frac{v_T}{\omega_0} = R_0,$$

we have:

$$\frac{dW}{dt} \sim \frac{\varepsilon^2}{R_0}\omega_0, \tag{47.19}$$

and since $\frac{\varepsilon^2}{R_0}$ represents the energy of the Debye polarization for the standard case, the last formula indicates the great intensity of radiation of polarization waves. The radiation per period is on the order of the energy of the Debye sphere of polarization, that is, for

$$\frac{dW}{dt} \sim \frac{\varepsilon^2}{R_0}\omega_0\left(\frac{R_0}{a}\right)^2, \tag{47.20}$$

that is, the intensity of radiation is respectively $\left(\frac{R_0^3}{a}\right)$ times less.

2. <u>Equations for transverse oscillations</u>. The initial system of equations for the transverse field in an electron plasma is

$$\operatorname{div} \boldsymbol{E} = 0, \quad \operatorname{div} \boldsymbol{H} = 0,$$
$$\operatorname{rot} \boldsymbol{E} = -\frac{1}{c}\dot{\boldsymbol{H}}, \ \operatorname{rot} \boldsymbol{H} = \frac{1}{a}\dot{\boldsymbol{E}} + \frac{4\pi}{c}\rho_0\frac{e}{m}\boldsymbol{u}, \ \dot{\boldsymbol{u}} = \frac{e}{m}\boldsymbol{E}. \tag{47.21}$$

For a transverse field it is sufficient to introduce only a single vector potential.

Let us set

$$\boldsymbol{E} = -\frac{1}{c}\dot{\boldsymbol{A}}, \quad \boldsymbol{H} = \operatorname{rot} \boldsymbol{A}, \quad \operatorname{div} \boldsymbol{A} = 0, \tag{47.22}$$

in which case the equations with divergences are satisfied. In order to satisfy the third and fourth equations, we must have

$$\Delta \boldsymbol{A} - \frac{1}{c^2}\ddot{\boldsymbol{A}} = -\frac{4\pi}{c}\rho_0\frac{e}{m}\boldsymbol{u},$$

but thanks to the equation of motion

$$\dot{\boldsymbol{u}} = \frac{e}{m}\boldsymbol{E} = -\frac{e}{mc}\dot{\boldsymbol{A}},$$

from which

$$\boldsymbol{u} = -\frac{e}{m}\boldsymbol{A} + \text{const}\,(x, y, z).$$

The undetermined constant takes into consideration the possibility of magnetostatic fields in the plasma. Assuming their absence (that is for $\underline{u} = 0$, $\underline{A} = 0$), we have:

$$u = -\frac{e}{mc}A. \qquad (47.23)$$

Thus, the initial system of equations is replaced by an equivalent system of equations for a single vector potential.

$$\ddot{A} + \omega_0^2 A - c^2 \Delta A = 0, \quad \operatorname{div} A = 0, \qquad (47.24)$$

where $\omega_0^2 = 4\pi \frac{e^2}{m^2} \cdot \rho_0$. For $\rho_0 \to 0$, the equations obtained go over into the usual system of equations for transverse fields in a vacuum.

The general solution satisfying system (47.24) may be represented in the form of shifting transverse waves, subject to dispersion law

$$\omega^2 = \omega_0^2 + c^2 k_\omega^2. \qquad (47.25)$$

This law of dispersion is analogous to the case of longitudinal waves and gives two forms of solution:

1. If $\omega < \omega_0$, $k_\omega^2 < 0$, then k_ω is a purely imaginary quantity; the functional dependence of the solutions on the coordinate will decrease or increase exponentially.

2. For $\omega > \omega_0$, $k_\omega^2 > 0$ the solution has a wave character. The general solution will be the sum of these two types of solutions. For example, for the one-dimensional case

$$A = \int_0^{\omega_0} A_\omega e^{i\omega t \pm \frac{\sqrt{\omega_0^2 - \omega^2}}{c} x} d\omega + \int_{\omega > \omega_0} A_\omega e^{i\omega t \pm i \frac{\sqrt{\omega^2 - \omega_0^2}}{c} x} d\omega \qquad (47.26)$$

the solutions of the wave type play the role of spatial transverse oscillations, and the exponential solutions, of surface oscillations. The velocity with which the field drops for surface oscillations may be characterized by the distance at which the field is decreased e times. This distance equals

$$\lambda_\omega = \frac{1}{k_\omega} = \frac{c}{\sqrt{\omega_0^2 - \omega^2}} \qquad (47.27)$$

in complete analogy with the case for superficial longitudinal oscillations, only with the velocity of light c appearing instead of the value of the velocity ν_T[1]

For $\omega = 0$

$$\lambda_0 = \frac{c}{\omega_0}. \qquad (47.28)$$

This quantity plays the role of a Debye length, but only for the transverse field. Since $E = -\frac{1}{c}\dot{A}$ and the electric field (when $\omega \to 0$) disappears, the length λ_0 characterizes the velocity with which the magnetostatic field increases in the plasma. The fact that magnetic fields in electron plasma do not penetrate into the electron plasma deeply, but only to a distance λ_0 from the surface, indicates the anomalously strong diamagnetic properties of electron plasma. Of course, the last conclusion will apply only if the frequencies concerned satisfy the postulates that lie at the basis of the initial equations (ignoring the connection of electron plasma with ion plasma and neutral particles, which is valid for frequencies that are not too low).

When the region $\sim \frac{c}{\omega_0}$ is ignored, the magnetic state of electron plasma up to frequencies $\omega < \omega_0$ may be described phenomenologically by saying that the magnetic field is absent in it.

We point to the weighty consequences to which this leads:

1. The slight dependence of the frequency spectrum of the volume oscillations on the boundary cnditions. For in the wave length region for which $c^2k^2 \ll \omega_0^2$, the term $c^2\Delta A$ in system (47.24) may be ignored, which leads to the independence of the solutions from the boundary conditions. In the region where $c^2k^2 < \omega_0^2$, obviously the influence of the boundary conditions is likewise small. In the case $c^2k^2 \sim \omega_0^2$ or $c^2k^2 < \omega_0^2 c$ the structure of the frequency spectrum is essentially dependent on the boundary conditions. The facts just cited can be easily illustrated in the case of volume oscillations of plasma in a cube with reflecting walls. In this case, as we know,

$$k_x = \frac{2\pi}{L} n_x, \quad k_y = \frac{2\pi}{L} n_y, \quad k_z = \frac{2\pi}{L} n_z,$$

where n_x, n_y, n_z are integers.

Consequently,

$$\omega^2_{n_1 n_2 n_3} = \omega_0^2 + c^2\left(\frac{2\pi}{L}\right)^2 (n_x^2 + n_y^2 + n_z^2).$$

For $c^2k^2 < \omega_0^2$ the frequencies practically are independent of the last term, which is determined by the dimensions of the region.

2. The change with time of the vortical perturbations in the plasma had an oscillatory character. That is, the perturbational formation, which fundamentally comprises wave lengths $> \frac{c}{\omega_0}$, perform oscillations with a frequency ω_0 and disperses only slowly as compared with the period of the oscillations.

We consider the case in which the region of the vortical perturbation set up in the plasma is large as compared with the characteristic distance $\frac{c}{\omega_0}$, that is, we assume that the wavelengths of the harmonics into which the given perturbation can be decomposed considerably exceeds $\frac{c}{\omega_0}$ (or $k \ll \frac{\omega_0}{c}$).

We can then find the general character of the change in the perturbation as time goes on, writing the dispersion equation in the form

$$\omega = \omega_0 \sqrt{1 + \frac{c^2k^2}{\omega_0}} \approx \omega_0 + \frac{1}{2}\frac{c^2}{\omega_0}k^2, \tag{47.29}$$

while for the velocity, we have:

$$u(x, y, z, t) = e^{i\omega_0 t} \int e^{\frac{ic^2}{2\omega_0}k^2 t + ikr} c(k)\, dk = e^{i\omega_0 t} f(x, y, z, t), \tag{47.30}$$

where c(k) is defined by the initial distribution of velocities (for t = 0). The integral in it coincides in form with the integral that appears in the well-known problem for the dispersion of wave packets for free motion in quantum mechanics, as is obvious from the fact that the function f (x, y, z, t) is the general solution of the equation

$$\Delta f = i\frac{2\omega_0}{c^2}\frac{\partial f}{\partial t}, \tag{47.31}$$

which is formally mathematically analogous to the wave equation for the free motion of a particle, merely substituting $\frac{m}{\hbar}$ for $\frac{\omega_0}{c^2}$. We therefore need not perform the calculations, but we make use directly in our problem of the well-known formula for the velocity of scattering of packets, substituting $\frac{\omega_0}{c^2}$ for $\frac{m}{\hbar}$ in it. Thus we have for the effective width of the region of perturbations at time t:

$$\delta_t^2 = \delta_0^2 + \left(\frac{c^2}{\delta_0 \omega_0}\right)^2 t^2, \tag{47.32}$$

where the δ_0 are the linear dimensions of inhomogeneity at the initial moment of time.

Consequently, the vortical perturbation set up in the plasma vibrates with the frequency ω_0 and, in addition, scatters. We can characterize the rate of scattering by the time in the course of which the dimensions are doubled. For this time, we have (substituting in (47.32) $\delta_t = 2\delta_0$):

$$\tau = \sqrt{3}\delta_0^2 \frac{\omega_0}{c^2}, \tag{47.33}$$

and since by assumption the linear dimensions of the region of perturbation are greater than $\frac{c}{\omega_0}$, we find that the time for the dispersion is greater than the period of the pulsations $T_0 = \frac{2\pi}{\omega_0}$:

$$\frac{\tau}{T_0} = \frac{\sqrt{3}}{2\pi} \cdot \frac{\omega_0^2}{c^2} \delta^2 \gg 1. \tag{47.34}$$

Hence, the perturbations under consideration vibrate with the frequency ω_0, and scatter slowly, as compared with the period of the oscillations.

A similar state of affairs holds true as well for the case of longitudinal oscillations.

Section 48. Method of Exciting Plasma Oscillations.

We desire to discuss two methods of exciting oscillations.

1. Excitation of oscillations by a moving charge. Condition for formation of oscillations and of spectra of radiation frequencies. We consider the change in the character of the polarization of plasma in the case when a point charge moves uniformly in a straight line through it with the velocity v_0. Let the motion occur along the z axis. The

distribution of the density of the charge of the moving particle may be described by means of Dirac's Delta Function, namely:

$$\rho_e(x, y, z, t) = \varepsilon\delta(x)\delta(y)\delta(z - vt). \tag{48.1}$$

The problem reduces to the solution of a system of equations

$$\left.\begin{aligned} \Delta\varphi &= -\frac{1}{\rho_0}\frac{\partial\sigma}{\partial t}, \\ \Delta\Phi &= 4\pi\frac{e}{m}\sigma - 4\pi\rho_e(x, y, z, t), \\ \frac{\partial\varphi}{\partial t} &+ \frac{e}{m}\Phi + \frac{v_T^2}{\rho_0}\sigma = 0 \end{aligned}\right\} \tag{48.2}$$

with the corresponding member ρ_e, which represents the given function of coordinate and time.

The general solution of system (48.2) will be sought, as previously, in the form of Fourier integrals for all the unknown functions:

$$\Phi = \int_{-\infty}^{+\infty}\Phi_\omega e^{i\omega t}\,d\omega,\quad \varphi = \int_{-\infty}^{+\infty}\varphi_\omega e^{i\omega t}\,d\omega,\quad \sigma = \int_{-\infty}^{+\infty}\sigma_\omega e^{i\omega t}\,d\omega. \tag{48.3}$$

We represent the given function $\rho_e(x, y, z, t)$ likewise in the form of the Fourier integral

$$\rho_e(x, y, z, t) = \int_{-\infty}^{+\infty}\rho_{e\omega}e^{i\omega t}\,d\omega. \tag{48.4}$$

The Fourier components of this function are

$$\rho_{e\omega} = \frac{\varepsilon}{2\pi}\int_{-\infty}^{+\infty}\delta(x)\cdot\delta(y)\cdot\delta(z - v_0 t)e^{-i\omega t}\,dt = \frac{\varepsilon}{2\pi v_0}\delta(x)\cdot\delta(y)e^{-i\frac{\omega}{v_0}z}. \tag{48.5}$$

Substituting the proposed solutions in our initial equations, we get a system of equations with respect to the Fourier components:

$$\left.\begin{aligned} \Delta\varphi_\omega &= -\frac{i\omega}{\rho_0}\sigma_\omega, \\ \Delta\Phi_\omega &= -4\pi\frac{e}{m}\sigma_\omega - \frac{2\varepsilon}{v_0}\delta(x)\delta(y)e^{-i\frac{\omega}{v_0}z}, \\ i\omega\varphi_\omega &+ \frac{e}{m}\Phi_\omega + \frac{v_T^2}{\rho_0}\sigma_\omega = 0. \end{aligned}\right\} \tag{48.6}$$

Because of the presence of a specified direction of an axis, we employ a cylindrical system of coordinates (ρ, φ, z). In these coordinates, expression (48.5) is

$$\rho_{e\omega} = \rho_{e\omega}(z, \rho) = \frac{e}{2\pi} \frac{\delta(\rho)}{v_0 \rho} e^{-i\frac{\omega}{v_0}z}. \tag{48.7}$$

The equation for the Fourier component for the change of density σ_ω is obtained from system (48.6) by eliminating φ_ω and Φ_ω, and is (outside of the singular point):

$$\Delta\sigma_\omega + k_\omega^2 \sigma_\omega = 0; \quad k_\omega^2 = \frac{\omega^2 - \omega_0^2}{v_T^2}, \tag{48.8}$$

or, in the cylindrical system of coordinates

$$\frac{\partial^2 \sigma_\omega}{\partial z^2} + \frac{\partial^2 \sigma_\omega}{\partial \rho^2} + \frac{1}{\rho} \cdot \frac{\partial \sigma_\omega}{\partial \rho} + k_\omega^2 \sigma_\omega = 0. \tag{48.9}$$

The solution of this equation is

$$\sigma_\omega = C_\omega Z_0 (\sqrt{k_\omega^2 - h_\omega} \rho) e^{-ihz}, \tag{48.10}$$

where Z_0 is a cylindrical function, as yet not specified, satisfying Bessel's equation

$$\frac{\partial^2 Z_0}{\partial \rho^2} + \frac{1}{\rho} \frac{\partial Z_0}{\partial \rho} + (k_\omega^2 - h_\omega^2) Z_0 = 0$$

To determine Φ_ω, we employ Poisson's equation and since the right side of this equation is a cylindrical function, then we obtain a solution, setting

$$\Phi_\omega = A_\omega Z_0 \cdot (\sqrt{k_\omega^2 - h_\omega^2} \rho) e^{ihz}, \tag{48.11}$$

in which case $\Delta\Phi_\omega = -k_\omega^2 \Phi_\omega$, and Poisson's equation reduces to the following equation:

$$k_\omega^2 \Phi_\omega = 4\pi \frac{e}{m} \sigma_\omega + \frac{2}{\pi} \frac{e}{v_0 \rho_0} \delta(\rho) e^{-i\frac{\omega}{v_0}z}. \tag{48.12}$$

In connection with expression (48.10) and (48.11), this equation determines h_ω and the connection between C_ω and A_ω, namely,

$$h_\omega = \frac{\omega}{v_0}, \qquad C_\omega = A_\omega \frac{k_\omega^2}{4\pi \frac{e}{m}}. \tag{48.13}$$

The Fourier amplitude φ_ω is determined from the continuity equation

$$\Delta\varphi_\omega = -\frac{1}{\rho_0} i\omega\sigma_\omega. \tag{48.14}$$

Setting here, too,

$$\varphi_\omega = B_\omega Z_0\left(\sqrt{k_\omega^2 - h_\omega^2}\,\rho\right) e^{-ihz}, \tag{48.15}$$

we get:

$$B_\omega = \frac{i\omega}{\rho_0 k_\omega^2} C_\omega = \frac{i\omega}{4\pi \frac{e}{m}\rho_0} A_\omega. \tag{48.16}$$

Hence, the solution of the initial system, expressed in cylindrical functions, is

$$\left.\begin{aligned}
\Phi_\omega &= A_\omega Z_0\left(\sqrt{k_\omega^2 - h_\omega^2}\,\rho\right) e^{-i\frac{\omega}{v_0}z},\\
\varphi_\omega &= A_\omega \frac{i\omega}{4\pi \frac{e}{m}\rho_0} Z_0\left(\sqrt{k_\omega^2 - h_\omega^2}\,\rho\right) e^{-i\frac{\omega}{v_0}z},\\
\sigma_\omega &= A_\omega \frac{k_\omega^2}{4\pi \frac{e}{m}} Z_0\left(\sqrt{k_\omega^2 - h_\omega^2}\,\rho\right) e^{-i\frac{\omega}{v_0}z}.
\end{aligned}\right\} \tag{48.17}$$

This expression should vanish at infinity, and at zero should have a singular point (presence of a point charge). Such properties are possessed by Hankel's functions of the first and second class

$$H_0^{(1)}(i\varkappa_\omega \rho) \qquad \text{and} \qquad H_0^{(2)}(\varkappa_\omega \rho), \tag{48.18}$$

where

$$\varkappa_\omega = \left|\sqrt{k_\omega^2 - h_\omega^2}\right|.$$

Since we have not made use as yet of the conditions for the behavior of the functions in the neighborhood of the singular point, the solution contains an undetermined constant. This constant may be determined in a suitable manner (see Tamm[+]). However, we conclude the exposition of the mathematical side of the problem in question at this point. For our purposes, in this part of the work (determination of the conditions for obtaining a radiation vector and its character), the result obtained is sufficient.

We consider the solution in the form (48.17).

1. The form of the equation enables us to establish at once the possibility of exciting the space oscillations of plasma in the form of radiation of longitudinal waves from a charge moving uniformly in a straight line, and in addition to this, the conditions necessary for the formation of this radiation. The nature of the solution expressed by the Hankel function is sharply different for the two cases:

$v_0 > v_T$ and $v_0 < v_T$, depending, that is, on whether the following quantity is greater or less than zero:

$$k_\omega^2 - h_\omega^2 = \frac{\omega^2}{v_T^2} \cdot \left(1 - \frac{v_T^2}{v_0^2}\right) - \frac{\omega_0^2}{v_T^2} = \varkappa_\omega^2. \tag{48.19}$$

Let us assume that $k_\omega^2 - h_\omega^2 < 0$; this always holds good if the velocity of the moving charge does not exceed the velocity of sound in the electron plasma. The solution in this case is expressed by means of a Hankel function of the first class.

For example, for Φ_ω:

$$\Phi = A_\omega \cdot H_0^{(1)}(i\varkappa_\omega \rho); \tag{48.20}$$

for $\varkappa_\omega \rho \gg 1$, the asymptotic form of this function is

$$\Phi_\omega = A_\omega e^{-\varkappa_\omega \rho}. \tag{48.21}$$

Thus, for velocities charge not exceeding the velocity of sound in the electron plasma, the polarization state of the plasma around the moving charge is similar to the Debye polarization for the static case.

[+] I. Tamm, Journ. of Phys. 1, 439 (1939).

We are dealing here with exponentially damped solutions.

The polarized region of the plasma, surrounding the charge, follows its motion, merely changing its form according to the velocity of the moving charge. There is no budding off of polarization, that is in this case there is no emission of radiation.

We consider another case $k_\omega^2 - h_\omega^2 > 0$; on the basis of (48.19), we see that this applies for the case $v_0 > v_T$, and only beginning with a minimum critical frequency. The solutions for Φ_ω in this case are given by Hankel functions of the second class

$$\Phi_\omega = -A_\omega i H_0^{(2)}(\varkappa_\omega \rho) \quad \text{if} \quad \omega > 0,$$

$$\Phi_\omega = A_\omega i H_0^{(2)}(\varkappa_\omega \rho) \quad \text{if} \quad \omega < 0. \tag{48.22}$$

For $\varkappa_\omega \rho \gg 1$, we may use the asymptotic representation of these functions

$$\Phi_\omega = -A_\omega i \sqrt{\frac{\pi}{2\varkappa_\omega \rho}}\, e^{-i\left(\varkappa_\omega \rho - \frac{\pi}{2}\right)} \quad \text{for} \quad \omega > 0 \tag{48.23}$$

and the corresponding complex-conjugate express for $\omega < 0$.

The solutions in this case have a wave character. The charge moving uniformly in a straight line is a source for the emission of polarized waves. The condition for the emision is the requirement that the velocity of the moving charge should exceed the velocity of sound in the electron plasma, v_T .

2. Starting from the radiation condition (48.19), we establish the characteristic property of dependence of the emitted spectrum of frequencies on the velocity of the moving particle.

For a given velocity $v_0 > v_T$ the quantity

$$\frac{\omega^2}{v_T^2}\left(1 - \frac{v_T^2}{v_0^2}\right) - \frac{\omega_0^2}{v_T^2} = S^2$$

has a positive value, beginning only with a certain minimum frequency.

We get the value of this minimum frequency by setting $S^2 = 0$, from which

$$\omega_{\min} = \frac{\omega_0}{\sqrt{1 - \frac{v_T^2}{v_0^2}}}. \tag{48.24}$$

There is no emission of lower frequencies, since in this case $S^2 < 0$ and we are concerned with exponential functions, and all the higher frequencies are comprised in the radiation. Accordingly, the emission spectrum of the longitudinal polarization waves has the peculiarity that it is limited on the side of lower frequencies. As the velocity v_0 increases, the minimum frequency of radiation approaches the frequency ω_0.

In this point, we have considered only the condition for the excitation of oscillations and the characteristics of the spectrum emitted by the moving charge. We shall return later to this problem, but we now consider it in a different aspect.

2. Excitation of longitudinal oscillations by transverse waves in an inhomogeneous plasma. In linear approximation, longitudinal and transverse fields in a homogeneous plasma are described independently of each other. Interaction between them occurs only, first in a uniform medium when nonlinear terms are taken into consideration and secondly, in the case of an inhomogeneous medium, when the plasma density is a function of the coordinates.

We consider this second case. We assume that we are dealing with a neutral plasma, that is, we assume that at every microscopic point of the plasma, in the stationary case, the positive charges are compensated by negative charges. We assume that in the stationary state the total density of the mass is given by the specified function of the coordinates.

The dynamic conditions that guarantee that the inhomogeneous plasma is stationary may be expressed by introducing the field of "side forces," which in the stationary case satisfy the relationship

$$\operatorname{grad} P(\rho_0) + F_{\text{стор}} = 0. \tag{48.25}$$

When there are perturbations in the electron plasma, we shall consider that the concentration of the ion plasma and of the neutral particles remains unchanged. This assumption is justifiable because of the great mass of these plasmas. In other terms, we assume that condition (48.25) is retained unchanged, even in the presence of perturbations in the electron plasma. Under this condition, perturbations of the density in the electron plasma are determined only by the electrical forces and the change of pressure in the plasma itself, due to variations of the density, and does not depend on side forces. Hence, in linear approximation, the equations of motion for electron plasma will have the same form as in a uniform medium, with the difference that the unperturbed density is a function of the coordinates.

We assume in the first approximation

$$\rho=\rho_0+\rho_1 \quad \text{и} \quad P_э=P_э(\rho_0)+\left(\frac{\partial P}{\partial \rho}\right)_{\rho=\rho_0}\cdot\rho_1=P_э(\rho_0)+v_T^2\rho_1, \tag{48.26}$$

where $P(\rho_0)$ is the pressure, relating only to the electron plasma. Under the conditions of a gaseous plasma, it may be held that the pressure of each of the gases (neutral, ion, electron) is proportional to its density, and therefore v_T^2 independent of the density

$$\left(P_э=\frac{kT}{m}\rho\right),$$

and in what is called isothermal plasma does not depend from the coordinates either.

The equation of motion will have the form

$$\rho_0\frac{\partial u_1}{\partial t}=-\operatorname{grad}P_и-\operatorname{grad}P_н-\operatorname{grad}P_{эл}+F_{стор}+\frac{e}{m}E_1, \tag{48.27}$$

where $P_и$, $P_н$, $P_{эл}$ are the partial pressures of the ion, neutral and electron gases.

Taking (48.26) and (48.25) into account, we get:

$$\rho_0\frac{\partial u_1}{\partial t}=-v_T^2\operatorname{grad}\rho_1+\frac{e}{m}\cdot E_1. \tag{48.28}$$

Field $\underline{E}_1$ is determined by the forming charges and currents

$\left(\rho_э=\frac{e}{m}\rho,\ j=\rho_0\frac{e}{m}u_1\right)$ by means of Maxwell's equations, from which, making

use of (48.28), we obtain the following system of equations for the longitudinal and transverse fields in electron plasma:

$$\left.\begin{aligned}\ddot{\rho}_1+\omega_0^2\rho_1-v_T^2\Delta\rho_1&=-E_1\operatorname{grad}\rho_0\frac{e}{m},\\ \ddot{\Omega}+\omega_0^2\Omega-c^2\Delta\Omega&=-\left[E_1\operatorname{grad}\rho_0\frac{e}{m}\right],\end{aligned}\right\} \tag{48.29}$$

where the quantity ρ_1 characterizes the longitudinal field

$$\rho_1 = \frac{1}{4\pi\frac{e}{m}} \operatorname{div} \boldsymbol{E}_1. \tag{48.30}$$

and the vector

$$\boldsymbol{\Omega} = \frac{1}{4\pi\frac{e}{m}} \operatorname{rot} \boldsymbol{E}_1 \tag{48.31}$$

characterizes the transverse field,

$$\omega_0^2 = 4\pi \frac{e^2}{m^2} \rho_0 (x, y, z). \tag{48.32}$$

Equations (48.29) show that in regions where ρ_0 = const., the connection between longitudinal and transverse fields is absent, in accordance with the previous results. When there is inhomogeneity, there is an interaction between these two types of field, that is, one of them induces the other. Since the total field is $\boldsymbol{E}_1 = \boldsymbol{E}_1^l + \boldsymbol{E}_1^{tr}$, the presence of only the transverse field $\boldsymbol{E}_1^{tr}$, as will be seen from the equation for ρ_1, entails the appearance of the longitudinal field as well. The equation for Ω shows that a similar situation occurs for the formation of a transverse field under the influence of the longitudinal field.

The solution of the problem interesting us, the formation of a longitudinal field in plasma under the influence of the transverse field, may be obtained in approximation, if we assume that for t = 0, only the transverse field is given, and then substitute it in the right side (48.29). For a harmonically variable $\boldsymbol{E}_1^{tr}$ (frequency ω), we get:

$$\Delta\rho_{10} + \frac{\omega^2 - \omega_0^2}{v_T^2}\rho_{10} = -\frac{1}{v_T^2} (\boldsymbol{E}_{10}^{tr} \operatorname{grad} \rho_0) \frac{e}{m}, \tag{48.33}$$

where $\boldsymbol{E}_{10}^{(tr)}$, ρ_{10} are the amplitudes of the corresponding quantitites $\boldsymbol{E}_1^{tr}$, ρ_1. Equation (48.33) also makes it possible, with a known inhomogeneity $\rho_0 = \rho_0 (x, y, z)$ and a known amplitude for the transverse field $\boldsymbol{E}_{10}^{(tr)}$ to define the fundamental characteristics of longitudinal field ρ_1.

We shall illustrate all that has been said above for the example of the formation of longitudinal oscillations under the influence of an electromagnetic wave, falling on the boundary of the plasma in the case where the plasma occupies the half-space.

Let ρ_0 be defined by the condition for

$$\rho=\rho_0=\text{const.} \qquad \text{for} \quad z>0, \qquad\qquad \rho=0 \qquad \text{for} \quad z<0. \tag{48.34}$$

Green's function for equation (48.33) is

$$\frac{e^{ik_\omega|r-r'|}}{|r-r'|}, \qquad k_\omega=\sqrt{\frac{\omega^2-\omega_0^2}{v_T^2}}. \tag{48.35}$$

Consequently, the solution may be written as:

$$\rho_1=-\frac{1}{4\pi}\frac{e}{m}\frac{(E_{10}^{tr}\,n_1)}{v_T^2}\int\frac{|\operatorname{grad}\,\rho_0|\,e^{ik_\omega|r-r'|}}{|r-r'|}\,dr' \tag{48.36}$$

($\underline{n}_1$ is the unit vector in the z direction), in which we have assumed that E^{α}_{10} is independent of the coordinates, which is perfectly permissible, if the length of the longitudinal wave in the direction of the z axis considerably exceeds the dimensions of the inhomogeneity in the same dimension.

The discontinuous function $\operatorname{grad}_z \rho_0$ may be expressed by means of Dirac's δ-function

$$\frac{\partial}{\partial z}\,\rho_0(z)=\rho_0\delta(z). \tag{48.37}$$

Substituting (48.37) in (48.36) and integrating, we have:

$$\rho_1(z)=-\frac{1}{4\pi}\cdot\frac{(E^{(tr)}\,n_1)\frac{e}{m}}{v_T^2}\,\rho_0\int\frac{\delta(z')\,e^{ik_\omega\sqrt{y'^2+x'^2+(z-z')^2}}}{\sqrt{x'^2+y'^2+z'^2}}\,dx'dy'dz'=$$

$$=a\int\frac{e^{ik_\omega\sqrt{y'^2+x'^2+z^2}}\,dx'\,dy'}{\sqrt{y'^2+x'^2+z^2}}=2\pi a\int\limits_0^\infty\frac{e^{ik_\omega\sqrt{r^2+z^2}}}{\sqrt{r^2+z^2}}\,rdr=$$

$$= \pi a \int_0^{\infty} \frac{e^{ik\sqrt{t+z^2}}}{\sqrt{t+z^2}}\,dt = 2\pi a \int_z^{\infty} e^{ik_\omega u}\,du \tag{48.38}$$

$$\left(a = -\frac{1}{4\pi}\frac{(E^{tr} n_1)}{v_T^2}\rho_0\frac{e}{m}\right).$$

The integral in (48.38) has a different meaning depending on whether $\omega < \omega_0$ or $\omega > \omega_0$.

In the first case k_ω is imaginary and we have:

$$\rho_1(z) = \frac{(E_{10}^{tr} n_1)\frac{e}{m}\rho_0}{4v_T^2}\,\frac{e^{-|k_\omega| z}}{|k_\omega|}, \quad |k_\omega| = \sqrt{\frac{\omega_0^2-\omega^2}{v_T^2}} \tag{48.39}$$

that is, for frequencies $\omega < \omega_0$, the transverse waves excite superficial longitudinal oscillations of plasma.

For $\omega > \omega_0$

$$\rho_1 = \frac{(E_1^{tr} n_1)\frac{e}{m}\rho_0}{4v_T^2}\,\frac{e^{-i|k_\omega| z}}{i|k_\omega|}; \quad |k_\omega| = \sqrt{\frac{\omega^2-\omega_0^2}{v_T^2}}, \tag{48.40}$$

in which case the longitudinal waves excited are space oscillations of the plasma.

The following are characteristic for both cases:

1. Resonance phenomena for $\omega = \omega_0$.

2. The dependence of the result on the direction of the electrical vector in the transverse wave with respect to the surface. If this vector is directed perpendicular to the surface, the effect is at a maximum.

We proceed to find the energy lost by the transverse wave by exciting the longitudinal wave in the example being considered. The loss of energy in unit time per unit of surface of the section is equal to the energy flow conditioned by the longitudinal wave excited.

We have for the energy flow in this case (see Sec. 46):

$$S_z = \rho_0 \dot{\varphi} \frac{\partial \varphi}{\partial z} + \frac{1}{4\pi} \Phi \frac{\partial \dot{\Phi}}{\partial z}. \tag{48.41}$$

The change in density $\sigma(z, t)$ is, on the basis of (48.40),

$$\rho_1 = \sigma = \frac{(E_{10}^{tr} n)}{4v_T^2} \frac{e}{m} \frac{\rho_0}{k_\omega} \sin(\omega t - k_\omega z) \tag{48.42}$$

(taking the real portion of (48.40), from which by means of

$$\Delta\Phi = -4\pi \frac{e}{m} \sigma, \quad \Delta\varphi = -\frac{1}{\rho_0} \dot{\sigma},$$

we find Φ and φ for $z > 0$.

$$\left.\begin{aligned} \Phi &= \frac{(E_1^{tr} n_1)\frac{e}{m}\rho_0}{4v_T^2} \cdot \frac{\omega_0^2}{k_\omega^2} \sin(\omega t - k_\omega z), \\ \varphi &= \frac{(E_1^{tr} n_1)\frac{e}{m}\rho_0}{4v_T^2} \cdot \frac{\frac{e}{m}\omega}{k_\omega^3} \cdot \cos(\omega t - k_\omega z). \end{aligned}\right\} \tag{48.43}$$

Inserting this in formula (48.41), we have, averaging for the time:

$$S_z = \frac{(E_{10}^{tr} \cdot n_1)^2}{32 v_T^4} \frac{\omega}{k_\omega^5} \frac{e^2}{m^2} \rho_0 (\omega^2 + \omega_0^2). \tag{48.44}$$

The expression obtained for the flow indicates the presence of resonance losses in energy as $\omega \to \omega_0$, and makes it possible to establish the condition for the applicability of the expression for S_z.

Formula (48.44) applies on the condition that the magnitude of the energy flow S_z does not exceed the magnitude of the decreasing flow of energy

$$S_0 = \frac{c}{4\pi} [EH],$$

that is,

$$\frac{S_0}{S_z} = \frac{\frac{32}{8\pi} \cdot (E_{10}^{tr})^2 \cdot c\, (\omega^2 - \omega_0^2)^{\frac{5}{2}}}{\frac{e^2}{m^2} \rho_0 (\omega^2 + \omega_0^2)\, \omega\, (E_\omega^{tr})^2 \cos^2\theta\, v_T} \simeq 2^{\frac{5}{2}} \frac{16c}{v_T} \left(\frac{\Delta\omega}{\omega}\right)^{\frac{5}{2}} \gg 1, \tag{48.45}$$

from which

$$\Delta\omega \sim \frac{2^{\frac{2}{5}}}{16}\left(\frac{v_T}{c}\right)^{\frac{2}{5}} \omega_0.$$

Section 49. Properties of Oscillations Under Various Physical Conditions.

1. Influence of temperature on fading of longitudinal space oscillations. We know that change in time of the deviation of the density from the equilibrium state is defined by the formula

$$\rho(x, y, z, t) = \int_{-\infty}^{+\infty} c(\mathbf{k})\, e^{i\omega t - i\mathbf{k}\mathbf{r}}\, d\mathbf{k}, \tag{49.1}$$

where

$$\omega^2 = \omega_0^2 + v_T^2 k^2, \quad v_T^2 = \frac{3kT}{m}, \quad d\mathbf{k} = dk_x\, dk_y\, dk_z. \tag{49.2}$$

The amplitudes c(k) are defined by the initial value for :

$$\rho(x, y, z, 0) = \int_{-\infty}^{+\infty} e^{-i\mathbf{k}\mathbf{z}}\, c(\mathbf{k})\, d\mathbf{k}. \tag{49.3}$$

Since the law of dispersion (49.2) applies under conditions when the dimensions of inhomogeneity exceed the Debye length, we may set

$$\omega = \omega_0 \sqrt{1 + \frac{v_T^2 k^2}{\omega_0^2}} \cong \omega_0 + \frac{1}{2}\,\frac{v_T^2}{\omega_0}\, k^2, \tag{49.4}$$

and consequently,

$$\rho(x, y, z, t) = e^{i\omega_0 t} \int_{-\infty}^{+\infty} e^{\frac{1}{2}\frac{v_T^2}{\omega_0} k^2 t - i\mathbf{k}\mathbf{r}}\, c(\mathbf{k})\, d\mathbf{k} = e^{i\omega_0 t} f(x, y, z, t). \tag{49.5}$$

The integral in this expression is analogous to the one we have in discussing the scattering of transverse perturbations, except that instead of the velocity of the velocity of light c, we have here

$v_T = \sqrt{\frac{3kT}{m}}$. Consequently, the spatial dimensions of the inhomogeneity at moment of time t will be

$$\delta_t^2 = \delta_0^2 + \left(\frac{v_T^2}{\delta_0 \omega_0}\right)^2 t^2, \qquad (49.6)$$

where δ_0 are the linear dimensions of the inhomogeneity at the initial moment.

Thus, the inhomogeneity of density of distribution of electrons set up vibrates with a frequency and in addition scatters. The velocity of scattering again can be characterized by the time in the course of which the dimensions are multiplied by two; for this time τ we have:

$$\tau = \sqrt{3}\delta_0^2 \frac{\omega_0}{v_T^2} = \delta_0^2 \frac{\sqrt{3}}{3} \frac{\sqrt{4\pi N c^2 m}}{kT}. \qquad (49.7)$$

The formula obtained shows the influence of temperature on the speed of fading of vortex-free perturbations.

Since by assumption, only perturbed regions are considered with dimensions greater than the Debye length, then automatically, the time of fading τ is greater than the period of dispersion $\tau_0 = \frac{2\pi}{\omega_0}$, since

$$\frac{\tau}{\tau_0} = \frac{\sqrt{3}}{2\pi} \frac{\omega_0^3}{v_T^2} \delta_0^2 = \frac{\sqrt{3}}{6\pi} \left(\frac{\delta_0}{d}\right)^2, \qquad (49.8)$$

where

$$D = \sqrt{\frac{kT}{4\pi N e^2}}$$

is the Debye length.

Consequently, the inhomogeneity of the electron density established in any way vibrates and slowly (compared to the period of oscillation) fades.

2. Presence of a double layer of fluctuational origin on the boundary of the plasma, and definition of its properties. a. Elementary considerations. We consider the displacement of a volume of electron plasma on the plasma-vacuum boundary. Let the displacement occur normally to the boundary; for simplicity's sake, we choose the form of the

displaced volume in the form of a parallelopiped with width d. Assuming the displacement to be small as compared with d, we have for the intensity of the electric field formed:

$$E = 4\pi\sigma = 4\pi eN\xi, \tag{49.9}$$

where N is the electron density in the plasma. The equation of motion will be

$$m\ddot{\xi} = eE,$$

or

$$\ddot{\xi} + \omega_0^2\xi = 0, \tag{49.10}$$

where

$$\omega_0^2 = 4\pi N \frac{e^2}{m}. \tag{49.11}$$

When there is thermal equilibrium

$$\omega_0^2\overline{\xi^2} = kT, \tag{49.12}$$

from which

$$\overline{\xi^2} = \frac{kT}{m}\frac{1}{\omega_0^2}. \tag{49.13}$$

The side of the space being translated remains undetermined. Obviously, to greater d there corresponds higher velocities for the propagation of perturbations in the plasma. However, as as result of taking into consideration the compressibility of the medium, the polarization actions

in electron plasma are propagated to a distance $\frac{v_0}{\omega_0}$, where

$$v_0 = \sqrt{\frac{kT}{m}} \tag{49.14}$$

because of Debye screening. We may therefore set

$$d = \frac{v_0}{\omega_0}. \qquad (49.15)$$

This fact leads to the idea of a fluctuational double layer with a thickness equal to the length of the Debye polarization, due to oscillation according to (49.10) with a frequency ω_0.

The sudden change in the potential energy upon passing through the double layer gives

$$eV = eEd = 4\pi e^2 N \frac{v_0}{\omega_0} \xi \sin(\omega_0 t + \varphi) = V_0 \sin(\omega_0 t + \varphi). \qquad (49.16)$$

On the basis of (49.13) and (49.14)

$$V_0 = kT. \qquad (49.17)$$

However, the elementary treatment given above requires more detailed consideration. The following questions remain unclear.

1. The quantity N, which appears in formula (49.9), plays the role of the density of moving electrons, which in the general case is obviously not equal to the density of all the electrons in the plasma. In connection with this, the frequency of oscillations ω_0 remains undetermined. Only its upper limit is determined.

2. The distance to which polarization perturbations are propagated depends on the frequency of the oscillations, as follows from the formula exp $\varkappa$ x,

$$\varkappa^2 = \frac{\omega_0^2 - \omega^2}{v_T^2}.$$

Consequently, the change in magnitude d in the previous treatment of the Debye length, corresponding to the static case, is not immediately obvious and requires foundation.

b. Fuller treatment. A more exact solution of the problem should start from the general solution of the equations for plasma corresponding to surface oscillations.

In the plane case, the solution giving the surface vortex-free motions in electron plasma has the form

$$u = \int_{-\omega_0}^{+\omega_0} \dot{x}_{\omega} e^{-k_{\omega} x} d\omega,$$

$$\sigma = \varepsilon_0 \int_{-\omega_0}^{+\omega_0} k_{\omega} x_{\omega} e^{-k_{\omega} x} d\omega, \tag{49.18}$$

$$E = -4\pi \frac{e}{m} \int_{-\omega_0}^{+\omega_0} x_{\omega} e^{-k_{\omega} x} d\omega,$$

where

$$x_{\omega} = -\frac{1}{4\pi \frac{e}{m}} E_{\omega}^{0} e^{i\omega t}, \tag{49.19}$$

and E_{ω}^{0} is the amplitude of the oscillatory fluctuations of frequency

$$k_{\omega} = \sqrt{\frac{\omega_0^2 - \omega^2}{v_0^2}}. \tag{49.20}$$

The total energy in the case of surface oscillations, as in the case of volume oscillations, is expressed in the form of energy of an aggregate of oscillators with frequencies from o to ω_0.

$$W = \frac{m}{2} \int_{\omega} \{ \dot{Z}_{\omega}^2 + (\omega_0^2 - v_T^2 k_{\omega}^2) Z_{\omega}^2 \} \rho_{\omega} \, d\omega, \tag{49.21}$$

where $Z_{\omega} = \omega_0 x_{\omega}$ has the dimensionality of a length, and ρ_{ω} is the density of surface oscillations

$$\rho_{\omega} = \frac{SN}{\omega_0 k_{\omega}}, \tag{49.22}$$

where S is the size of the surface; then

$$\frac{m}{2} (\dot{Z}_{\omega}^2 + \omega^2 Z_{\omega}^2)$$

represents the energy of an oscillator corresponding to a surface oscillation of the plasma with frequency $\omega = \sqrt{\omega_0^2 - v_T^2 k^2}$.

For the case of thermal equilibrium, the mean energy for each oscillator is kT.

Accordingly,

$$m\omega^2 \overline{Z_\omega^2} = m\omega_0^2 \overline{x_\omega^2}\, \omega^2 = kT. \tag{49.23}$$

The change in the potential energy upon passing through the region of surface oscillations will be

$$eV = e\int_0^\infty E\,dx \tag{49.24}$$

or, employing the expression for E in (49.18),

$$eV = -4\pi \frac{e^2}{m} \rho_0 \int_{-\omega_0}^{+\omega_0} x_\omega \frac{d\omega}{k_\omega} = -m\omega_0^2 \int_{-\omega_0}^{+\omega_0} \frac{d\omega}{k_\omega} x_\omega. \tag{49.25}$$

The region $\omega \sim \omega_0$ is of greatest importance for the magnitude of the integral because of the presence of the factor

$$\frac{1}{k_\omega} = \frac{v_T}{\sqrt{\omega_0^2 - \omega^2}}. \tag{49.26}$$

It is this fact that justifies the assumptions made above. We obtain the right order of magnitude for the amplitude of change of the potential energy eV, if we set instead of amplitude x_ω, the value of this quantity, corresponding first to the case of thermal equilibrium and secondly, taken for $\omega = \omega_0$.

In this way we have, setting $x_\omega = x_{\omega}^0 e^{i\omega t + i\varphi}$:

$$eV(t) = -m\omega_0^2 v_T x_\omega^0 \left\{ e^{i(\omega_0 t + \varphi)} \int_0^{\omega_0} \frac{d\omega}{\sqrt{\omega_0^2 - \omega^2}} + e^{-i(\omega_0 t + \varphi)} \int_{-\omega_0}^{0} \frac{d\omega}{\sqrt{\omega_0^2 - \omega^2}} \right\}. \tag{49.27}$$

or since

$$\int_0^{\omega_0} \frac{d\omega}{\sqrt{\omega_0^2-\omega^2}} = \arcsin\frac{\omega}{\omega_0}\Big|_0^1 = \frac{\pi}{2} = \int_{-\omega_0}^{0} \frac{d\omega}{\sqrt{\omega_0^2-\omega^2}},$$

$$eV(t) = -\pi m\omega_0^2 v_T x_\omega^0 \cos(\omega_0 t+\varphi) = -eV_0\cos(\omega_0 t+\varphi). \tag{49.28}$$

We find the magnitude of amplitude x_ω^0 by employing (49.23). In this way,

$$eV_0 = \pi m\omega_0^2 v_T \sqrt{\frac{kT}{m\omega_0^4}} = \pi kT,$$

since

$$v_T = \sqrt{\frac{kT}{m}} \tag{49.29}$$

Comparison with formula (49.17) shows that this differs from the corresponding result obtained from elementary considerations only in the numerical coefficient.

The above description furnishes three characteristics of the double layer. The thickness of the double layer can be determined. The formula for the potential difference upon passing through the region of surface oscillations may be written in the form V = E.d, from which d also is determined

$$E = -4\pi \frac{e}{m}\rho_0 \int_{-\omega_0}^{+\omega_0} x_\omega e^{-k_\omega x}\, d\omega; \tag{49.30}$$

for the frequency region $\omega \sim \omega_0$, which is the most important, according to (49.26):

$$E \cong -4\pi \frac{e}{m}\rho_0 x_\omega^0 2\cos(\omega_0 t+\varphi)\,\omega_0 = -\frac{m}{e}\omega_0^3 x_\omega^0 2\cos(\omega_0 t+\varphi). \tag{49.31}$$

The comparison with (49.28) give

$$eV \cong \frac{\pi}{2}\frac{v_T}{\omega_0} eE. \tag{49.32}$$

Accordingly

$$d \cong \frac{\pi}{2} \frac{v_T}{\omega_0}, \tag{49.33}$$

which differs from the result obtained above only in the nonessential numerical coefficient. The field strength in the double layer oscillates with a frequency equal to ω_0. The magnitude of the sudden change of potential upon passing through the double layer is proportional to KT.

3. Longitudinal oscillations in the flow. Change in vibration spectrum. Spatially periodic structure. We assume that in a certain region of the electron plasma, there is a stationary flow of electrons, with a certain velocity relative to the ions, and we consider the nature of the behavior of the perturbations in this case. We confine ourselves here to "linear" treatment of the problem, that is, we assume that the perturbations of density and velocities are so small as compared to their stationary values that we may limit ourselves to their first powers. The fundamental equations determining the character of the propagation of perturbation in the flow is obtained, as above, on the basis of three equations: the equation of continuity, the equation of of fluid motion and Poisson's equation. Setting $\rho = \rho_0 + \rho_1$ and $u = u_0 + u_1$, we have the following system of equations for the first approximation:

$$\left.\begin{aligned} \operatorname{div}(\rho_0 u_1 + u_0 \rho_1) &= -\frac{\partial \rho_1}{\partial t}, \\ \frac{\partial u_1}{\partial t} + (u_0 \operatorname{grad}) u_1 &= -\frac{v_T^2}{\rho_0} \operatorname{grad} \rho_1 + \frac{e}{m} E_1, \\ \operatorname{div} E_1 &= 4\pi \frac{e}{m} \rho_1. \end{aligned}\right\} \tag{49.34}$$

For the case in which the direction of u_0 coincides with the x axis, we have, elimination E_1 and u_1, the equation for ρ_1:

$$\frac{\partial^2 \rho_1}{\partial t^2} + \omega_0^2 \rho_1 + 2u_0 \frac{\partial^2 \rho_1}{\partial t \partial x} + (u_0^2 - v_T^2) \frac{\partial^2 \rho_1}{\partial x^2} - \\ - v_T^2 \left(\frac{\partial^2 \rho_1}{\partial y^2} + \frac{\partial^2 \rho_1}{\partial z^2} \right) = 0. \tag{39.35}$$

We seek a solution, as usual, in the form of plane waves, and we find that in the case of a longitudinal wave, being propagated along the x axis, the law of dispersion will have the form

$$\omega = u_0 k \pm \sqrt{v_T^2 k^2 + \omega_0^2}. \tag{49.36}$$

For propagation in the perpendicular direction

$$\omega = \pm \sqrt{v_T^2 k^2 + \omega_0^2}. \tag{49.37}$$

Two cases must be distinguished $u_0 < v_T$ and $u_0 > v_T$, the form of the solutions for them differing essentially. For the first case, the difference between formulas (49.36) and (49.37) is not essential. The general structure of the vibration structure is the same as when there is no flow. Solutions differing in sign lead to results qualitatively the same. It is easy to see that the region from 0 to ω_0 corresponds to imaginary values of k, that is, in this field there are only exponential solutions. In the region $\omega > \omega_0$, k is real, and the solutions already have a wave character. On the other hand, if $u_0 > v_T$, then the solution with the (+) corresponds to the previous result, while the solution with the (-) has the property that, beginning from a certain real k, ω has positive values from zero up, that is, in this case, the real character of the solutions applies in the domain of low frequencies as well. The group and phase velocities of the perturbations are

$$\left.\begin{aligned} v_{rp} &= \frac{d\omega}{dk} = u_0 - v_T \frac{v_T k}{\sqrt{\omega_0^2 + v_T^2 k^2}}, \\ v_\phi &= \frac{\omega}{k} = u_0 - \sqrt{v_T^2 + \left(\frac{\omega_0}{k}\right)^2}. \end{aligned}\right\} \tag{49.38}$$

The connection between them is defined by the formula

$$(v_{rp} - u_0)(v_\phi - u_0) = v_T^2. \tag{49.39}$$

To explain the character of the propagation of perturbations in the case of a stream, we write a more general solution for equation (49.35). We confine ourselves to consideration of spatially extended perturbations for which it is possible to put $\lambda \gg \lambda_D$, where λ is wave length of the several harmonics, into which the perturbation can be decomposed, and λ_D is the Debye distance. With this condition, we can drop $v_T^2 k^2$ in the dispersion formula (49.36), and write approximately

$$\omega = u_0 k \pm \omega_0. \qquad (49.40)$$

In that case, the solution will have the following appearance:

$$\rho_1(x, t) = \int a(k)\, e^{i\omega t - ikx}\, dk = e^{\pm i\omega_0 t} \int a(k)\, e^{iku_0\left(t - \frac{x}{u_0}\right)} = e^{\pm i\omega_0 t} F\left(t - \frac{x}{u_0}\right). \qquad (49.41)$$

This function shows that the perturbation spreads in the direction of a flow with the velocity u_0, vibrating with frequency ω . Another form as well may be given to solution (49.41). Replacing k by ω in the first integral, by means of (49.40), we get:

$$\rho_1(x, t) = \int a(k)\, e^{i\omega t - ikx}\, dk = e^{\pm i \frac{\omega_0}{u_0} x} \int a\left(\frac{\omega \pm \omega_0}{u_0}\right) e^{i\omega\left(t - \frac{x}{u_0}\right)} d\left(\frac{\omega}{\omega_0}\right) = e^{\pm i \frac{\omega_0}{u_0} x} g\left(t - \frac{x}{u_0}\right).$$

$$(49.42)$$

Both formulas, obviously, are equivalent and the choice of one or the other is determined merely by convenience in calculations. The first formula is convenient for the case in which the density distribution at the initial moment of time is given as a function of the coordinates. Then in solution (49.41), we get vibrations with frequency ω_0 and their removal by the flow. If the perturbation is given as a function of t for a definite x, then (49.42) immediately gives a solution satisfying this condition, for a suitably chosen function g(t). In this case, there appears the characteristic periodicity in the space distribution of the perturbation with a wave length equal to $\frac{2\pi}{\omega_0} u_0$, defined by the concentration of the electron plasma and the flow velocity. It is of importance to note that this periodicity may be stationary, if the action of the source of perturbation is independent of the time. The cause for the appearance of such a periodicity once again consists in the presence of the vibration of frequency ω_0, which is visible from the fact, for example, that the periodic character of the solution is conditioned by the term with ω_0 in the law of dispersion (49.40).

Section 50. Change in Debye Polarization With Motion of Charge.

In this section, we consider the change of the polarization of plasma around a charge moving through it uniformly along a straight line. We saw in Sec. 47 that in the case of a charge at rest, the polarization in the electron plasma is of Debye type. When there is uniform rectilinear motion, the picture of the change in polarization was established qualitatively in Sec. 48. It was shown there that for velocities not exceeding the velocity of sound, the polarized region in an electron plasma moves along with the particle, merely changing its form. At velocities exceeding the velocity of sound, the character of polarization changes sharply: the emission of longitudinal rays begins, that is, perturbation of the space oscillations of the plasma. However, a complete solution of the problem of the change in polarization was not given in the section in question. We now desire to obtain a detailed picture of the change in polarization depending on the velocity of the moving charge.

The initial system of equations will be as in Sec. 48:

$$\begin{gathered}\Delta\varphi = -\frac{1}{\rho_0}\frac{\partial\sigma}{\partial t},\\ \Delta\Phi = -4\pi\frac{e}{m}\sigma - 4\pi\rho_e(x, y, z, t),\\ \frac{\partial\varphi}{\partial t} + \alpha\varphi + \frac{e}{m}\Phi + \frac{v_T^2}{\rho_0}\sigma = 0,\end{gathered} \tag{50.1}$$

where, for a point charge moving uniformly along the z axis

$$\rho_e(x, y, z, t) = \varepsilon\delta(x)\delta(y)\delta(z - v_0 t). \tag{50.2}$$

Once again, we shall seek solutions in the form

$$\Phi = \int_{-\infty}^{+\infty}\Phi_\omega e^{i\omega t}\,d\omega;\quad \varphi = \int_{-\infty}^{+\infty}\varphi_\omega e^{i\omega t}\,d\omega;\quad \sigma = \int_{-\infty}^{+\infty}\sigma_\omega e^{i\omega t}\,d\omega, \tag{50.3}$$

and likewise represent the given function ρ_e by a Fourier integral:

$$\rho_e = \int_{-\infty}^{+\infty}\rho_{e\omega}e^{i\omega t}\,d\omega, \tag{50.4}$$

where

$$\rho_{e\omega}=\frac{1}{2\pi}\int_{-\infty}^{+\infty}\rho_e e^{-i\omega t}\,dt=\frac{\varepsilon}{2\pi}\int_{-\infty}^{+\infty}\delta(x)\,\delta(y)\,\delta(z-v_0t)\,e^{-i\omega t}\,dt=$$

$$=\frac{\varepsilon}{2\pi}\delta(x)\,\delta(y)\frac{1}{v_0}e^{-i\omega\frac{z}{v_0}}. \tag{50.5}$$

Substituting the assumed solution in our initial equation, we obtain the following system of equations with respect to the Fourier components:

$$\begin{aligned}\Delta\varphi_\omega&=-\frac{i\omega}{\rho_0}\sigma_\omega,\\ \Delta\Phi_\omega&=-4\pi\frac{e}{m}\sigma_\omega-4\pi\frac{\varepsilon}{2\pi v_0}\delta(x)\,\delta(y)\,e^{-i\frac{\omega z}{v_0}},\\ i\omega\varphi_\omega+\alpha\varphi_\omega&+\frac{e}{m}\Phi_\omega+\frac{v_T^2}{\rho_0}\sigma_\omega=0.\end{aligned} \tag{50.6}$$

Naturally, here too, we seek solutions in the form of Fourier integrals:

$$\begin{aligned}\sigma_\omega(x,y,z)&=e^{-i\frac{\omega}{v_0}z}\iint_{-\infty}^{+\infty}\sigma_\omega(\mathbf{k}_\perp)\,e^{i\mathbf{k}_\perp\mathbf{r}}\,d\mathbf{k}_\perp,\\ \Phi_\omega(x,y,z)&=e^{-i\frac{\omega}{v_0}z}\iint_{-\infty}^{+\infty}\Phi_\omega(\mathbf{k}_\perp)\,e^{i\mathbf{k}_\perp\mathbf{r}}\,d\mathbf{k}_\perp,\\ \varphi_\omega(x,y,z)&=e^{-i\frac{\omega}{v_0}z}\iint_{-\infty}^{+\infty}\varphi_\omega(\mathbf{k}_\perp)\,e^{i\mathbf{k}_\perp\mathbf{r}}\,d\mathbf{k}_\perp,\end{aligned} \tag{50.7}$$

where $\mathbf{k}_\perp$ is a vector with the components k_x and k_y. Similarly for the known function $\rho_{e\omega}$:

$$\rho_{e\omega}=e^{-i\frac{\omega}{v_0}z}\int_{-\infty}^{+\infty}\rho_{e\omega}(\mathbf{k}_\perp)\,e^{i\mathbf{k}_\perp\mathbf{r}}\,d\mathbf{k}_\perp, \tag{50.8}$$

$$\rho_{e\omega}(\mathbf{k}_\perp)=\frac{1}{(2\pi)^2}\int_{-\infty}^{+\infty}\frac{\varepsilon}{2\pi v_0}\delta(x)\,\delta(y)\,e^{-i\mathbf{k}_\perp\mathbf{r}}\,d\mathbf{r}=\frac{\varepsilon}{(2\pi)^3v_0}. \tag{50.9}$$

Substituting these expressions in the preceding system of differential equations, we now obtain algebraic equations for the components, the solution of which is elementary:

$$\sigma_\omega(k_\perp) = \frac{z}{2\pi^2 v_0}\frac{e}{m}\rho\frac{1}{\omega^2\left(1-\frac{v_T^2}{v_0^2}\right) - i\omega\alpha - v_T^2 k_\perp^2 - \omega_0^2},$$

$$\Phi_\omega(k_\perp) = \frac{z}{2\pi v_0}\frac{1}{k_\perp^2 + \frac{\omega^2}{v_0^2}}\frac{\omega^2\left(1+\frac{v_T^2}{v_0^2}\right) - i\omega\alpha - v_T^2 k_\perp^2}{\omega^2\left(1-\frac{v_T^2}{v_0^2}\right) - i\omega\alpha - v_T^2 k_\perp^2 - \omega_0^2}, \tag{50.10}$$

$$\varphi_\omega(k_\perp) = \frac{z}{2\pi^2 v_0}\frac{e}{m}\frac{1}{k_\perp^2 + \frac{\omega^2}{v_0^2}}\frac{i\omega}{\omega^2\left(1-\frac{v_T^2}{v_0^2}\right) - i\omega\alpha - v_T^2 k_\perp^2 - \omega_0^2}.$$

Here

$$\omega_0^2 = 4\pi\frac{e^2}{m^2}N \qquad k_\perp^2 = k_x^2 + k_y^2.$$

Thus, the solution of the initial system of equations has the following form:

$$\sigma = \frac{z}{2\pi^2 v_0}\frac{e}{m}\rho_0\int\int_{-\infty}^{+\infty}\int d\omega e^{i\omega\left(t-\frac{z}{v_0}\right)}\frac{e^{ik_\perp r}dk_\perp}{\omega^2\left(1-\frac{v_T^2}{v_0^2}\right) - i\omega\alpha - v_T^2 k_\perp^2 - \omega_0^2},$$

$$\Phi = \frac{z}{2\pi^2 v_0}\int\int_{-\infty}^{+\infty}\int d\omega e^{i\omega\left(t-\frac{z}{v_0}\right)} \times$$

$$\times\frac{\omega^2\left(1-\frac{v_T^2}{v_0^2}\right) - i\omega\alpha - v_T^2 k_\perp^2}{\left(k_\perp^2 + \frac{\omega^2}{v_0^2}\right)\left\{\omega^2\left(1-\frac{v_T^2}{v_0^2}\right) - i\omega\alpha - v_T^2 k_\perp^2 - \omega_0^2\right\}}e^{ik_\perp r}dk_\perp, \tag{50.11}$$

$$\varphi = \frac{z}{2\pi^2 v_0}\frac{e}{m}\int\int_{-\infty}^{+\infty}\int d\omega e^{i\omega\left(t-\frac{z}{v_0}\right)} \times$$

$$\times\frac{i\omega}{\left(k_\perp^2 + \frac{\omega^2}{v_0^2}\right)\left\{\omega^2\left(1-\frac{v_T^2}{v_0^2}\right) - i\omega\alpha - v_T^2 k_\perp^2 - \omega_0^2\right\}}e^{ik_\perp r}dk_\perp.$$

The solutions written down in this form are not intuitive, and it will therefore be our next task to calculate and investigate these integrals, so far as it is possible.

We consider the dependence of the solutions obtained on the coordinates and the time. Let the particle move with the velocity less than the velocity of sound in the electron plasma $u_0 < v_T$. The expression for σ will then have the form

$$\sigma = -\frac{se}{2\pi^2 v_0}\frac{p_0}{m}\int\int_{-\infty}^{+\infty}\int d\omega e^{i\omega t'}\frac{1}{\mu^2\omega^2 + i\omega a + v_T^2 k_\perp^2 + \omega_0^2}e^{ik_\perp r}dk_\perp, \qquad (50.12)$$

where $t' = t - \frac{z}{v_0}$, $\mu^2 = \frac{v_T^2}{v_0^2} - 1$. With the aid of the theory of residues, the integral is taken for ω. For $t' > 0$ the poles must be considered in the upper half-plane of the complex variable, and for $u_0 < v_T$ in the lower half.

The poles of the function under the integral sign are

$$\omega_{1,2} = -i\frac{a}{2\mu^2} \pm \sqrt{\frac{\omega_0^2}{\mu^2} + \frac{v_T^2}{\mu^2}k_\perp^2 + \frac{a^2}{4\mu^4}}. \qquad (50.13)$$

Consequently, the lower and upper half planes will each have one pole. Assuming $t' > 0$, we have:

$$\sigma = -\frac{sep_0}{2\pi v_0 m\mu^2}e^{-\frac{a}{2\mu^2}t'}\int_{-\infty}^{+\infty}\frac{e^{-t'\sqrt{\frac{\omega_0^2}{\mu^2} + \frac{v_T^2}{\mu^2}k_\perp^2 + \frac{a^2}{4\mu^4}}}}{\sqrt{\frac{\omega_0^2}{\mu^2} + \frac{v_T^2}{\mu^2}k_\perp^2 + \frac{a^2}{4\mu^4}}}e^{ik_\perp r}dk_\perp. \qquad (50.14)$$

Introducing the new variables $k_x = \tau\cos\varphi$, $k_y = \tau\sin\varphi$ and taking advantage of the fact that

$$\int_0^{2\pi} e^{i\tau r\cos\varphi}d\varphi = 2\pi I_0(\tau r), \qquad (50.15)$$

where I_0 is a Bessel function of zero order, we obtain:

$$\varphi = -\frac{\varepsilon e \rho_0}{v_0 m \mu^2} e^{-\frac{\alpha}{2\mu^2} t'} \int_0^\infty I_0(\tau r) \frac{e^{-t'\sqrt{\frac{\omega_0^2}{\mu^2} + \frac{v_T^2}{\mu^2} k_\perp^2 + \frac{\alpha^2}{4\mu^4}}}}{\sqrt{\frac{\omega_0^2}{\mu^2} + \frac{v_0^2}{\mu^2} k_\perp^2 + \frac{\alpha^2}{4\mu^4}}} \tau\, d\tau. \tag{50.16}$$

This integral coincides with Sommerfeld's integral

$$\int_0^\infty I_0(bt) \frac{e^{-c\sqrt{t^2+y^2}}}{\sqrt{t^2+y^2}} t\, dt = \frac{e^{-y\sqrt{b^2+c^2}}}{\sqrt{b^2+c^2}}. \tag{50.17}$$

Taking advantage of this, we get:

$$\varphi = -\frac{\varepsilon e \rho_0}{v_0 m \mu} e^{-\frac{\alpha}{2\mu^2}\left(t - \frac{z}{v_0}\right)} \frac{e^{-a\sqrt{\left(\frac{t - \frac{z}{v_0}}{\mu}\right)^2 + \frac{r^2}{v_T^2}}}}{\sqrt{\left(\frac{t - \frac{z}{v_0}}{\mu}\right)^2 + \frac{r^2}{v_T^2}}}, \tag{50.18}$$

where $a^2 = \omega_0^2 + \frac{\alpha^2}{4\mu^2}$. For $t' < 0$, we obtain the same equation, only with the + in the index of the factor

$$e^{\frac{\alpha}{2\mu^2}\left(t - \frac{z}{v_0}\right)}.$$

We write formula (50.18) in its final form as follows:

$$\varphi = -\frac{m\varepsilon}{4\pi e} \frac{\omega_0^2}{v_T^2} \frac{1}{\beta} \frac{1}{\sqrt{r^2 + \frac{1}{\beta^2}(z - v_0 t)^2}} e^{-a_0\sqrt{r^2 + \frac{1}{\beta^2}(z - v_0 t)^2}} e^{\pm \frac{\alpha}{2v_0 T \beta^2}(z - v_0 t)} \tag{50.19}$$

where

$$\beta^2 = 1 - \frac{v_0^2}{v_T^2}, \qquad a_0^2 = \frac{\omega_0^2}{v_T^2}\left(1 + \left(\frac{\alpha}{2\omega_0 \mu}\right)^2\right).$$

The $\pm$ is for $t - \frac{z}{v_0} \gtrless 0$, respectively. Formula (50.19) gives the change in polarization depending on the velocity of the moving charge. For $v_0 \to 0$ (50.19) goes over into the well-known formula for polarization in the Debye-Hückel theory:

$$\rho = -\frac{m\varepsilon}{4\pi e}\frac{\omega_0^2}{v_T^2}\frac{1}{R}e^{-\frac{\omega_0}{v_T}R}. \qquad (50.20)$$

As the velocity rises, if we neglect friction, the surfaces of equal densities represent ellipses of revolution flattened in the direction of the motion of the charge, with the ratio of the axes

$$1 : \frac{1}{\sqrt{\frac{v_0^2}{v_T^2} - 1}}. \qquad (50.21)$$

Since under ordinary conditions in the plasma $\alpha \ll \omega_0$, the role of friction is unimportant. As will be seen from (50.19), it introduces an asymmetry in the density distribution with respect to the plane $t - \frac{z}{v_0} = 0$. The friction only plays a notable role for velocities in the immediate vicinity of the critical velocity v_T; it must be taken into account if the terms containing α are comparable with the others. This gives the following condition:

$$\frac{\alpha}{2\omega_0\mu} \sim 1 \quad \text{или} \quad \frac{\alpha}{2\omega_0} \sim \sqrt{\frac{v_0^2}{v_T^2} - 1}, \qquad (50.22)$$

from which we find the order of width of the velocities interval in the neighborhood of v_T, for which it is necessary to take the friction into account

$$\Delta v_0 \sim \frac{\alpha^2}{8\omega_0^2} v_T. \qquad (50.23)$$

Since $\alpha \ll \omega_0$, the width of this interval of velocities is very small: $\Delta v_0 \ll v_T$. In the rest of the region, the effects connected with friction may be ignored. We now investigate the case of high particle velocities, that is, velocities exceeding critical velocity v_T. The expression for σ then has the form

$$\sigma = \frac{\iota e \rho_0}{2\pi^2 v_0 m} \int_{-\infty}^{+\infty} d\omega e^{i\omega t'} \frac{1}{\mu_1^2 \omega^2 - i\omega\alpha - v_T^2 k_\perp^2 - \omega_0^2} e^{ik_\perp r} dk_\perp ; \mu_1^2 = 1 - \frac{v_T^2}{v_0^2}. \quad (50.24)$$

In this case, the poles of the function under the integral sign are located highly unsymmetrically with respect to the plane $t - \frac{z}{v_0} = 0$. Actually, we obtain two poles in the upper plane and none in the lower

$$\omega_{1,2} = i \frac{\alpha}{2\mu_1^2} \pm \sqrt{\frac{\omega_0^2}{\mu_1^2} - \frac{\alpha^2}{4\mu_1^4} + \frac{v_T^2}{\mu_1^2} k_\perp^2}. \quad (50.25)$$

Here we must distinguish the cases of low and high friction:

$$1) \frac{\omega_0^2}{\mu_1^2} - \frac{\alpha^2}{4\mu_1^2} > 0 \quad \text{and} \quad 2) \frac{\omega_0^2}{\mu_1^2} - \frac{\alpha^2}{4\mu_1^4} < 0.$$

We consider the first case. Applying the theorem of residues for $t' > 0$, that is, in the region back of the particle, we have (similarly to the case $v_0 < v_T$)

$$\sigma = -\frac{\iota e \rho_0}{m v_0 v_T^2} e^{-\frac{\alpha}{2\mu_1^2} t'} \int_0^{\infty} I_0(\tau r) \frac{\sin \frac{t'}{\mu_1} \sqrt{a^2 + v_T^2 \tau^2}}{\sqrt{a^2 + v_T^2 \tau^2}} \tau \, d\tau ; \; a^2 = \omega_0^2 - \frac{\alpha^2}{4\mu_1^4}. \quad (50.26)$$

This integral is a particular case of the discontinuous integral studied by Sonin, if we make the change in (50.26)

$$\sin x = \sqrt{\frac{\pi x}{2}} I_{\frac{1}{2}}(x), \quad (50.27)$$

where $I_{\frac{1}{2}}$ is the corresponding Bessel function. Since in the lower half-plane $(t'<0)$, the denominator has no zeros, then for $t'<0$ (that is, in front of the particle) $\sigma=0$, so that we can write the result in a single formula:

$$\sigma=\begin{cases} 0, \text{ if } \quad t-\frac{z}{v_0}<0; \quad 0, \quad \textit{if} \quad \frac{r}{v_T}<\frac{t-\frac{z}{v_0}}{\mu_1}<0, \\ -\frac{2\varepsilon e\rho_0}{v_0 m\mu_1 v_T^2} e^{-\frac{\varepsilon}{2\mu_1^2}\left(t-\frac{z}{v_0}\right)} \frac{\cos a\sqrt{\left(\frac{t-\frac{z}{v_0}}{\mu_1}\right)^2-\frac{r^2}{v_T^2}}}{\sqrt{\left(\frac{t-\frac{z}{v_0}}{\mu_1}\right)^2-\frac{r^2}{v_T^2}}} \\ \frac{t-\frac{z}{v_0}}{\mu_1}>\frac{r}{v_T}>0. \end{cases} \tag{50.28}$$

Solution (50.28) indicates two characteristic properties of the polarization state of the plasma around a charge moving with the velocity exceeding the critical velocity.

1. The quantity σ is a discontinuous function in this case. It differs from zero only within a cone whose equation is

$$r^2=x^2+y^2=\frac{v_T^2}{\mu_1^2}\left(t-\frac{z}{v_0}\right)^2. \tag{50.29}$$

The half-vertex angle θ of the cone is given by the formula

$$\operatorname{tg}^2\theta=\frac{x^2+y^2}{(z-v_0t)^2}, \quad \operatorname{tg}\theta=\frac{\frac{v_T}{v_0}}{\mu_1^2}=\frac{1}{\sqrt{\frac{v_0^2}{v_T^2}-1}}. \tag{50.30}$$

The discontinuous relationships arising here are analogous to those that apply in discussing the effect of the formation of shock waves in the aerodynamics of supersonic velocities, as well the relations appearing in the radiation from a charge moving uniformly and rectilinearly in dielectric media (Cherenkov effect +), with the difference

+ I. Tamm, Journ. of Phys. 1, 439 (1939)

in principle from the second case, however, that here it is longitudinal and not tranverse perturbations that are involved ++,

2. The second characteristic property of the formula that has been given is that in contrast to the case $v_0 < v_T$, polarization for $v_0 > v_T$ is variable in sign and has a periodic structure in the cone.

We rewrite formula (50.28) in a more convenient form:

$$\sigma = -m \frac{\varepsilon}{2\pi e} \cdot \frac{\omega_0^2}{v_T^2} \frac{1}{\beta_1} \frac{\cos a_1 \sqrt{\frac{1}{\beta_1^2}(z - v_0 t)^2 - r^2}}{\sqrt{\frac{1}{\beta_1^2}(z - v_0 t)^2 - r^2}} e^{-\frac{\alpha v_0}{2 v_T^2 \beta_1^2}(z - v_0 t)}$$

$$\left\{ \beta_1^2 = \frac{v_0^2}{v_T^2} - 1, \quad a_1^2 = \frac{\omega_0^2}{v_T^2}\left[1 - \left(\frac{\alpha}{2\omega_0 \mu_1}\right)^2\right]\right\}. \tag{50.31}$$

Setting $r = 0$ and $\alpha \ll \omega_0$ (see above), we have for the spatial period:

$$L = \frac{2\pi}{a_1}\beta_1 = \frac{2\pi}{\omega_0}\sqrt{v_0^2 - v_T^2}. \tag{50.32}$$

From the point of view of an observer at rest, the periodic structure of the polarization is perceived in the form of a field periodically varying in the course of time, with the cyclical frequency

$$\omega^2 = \left(\frac{2\pi v_0}{L}\right)^2 = \frac{\omega_0^2}{1 - \frac{v_T^2}{v_0^2}}. \tag{50.33}$$

If $\frac{\alpha}{2\omega_0 \beta_1} \ll 1$, then the lines σ = const. coincide with the lines

$$\left(\frac{t - \frac{z}{v_0}}{\mu_1}\right)^2 - \frac{r^2}{v_T^2} = \text{const.}$$

++ There is no radiation of transverse waves in electron plasma, since the phase velocity of transverse waves in plasma always exceeds the velocity of light.

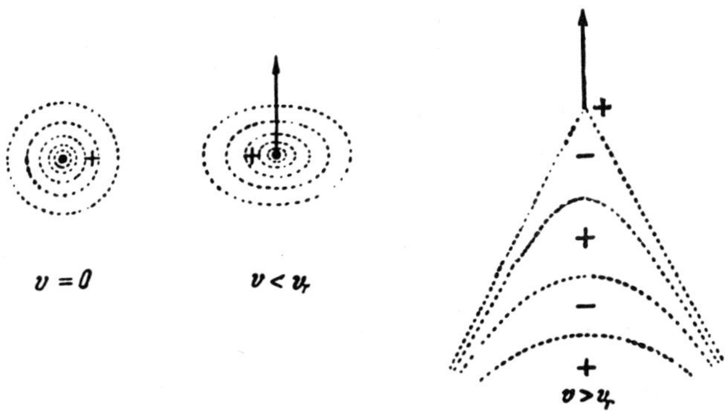

Fig. 20. Changes in the state of polarization of plasma in dependence on the velocity of the moving charge.

The last equation is the equation of the family of hyperboloids with a common asymptotic cone (50.29).

We now consider the case of high friction, that is, $\frac{\omega_0^2}{\mu_1^2} - \frac{a^2}{4\mu_1^4} < 0.$

All the calculations are similar and the result has the form

$$
\sigma = \begin{cases} 0, \text{ if } t - \dfrac{z}{v_0} < 0, \\[2ex] 0, \text{ if } \dfrac{r}{v_0} < \dfrac{t - \frac{z}{v_0}}{\mu_1}, \\[2ex] -\dfrac{2e\rho_0}{v_0 m \mu_1 v_T^2} \cdot e^{-\frac{a}{2\mu_1^2}\left(t - \frac{z}{v_0}\right)} \dfrac{\operatorname{ch} a \sqrt{\left(\dfrac{t - \frac{z}{v_0}}{\mu_1}\right)^2 - \dfrac{r^2}{v_T^2}}}{\sqrt{\left(\dfrac{t - \frac{z}{v_0}}{\mu_1}\right)^2 - \dfrac{r^2}{v_T^2}}}, \\[2ex] \dfrac{t - \frac{z}{v_0}}{\mu_1} > \dfrac{z}{v_T} > 0. \end{cases} \qquad (50.34)
$$

In this case, σ does not change sign, the periodicity vanishes, but the cone of perturbation remains.

The presence of a strong spatial asymmetry (the cone) in the polarization distribution even in the absence of friction creates behind the moving charge a spatial density of charge contrary (in the nearest zone) to the charge of the particle itself. This is visible from the fact that σ has a sign opposite to the sign of the product εe. This should lead to the appearance of a retarding force acting on the charge in the direction opposite to its motion.

We shall take up the calculation of this force somewhat later.

Section 51. Theory of Nonisothermal Plasma

Under the conditions of a gas-electron plasma, as is known, great temperature differences are set up between the electrons, on the one hand, and the ions and neutral particles, on the other, reaching as high as several tens of thousands of degrees. The literature does not give any theory for this phenomenon. There is merely the qualitative explanation consisting in the fact that the establishment of statistical equilibrium in the electron gas goes forward more rapidly than the energy exchange between the electrons and the other partners, which is the condition for the possibility of accumulation of a high internal energy in the electron gas.

We desire to show that even from purely energetic considerations, it is possible to explain the origin of temperature differences in the presence of a flow, and likewise even the fundamental laws that appear with this phenomenon. We make the following physical assumptions:

a) We require for what follows that the time for establishing the equilibrium in the electron gas and correspondingly the time connected with the establishment of equilibrium in the ion gas and the gas consisting of the neutral particles be considerably less than the time for establishing equilibrium between the electron gas and the other partners. In this case, the state of the plasma for the nonequilibrium processes may be described by two temperatures, for the electrons T_ε and for the ions T_p, with a finite difference of these temperatures at each macroscopic point of the plasma. Introducing two temperatures will be correct, obviously, only if processes are considered for which the characteristic periods of time considerably exceed the times for the establishment of equilibrium in each of the gases separately.

b) For what follows, there is necessary a second postulate as well, concerning the nature of the temperature equalization between the electron and the ion gases. On the basis of ideas as to the theory of heat conductivity, we can say that this equalization is realized by means of a process of heat conduction taking place at each macroscopic point of the plasma (for brevity, and to distinguish it from ordinary heating conductivity, we shall designate this process as "heat conduction at a point"). In a first approximation, we may consider the magnitude of the flow of heat in a unit volume as proportional to the temperature difference, as in the case of ordinary heat conductivity

$$dQ = \alpha \ (T_e - T_p)\,dt, \qquad (51.1)$$

where α is a phenomenologically introduced coefficient.

These two suppositions are sufficient to set up the basic equations.

We express the energy balance in the system being considered when an external field E is present. Designating by c_ε and c_p the specific heats of the electron and ion gas, respectively, we have two equations for the energy balance, for electrons and ions, in a plasma when there is an electric current of density j:

$$\left.\begin{aligned} c_e dT_e &= jEdt - dQ - \Delta', \\ c_p dT_p &= dQ - \Delta'', \end{aligned}\right\} \qquad (51.2)$$

where dQ is the amount of heat given off by the electrons /and/ ions in the heat conduction process, according to formula (51.1), and Δ' and Δ'' designate the flows of heat into outside space, which we shall not take into account here, E is the strength of the electric field connected with a current density j by Ohm's law:

$$j = \sigma E, \qquad (51.3)$$

where σ is the coefficient of the electric conductivity of the plasma.

Making use of (51.1), (51.2), and (51.3), we obtain the initial system of equations defining the change of T_e and T_p in the presence of a current depending on the time:

$$\left.\begin{aligned} \frac{dT_e}{dt} &= -\frac{\alpha}{c_e}(T_e - T_p) + \frac{1}{c_e\sigma}\cdot j^2, \\ \frac{dT_p}{dt} &= \frac{\alpha}{c_p}(T_e - T_p). \end{aligned}\right\} \qquad (51.4)$$

For a gas plasma c_ε and c_p do not depend on the temperature. Here, to evaluate the order of magnitudes and the appearance of the nature of the phenomenon, we shall also take it that the other constants α and σ likewise do not depend on the temperature. Then, the equation for the temperature difference in which we are interested,

$\vartheta = T_\varepsilon - T_p$ follows at once from system (51.4)

$$\frac{d\vartheta}{dt} + \frac{1}{\tau}\vartheta = \frac{1}{c_\varepsilon \sigma} \cdot j^2, \tag{51.5}$$

where

$$\frac{1}{\tau} = \alpha\left(\frac{1}{c_\varepsilon} + \frac{1}{c_p}\right). \tag{51.6}$$

The general solution of (51.5) for a continuous current is

$$\vartheta = \vartheta_0 e^{-\frac{t}{\tau}} + \frac{\tau}{c_\varepsilon \sigma} \cdot j^2. \tag{51.7}$$

In this way we see that the existence of a "finite thermal conductivity at a point" leads by (51.7) to the establishment of a stationary, (for $j = \text{const.}$), finite temperature difference between the electron and ion gases.

This temperature difference is proportional to the input power of the current, proportional to τ and inversely proportional to the heat capacity of the electron gas. The time τ plays the role of the time for establishing thermal equilibrium between the electron and ion gases.

The temperature changes T_ε and T_p with time are obtained from (51.4), assuming that at $t = 0$ a continuous current is applied with a density j and $T_\varepsilon(0) = T_p(0) = T_0$,

$$\left.\begin{aligned} T_\varepsilon(t) - T_0 &= \frac{j^2}{\sigma} \cdot \frac{1}{c_p + c_\varepsilon}\left[t + \tau\frac{c_p}{c_\varepsilon}\left(1 - e^{-\frac{t}{\tau}}\right)\right], \\ T_p(t) - T_0 &= \frac{j^2}{\sigma} \cdot \frac{1}{c_p + c_\varepsilon}\left[t - \tau\left(1 - e^{-\frac{t}{\tau}}\right)\right]. \end{aligned}\right\} \tag{51.8}$$

Formulas (51.7) and (51.8) give the solution to the problem raised as to the origin of the finite temperature difference in the presence of an electric current, and likewise the fundamental connections between the magnitudes in this phenomenon.

It will be seen from (51.8) that a finite conditions a delay in raising the temperature of the ion gas as compared with the time required for the rise to the same temperature in the case of equilibrium heating of the plasma as an entire system, that is, in comparison with a heating process in which $a \to \infty$, which means by (51.6) $\tau \to 0$. The retardation, as will be seen from (51.8) will be of the order τ for $t > \tau$, and therefore does not depend on the current density.

For an alternating current, the general solution of (51.5), assuming $j = j_0 \sin \omega t$, is

$$\vartheta = \vartheta_0 e^{-\frac{t}{\tau}} + \frac{\tau}{c_v \sigma} j_0^2 \left[1 - \frac{1}{1+(2\omega\tau)^2} \cdot (2\omega\tau \cos 2\omega t + \sin 2\omega t)\right]. \qquad (51.9)$$

In this case, the temperature difference is made up of two components, one of which is alternating. The harmonic part gives the fluctuations of the double frequency as compared with the current frequency. It is characteristic that the amplitude of the alternating component depends on the frequency; it has a form similar to the case of what is called the relaxation dispersion.

We reevaluate the order of temperature difference for the usual conditions in a gas electron plasma. For a stationary temperature difference, we have:

$$\vartheta = \frac{\tau}{c_v \sigma} \cdot j^2. \qquad (51.10)$$

The value of the quantity that plays the role of the time for establishing energy equilibrium between the electron and ion gases, as well as the neutral particles, is obtained if we take into account the energy exchange between the electrons and the heavy particles. For a not excessively large velocities interval, it is possible, as a first approximation, to take into account only the exchange energy upon elastic impact between the electrons and the heavy particles. The energy loss of an electron of mass m upon collision with a heavy particle of mass M upon one elastic impact is of the order of magnitude

$$\Delta E = \frac{4m}{M} \cdot E, \qquad (51.11)$$

where Δ E is the initial energy of the electron. This follows from the laws of conservation of energy and momentum.

The energy loss in unit time is obtained by multiplying ΔE by the number of collisions in unit time:

$$\frac{dE}{dt} = \Delta E N_{st} = \frac{4mE}{M} \frac{1}{\tau^{st}}, \tag{51.12}$$

where τ^{st} is the time of free travel of the electron between heavy particles.

The magnitude sought, τ, is defined as to the order of magnitude by the following formula:

$$\frac{1}{\tau} = \frac{1}{E} \frac{dE}{dt} = \frac{4m}{M} \frac{1}{\tau^{st}}. \tag{51.13}$$

Thus, the time of energy exchange and τ^{st}, the time of the free path are connected by the simple relationship

$$\tau = \frac{M}{m} \tau^{st}, \tag{51.14}$$

that is, $\tau \gg \tau^{st}$.

To determine θ from formula (51.10), it is necessary to assign the value of σ. But

$$\sigma = \alpha \frac{N_e e^2}{m} \frac{1}{\tau^{st}}, \tag{51.15}$$

for the order of magnitude of α is 1, and N_ϵ denotes the concentration of the electron gas.

Substituting (51.13) and (51.15) in the formula for θ, we have:

$$\theta = \frac{\tau}{c_e \sigma} j^2 \approx \frac{\frac{1}{4} M}{c_e N_e e^2} j^2. \tag{51.16}$$

Hence, the temperature difference is determined by the current, the mass of the heavy particle, the concentration of electrons $c_e = \frac{3}{2} N_e k$.

It is of interest to note that τ^{st} has disappeared from the final result.

In ordinary neon tubes $N_e \sim 5.10^{12}$ эл/см^3 $j \sim 1$ amp. $M = 20 \times 1.6 \times 10^{-24}$ g. Substituting in (51.16), we have:

$$\theta \approx 1{,}2 \cdot 10^5 \ {}^\circ\mathrm{K}.$$

The observed temperature difference for the conditions under consideration, obtained by the method of sondes, is equal to 25,000°K.

Section 52. Spectrum of Volume and Surface Oscillations for General Case of Nonisothermal Plasma

In this section, we consider the influence of the motion of ions on the spectrum of oscillations in plasma, as well as the influence of the temperature difference between electron and ion gas.

For a mixture of gases, the hydrodynamic equations of motion may be written down separately for each gas, taking into account the interaction between them.

We designate by e_1, m_1, e_2, m_2, the charge and mass particles of the electron and ion gases, respectively. We also write

$$\left.\begin{aligned} P_1(\rho_{01}+\rho_1) &= P_1(\rho_{01})+\left(\frac{\partial P_1}{\partial \rho}\right)_{\rho_{01}}\cdot\rho_1 = P_1+v_{T_1}^2\rho_1, \\ P_2(\rho_{02}+\rho_2) &= P_2(\rho_{02})+\left(\frac{\partial P_2}{\partial \rho}\right)_{\rho_{02}}\cdot\rho_2 = P_2+v_{T_2}^2\rho_2 \end{aligned}\right\} \tag{52.1}$$

their equations of state in linear approximation. For an ideal gas

$$v_{T_1}^2=\frac{kT_1}{m_1}, \qquad v_{T_2}^2=\frac{kT_2}{m_2}. \tag{52.2}$$

In a first approximation, the initial system of equations, describing the possible potential movements in a mixture of the two gases under consideration, has the following form:

$$\begin{gathered} \frac{\partial \boldsymbol{u}_1}{\partial t}=-\frac{v_{T_1}^2}{\rho_{01}}\operatorname{grad}\rho_1+\frac{e_1}{m_1}\boldsymbol{E}_1, \quad \operatorname{div}\boldsymbol{u}_1=-\frac{1}{\rho_{01}}\cdot\frac{\partial\rho_1}{\partial t}, \\ \operatorname{div}\boldsymbol{E}_1=\frac{4\pi e_1}{m_1}\rho_1+\frac{4\pi e_2}{m_2}\rho_2, \\ \frac{\partial \boldsymbol{u}_2}{\partial t}=-\frac{v_{T_2}^2}{\rho_{02}}\operatorname{grad}\rho_2+\frac{e_2}{m_2}\boldsymbol{E}_1, \quad \operatorname{div}\boldsymbol{u}_2=-\frac{1}{\rho_{02}}\cdot\frac{\partial\rho_2}{\partial t}. \end{gathered} \tag{52.3}$$

Taking the divergence of the equations of motion and making use of the continuity and field equations, we obtain:

$$\left.\begin{aligned}\ddot{\rho}_1+\omega_{01}^2\rho_1+\alpha\rho_2=v_{T_1}^2\Delta\rho_1,\\ \ddot{\rho}_2+\omega_{02}^2\rho_2+\beta\rho_1=v_{T_2}^2\Delta\rho_2.\end{aligned}\right\} \tag{52.4}$$

where

$$\omega_{01}^2=\frac{4\pi e_1^2}{m_1^2}\rho_{01},\quad \omega_{02}^2=\frac{4\pi e_2^2}{m_2^2}\cdot\rho_{02},\quad \alpha=\frac{4\pi e_1 e_2}{m_1 m_2}\cdot\rho_{01},\quad \beta=\frac{4\pi e_1 e_2}{m_1 m_2}\rho_{02}.$$

The general solution of system (52.4) may be represented in the form of a superposition of longitudinal plane waves

$$\rho_1=\rho_1^0 e^{i\omega t-ikx},\quad \rho_2=\rho_2^0 e^{i\omega t-ikx}.$$

Substitution in the initial system (52.4) leads to a simple system of equations for defining the amplitudes

$$\left.\begin{aligned}(\omega_{01}^2-\omega^2+v_{T_1}^2k^2)\rho_1^0+\alpha\rho_2^0=0,\\ \beta\rho_1^0+(\omega_{02}^2-\omega^2+v_{T_2}^2k^2)\rho_2^0=0.\end{aligned}\right\} \tag{52.5}$$

The solvability condition for this sytem gives the following dispersion law for longitudinal waves:

$$\omega^2=\frac{\omega_{01}^2+\omega_{02}^2+(v_{T_1}^2+v_{T_2}^2)k^2}{2}\pm$$
$$\pm\sqrt{\left(\frac{\omega_{01}^2+\omega_{02}^2+(v_{T_1}^2+v_{T_2}^2)k^2}{2}\right)-(\omega_{01}^2v_{T_2}^2+\omega_{02}^2v_{T_1}^2)k^2-v_{T_1}^2v_{T_2}^2k^4}. \tag{52.6}$$

This formula applies for the general case of a nonisothermal plasma, and includes within it two different types of oscillations of plasma corresponding to the two values of the root.

Formula (52.6) can be simplified. That is, since over the entire interval of k^2, the quantity $\eta\ll 1$, where

$$\eta=\frac{4\left[(\omega_{01}^2v_{T_2}^2+\omega_{02}^2v_{T_1}^2)k^2+v_{T_1}^2v_{T_2}^2k^4\right]}{\left[\omega_{01}^2+\omega_{02}^2+(v_{T_1}^2+v_{T_2}^2)k^2\right]^2}, \tag{52.7}$$

then the square root in the law of dispersion may be decomposed in a series of powers of η.

We obtain:

$$\omega^2 = \frac{\omega_{01}^2 + \omega_{02}^2 + (v_{T_1}^2 + v_{T_2}^2)\,k^2}{2} \times$$
$$\times \left\{ 1 \pm \left(1 - \frac{1}{2} \cdot \frac{4\,[(\omega_{01}^2 v_{T_2}^2 + \omega_{02}^2 v_{T_1}^2)\,k^2 + (v_{T_1}^2 v_{T_2}^2)\,k^4]}{[\omega_{01}^2 + \omega_{02}^2 + (v_{T_1}^2 + v_{T_2}^2)\,k^2]^2} \right) \right\}, \qquad (52.8)$$

or for the root with the plus sign (denoting the frequency corresponding to this root by the index d):

$$\omega_e^2 = \omega_{01}^2 + \omega_{02}^2 + (v_{T_1}^2 + v_{T_2}^2)\,k^2, \qquad (52.9)$$

for the root with the minus sign (introducing the index a):

$$\omega_a^2 = \frac{(\omega_{01}^2 v_{T_2}^2 + \omega_{02}^2 v_{T_1}^2)\,k^2 + v_{T_1}^2 v_{T_2}^2 k^4}{\omega_{01}^2 + \omega_{02}^2 + (v_{T_1}^2 + v_{T_2}^2)\,k^2}. \qquad (52.10)$$

We consider the two solutions.

1. Electron oscillations. We consider first the case of volume oscillations, that is, we assume $k^2 > 0$. Formula (52.9) corresponds to the dispersion law for an electron gas. Now, instead of

we shall take $\omega_0^2 = 4\pi \frac{e^2}{m^2} \rho_0$

$$\omega_{01}^2 + \omega_{02}^2 = 4\pi e^2 \left(\frac{1}{m_1} + \frac{1}{m_2} \right) N \quad (\text{for } e_1 = e_2).$$

In this way, a change in the frequency reduces to the fact that instead of the mass of the electron m, there appears the effective mass: $\frac{1}{\mu} = \frac{1}{m_1} + \frac{1}{m_2}$

The change in velocity for the case of an isothermal plasma likewise reduces merely to the introduction of the effective mass, since $v_{T_1}^2 + v_{T_2}^2 = kT\left(\frac{1}{m_1} + \frac{1}{m_2}\right)$. For a nonisothermal plasma, the role of a square of a velocity of sound is played by the quantity

$$k\left(\frac{T_1}{m_1} + \frac{T_2}{m_2}\right).$$

On the basis of the dispersion (52.9) and equation (52.5) for determining the amplitude, we usually find the following relationship between ρ_1^0 and ρ_2^0:

$$\frac{\rho_1^0}{\rho_2^0} = \frac{4\pi \frac{e_1 e_2}{m_1 m_2} \rho_{01}}{\omega_{02}^2 + v_{T_2}^2 k^2} = -\frac{\omega_{02}^2}{\omega_{02}^2 + v_{T_2}^2 k^2} \tag{52.11}$$

(since e_1 and e_2 have different signs), or for the change in concentration ($\rho_1^0 = m_1 \Delta N_1$, $\rho_2^0 = m_2 \Delta N_2$):

$$\frac{\Delta N_1}{\Delta N_2} = -\frac{m_2}{m_1} \cdot \frac{\omega_{02}^2}{\omega_{02}^2 + v_{T_2}^2 k^2}. \tag{52.12}$$

For rays that are not too short $\left(k^2 \leqslant \frac{\omega_{02}^2}{v_{T_2}^2}\right)$ in view of the fact that $m_2 \gg m_1$, we have $\Delta N_1 \gg \Delta N_2$. Hence, it is primarily the electron gas that is excited in the oscillations under considerations and not the ion gas, which justifies calling this type of oscillations electronic oscillations.

If we compare this type of oscillations with other types (see below), then one of the essential distinguishing features of electronic oscillations is that they are accompanied by the appearance of space charges, so that the charge density

$$\rho_e = -e(\Delta N_1 - \Delta N_2) \neq 0.$$

In this feature, and also in the similarity of the dispersion laws, electronic oscillations are similar to the well-known Born branch in ionic crystals.

The law of dispersion for electronic oscillations permits solutions with $\omega^2 > 0$, even for purely imaginary values of $k\,(k^2 < 0)$, which corresponds to an exponentially decreasing function, having a noticeable value only near the boundary of the plasma. We have called these solutions surface oscillations. Setting $k^2 = k_s^2$, in (52.0), we obtain the following equation, which determines the possible frequencies ω_s of the surface oscillations

$$\omega_s^2 = \omega_{01}^2 + \omega_{02}^2 - (v_{T_1}^2 + v_{T_2}^2) k_s^2. \tag{52.13}$$

Consequently, the spectrum of surface oscillations extends from zero up to

$$\omega_{01}^2 + \omega_{02}^2.$$

2. <u>Acoustic oscillations</u>. We consider the second type of oscillations, corresponding to the second value of the root in (52.6), with the (-). The characteristic difference of this spectrum from the preceding one consists in the fact that the spectrum of the volume oscillations extends from $\omega = 0$ up. Long wave lengths (small k) correspond to low frequencies.

We confine ourselves at first to the case of small values of k, that is, we assume in the dispersion law (52.10):

$$(v_{T_1}^2 + v_{T_2}^2)\, k^2 < \omega_{01}^2 + \omega_{02}^2. \tag{52.14}$$

Therefore, a fortiori

$$v_{T_1}^2\, v_{T_2}^2 k^2 < \omega_{01}^2 v_{T_2}^2 + \omega_{02}^2 v_{T_1}^2, \tag{52.15}$$

in which case

$$\omega_a^2 = \frac{\omega_{01}^2 v_{T_2}^2 + \omega_{02}^2 v_{T_1}^2}{\omega_{01}^2 + \omega_{02}^2}\, k^2 = \frac{kT_1 + kT_2}{m_1 + m_2}\, k^2, \tag{52.16}$$

that is, in this case, there is no wave dispersion.

We seek a relationship between the amplitudes ρ_1^0 and ρ_2^0 with the aid of (52.5) and (52.16), taking into account that $m_1 \ll m_2$, and we have:

$$\frac{\rho_1^0}{\rho_2^0} = \frac{\omega_{01}^2}{\omega_{01}^2 + v_{T_1}^2 k^2}\, \frac{m_1}{m_2}, \tag{52.17}$$

or for the concentrations:

$$\frac{\Delta N_1}{\Delta N_2} = \frac{\omega_{01}^2}{\omega_{01}^2 + v_{T1}^2 \cdot k^2}. \tag{52.18}$$

Thus, for long waves $\Delta N_1 \sim \Delta N_2$ in contradistinction to the case of electronic oscillations. It follows from this that for oscillations of the form described, the complete density of the electric charge $e(\Delta N_1 - \Delta N_2)$ is small (compare the preceding case) and for $k \to 0$ likewise tends to zero. This, combined with the absence of dispersion, indicates the acoustic character of these oscillations.

We consider the question of the possibility of surface acoustic oscillations. We see from the initial dispersion formula (50.6) that the solution with the (-) does not give positive values for $\omega^2 > 0$ as $k^2 < 0$, from which we conclude, to the absence of surface oscillations of the acoustic type.

3. Ionic oscillations. We pass to the case of a strongly non-isothermal plasma, that is, $T_1 \gg T_2$. The electronic and acoustic oscillations do not lose their importance in this case as well, what changes is only the characteristic parameter in the loss of their dispersion. But with this, there is also another type of oscillations (ionic oscillations).

We shall begin with the dispersion equation for a strongly non-isothermal plasma $v_{T_1}^2 \gg v_{T_2}^2$. Then (52.10) may be rewritten approximately in the form

$$\omega_u^2 = \frac{\omega_{02}^2 + \omega_{01}^2 \dfrac{v_{T_2}^2}{v_{T_1}^2} + v_{T_2}^2 k^2}{1 + \dfrac{\omega_{01}^2 + \omega_{02}^2}{v_{T_1}^2 k^2}}. \tag{52.19}$$

We see from this that for $v_{T_1}^2 \gg v_{T_2}^2 \frac{\omega_{01}^2}{\omega_{02}^2}$ and $v_{T_1}^2 \ll \frac{\omega_{01}^3 + \omega_{02}^3}{k_2}$ (52.20)

we have:

$$\omega_u^2 \to \omega_{02}^2 + v_{T_2}^2 k^2. \tag{52.21}$$

Hence, in a strongly nonisothermal plasma, the law of dispersion has a form similar to what we obtain considering the oscillations in an electron gas, but only with this essential difference, that the role of ω_{01}^2 is played by ω_{02}^2, and $v_{T_1}^2$ is replaced by $v_{T_2}^2$.

We define the relationship between the amplitudes for the case under consideration.

From (52.5) by means of (52.21), we have:

$$\frac{\rho_1^0}{\rho_2^0} = \frac{\omega_{02}^2}{\omega_{01}^2 + v_{T_1}^2 k^2} \tag{52.22}$$

or for the concentrations

$$\frac{\Delta N_1}{\Delta N_2} = \frac{m_2}{m_1} \frac{\omega_0^2}{\omega_{01}^2 + v_{T_1}^2 k^2} = \frac{\omega_{0_1}^2}{\omega_{01}^2 + v_{T_1}^2 k^2} \tag{52.23}$$

in connection with conditions (52.20), we get $\Delta N_2 \gg \Delta N_1$. This means that in oscillations of the type being considered ions and not electrons participate chiefly, which enables us to call them ionic oscillations.

Langmuir[+] discussed ionic oscillations. However, Langmuir starts from more particular presuppositions in his method. Langmuir's formula is as follows:

$$\omega_u^2 = \frac{\omega_{02}^2}{1 + \dfrac{\omega_{02}^2}{v_{T_1}^2 k^2}} . \tag{52.24}$$

In deducing this formula, no account was taken of: a) the compressibility of the ion gas; b) the acceleration in the electron gas. Formula (52.19) goes over into Langmuir's formula when the factors just mentioned are ignored.

Neglecting these factors, as Langmuir does, is valid for a narrow

[+] I. Langmuir and Tonks, Phys. Rev. 33, 195 (1929).

region of wave lengths; that is, it is required that $k^2 \ll \frac{\omega_{02}^2}{v_{T_2}^2}$ (and at the same time since we are speaking of ionic oscillations) $k^2 \gg \frac{\omega_{01}^2}{v_{T_1}^2}$. These two conditions determine the region within which the Langmuir approximation holds good.

4. <u>The static case in isothermal plasma</u>. The initial dispersion formula (52.6) includes the static case as well. Setting $\omega = 0$, we have:

$$\omega_{01}^2 v_{T_2}^2 + \omega_{02}^2 v_{T_1}^2 = - v_{T_1}^2 v_{T_2}^2 k^2 \tag{52.25}$$

or setting $v_{T_1}^2 = \frac{kT}{m_1}$, $v_{T_2}^2 = \frac{kT}{m_2}$, we have:

$$k^2 = - \frac{8\pi Ne^2}{kT}, \tag{52.26}$$

so that, consequently, k is an imaginary number. Hence, the solutions in this case coincide exactly with Debye's solutions for static polarization in electrolytes.

From (52.5), the relationship between ρ_1^0 and ρ_2^0 is

$$\frac{\rho_1^0}{\rho_2^0} = \frac{4\pi \frac{e_1 e_2}{m_1 m_2} \rho_{01}}{-\omega_{01}^2} = - \frac{m_1}{m_2}, \tag{52.27}$$

from which

$$\frac{\Delta N_1}{\Delta N_2} = -1, \tag{52.28}$$

that is, changes in the concentration for the static case are conditioned equally both by the ions and the electrons.

Section 53. Phenomenon of Anomalous-Violent Redistribution of Velocities of Beam of Electrons in Plasma, and Excitation of Oscillations

1. Retardation of charged particle, due to excitation of oscillations. The problem of the retardation of charged particles in gaseous-electron plasma has already been frequently discussed in the literature*. However, all the theories up to the present time have been based on conceptions of paired collisions, and, hence, the vibrational properties of the electron gas could not be taken into consideration; the question of the losses of energy in exciting the oscillations of the plasma remained essentially open. In the theory of paired collisions, calculation of the magnitude of energy losses leads to a divergent integral. The losses of energy over a unit of path are

$$-\frac{d\mathcal{E}^{st}}{dz}=\frac{E^2\omega_0^2}{v_0^2}\ln\frac{p}{a}, \qquad a=\frac{Ee(M+m)}{Mmv_0^2}. \tag{53.1}$$

The divergence of the total magnitude of the energy losses is conditioned by the interaction with electrons at great distances (greater than p). The presence of the divergence indicates, as we have already said, the important part played by distant interactions.

The theory of retardation, based on ideas of paired collisions, cannot be considered as satisfactory, since it:

1) Does not take into account the energy losses in exciting vibrations;

2) In general, does not take into account the action of distant forces;

3) Leads to indefiniteness in the final formula because of indefiniteness in choice of the maximum range distance.

We shall see below that solving the first difficulty will lead automatically to elimination of the last two.

Our problem consists in calculating the energy losses of a moving particle due to excitation of plasma oscillations. We have already seen that for velocities less than the velocity of sound, uniform rectilinear motion in plasma does not involve energy losses, because of the absence of emission of longitudinal waves. For higher velocities,

---

* See literature referred to in Sec. 1.

the moving charge is a source of longitudinal waves and, consequently, its motion involves energy losses.

Calculation of the energy losses may be conducted in two ways: either directly, by calculating the energy flow through the lateral surface of a cylinder whose axis coincides with the direction of motion, or by calculating the force with which the field of a moving charge acts on the charge itself. Both methods lead to the same result, as we showed. We confine ourselves here to conducting the calculations for the retarding force by the second method. For the calculation, we make use of the results obtained in Sec. 50 for calculating the potential in plasma for the case of a uniformly moving charge. To this end, we break up the expression for the potential into two parts

$$\Phi = \Phi_{\text{к}} + \Phi_{\text{п}},$$

where $\Phi_{\text{к}}$ is the Coulomb potential of the charge moving in a vacuum, and $\Phi_{\text{п}}$ is the potential arising out of the effect of polarization of the plasma by the moving charge. We then have for the force of retardation:

$$F_{\text{тор}} = -e \frac{\partial \Phi_{\text{п}}}{\partial z}\Big|_{r \to 0,\ t - \frac{z}{v_0} \to 0}. \tag{53.2}$$

The potential of the entire system has the form:

$$\Phi = \frac{e}{2\pi^2 v_0} \iiint_{-\infty}^{+\infty} d\omega e^{i\omega t'} \times$$

$$\times \frac{\omega^2\left(1 - \frac{v_T^2}{v_0^2}\right) - i\omega\alpha - v_T^2 k_\perp^2}{\left(k_\perp^2 + \frac{\omega^2}{v_0^2}\right)\left[\omega^2\left(1 - \frac{v_T^2}{v_0^2}\right) - i\alpha\omega - v_T^2 k_\perp^2 - \omega_0^2\right]} e^{ik_\perp r} dk_\perp. \tag{53.3}$$

Adding $\omega_0^2 + \omega_0^2$, to the numerator, we break up Φ into two integrals, $\Phi_{\text{к}} + \Phi_{\text{п}}$, where

$$\Phi_{к} = \frac{\varepsilon}{2\pi^2 v_0} \cdot \int\int\int\limits_{-\infty}^{+\infty} d\omega e^{i\omega t'} \frac{1}{k_\perp + \frac{\omega^2}{v_0^2}} e^{ik_\perp r} dk_\perp \tag{53.4}$$

and

$$\Phi_{п} = \frac{\varepsilon\omega_0^2}{2\pi^2 v_0} \int\int\int\limits_{-\infty}^{+\infty} d\omega e^{i\omega t'} \times$$
$$\times \frac{e^{ikr} dk_\perp}{\left(k_\perp^2 + \frac{\omega^2}{v_0^2}\right)\left[\omega^2\left(1 - \frac{v_T^2}{v_0^2}\right) - i\omega\nu - v_T^2 k_\perp^2 - \omega_0^2\right]}. \tag{53.5}$$

The first integral can be exactly calculated. For this, we introduce new variables $k_x = r\cos\varphi$ and $k_y = r\sin\varphi$. Integrating for φ and making use of (50.15), we have:

$$\Phi_{к} = \frac{\varepsilon}{\pi v_0} \int\limits_{-\infty}^{+\infty}\int\limits_0^{\infty} I_0(\tau r) \frac{e^{i\omega t'}}{\tau^2 + \frac{\omega^2}{v_0^2}} \tau\, d\tau\, d\omega, \tag{53.6}$$

and with the use of the theorm of residues, we have:

$$\int\limits_{-\infty}^{+\infty} \frac{e^{i\omega t'}}{\tau^2 + \frac{\omega^2}{v_0^2}} d\omega = \begin{cases} v_0 \pi e^{-v_0 t' \tau}, & t' > 0, \\ v_0 \pi e^{-v_0 |t'| \tau}, & t' < 0. \end{cases} \tag{53.7}$$

Consequently,

$$\Phi_{к} = \varepsilon \int\limits_0^{\infty} I_0(\tau r)\, e^{-v_0 t' \tau} d\tau, \tag{53.8}$$

This integral is well known in the theory of Bessel functions (Lipschitz integral):

$$\Phi_{\kappa} = \frac{e}{\sqrt{(v_0 t - z)^2 + r^2}}. \tag{53.9}$$

Hence, the Coulomb field of the charge is severed from the field due to polarization of the plasma.

To calculate the retarding force, we make use of the expression Φ_n . We consider at first the region of velocities less than the critical velocity of sound: $v_0 < v_T$. For ordinary conditions, in plasma $\omega_0 \gg \alpha$, so that there is no friction and we may set $\alpha = 0$. The expression under the integral in (53.3) then has a symmetrical distribution of its poles in the complex plane of (two poles in the upper half-plane and two poles in the lower). We may therefore expect a symmetrical distribution of the field in front of and behind the particle, and as a result, a zero value for the force of retardation.

Integrating with the help of the theorm of residues for ω , and making the substitutions $k_x = \tau \cos\varphi$ and $k_y = \tau \sin\varphi$, and integrating with respect to φ (similarly to what has preceded), we reduce the triple integral to a single one:

$$\Phi_n = -\varepsilon\omega_0^2 \int_0^{+\infty} \frac{I_0(\tau r)\, e^{-v_0 |t'| \tau}}{v_0^2\tau^2 + \omega_0^2}\, d\tau - \\ - \varepsilon\omega_0^3 \int_0^{\infty} I_0(\tau r) \frac{e^{-|t'| \sqrt{\frac{\omega_0^2}{\mu^2} + \frac{v_T^2}{\mu^2}\tau^2}}}{(v_0^2\tau^2 + \omega_0^2)\cdot \sqrt{\frac{\omega_0^2}{\mu^2} + \frac{v_T^2}{\mu^2}\tau^2}}\, \tau\, d\tau, \tag{53.10}$$

both for $t' = t - \frac{z}{v_0} > 0$, and for $t' < 0$, where $\mu^2 = \frac{v_T^2}{v_0^2} - 1$,

from which we see that for the force of retardation we actually obtain exactly zero

$$F_{\text{тор}} = -\varepsilon \frac{\partial \Phi_n}{\partial z}\Bigg|_{\substack{t - \frac{z}{v_0} \to 0 \\ r \to 0}} = 0. \tag{53.11}$$

We now pass to the case $v_0 > v_T$. Designating $1 - \frac{v_T^2}{v_0^2} = \mu_1^2$, we have:

$$\Phi = \frac{\varepsilon\omega_0^2}{(2\pi)^2 v_0} \int\int\int_{-\infty}^{+\infty} d\omega e^{i\omega t'} \frac{1}{\left(k_\perp^2 + \frac{\omega^2}{v_0^2}\right)(\mu_1^2\omega^2 - v_T^2 k_\perp^2 - \omega_0^2)} e^{ik r} dk. \tag{53.12}$$

Unlike the case $v_0 < v_T$, the expression under the integral now has an asymmetrical distribution of the poles in the complex plane: three poles in the upper half-plane and only one in the lower. Consequently, the fields in front of and behind the particle differ sharply, and as a result, a retarding force different from zero appears.

Integrating with the aid of the theory of residues in ω, introducing the change $k_x = \tau\cos\varphi$ and $k_y = \tau\sin\varphi$, and integrating with respect to φ, we obtain:

$$\Phi_n = -\varepsilon\omega_0 \int_0^\infty I_0(\tau r) e^{-v_0|t'|\tau} \frac{1}{v_0^2\tau^2 + \omega_0^2} d\tau, \qquad \left(\text{if} \quad t' = t - \frac{z}{v_0} < 0\right). \tag{53.13}$$

$$\Phi_n = -\varepsilon\omega_0 \int_0^\infty I_0(\tau r) \frac{e^{-v_0 t'\tau}}{v_0^2\tau^2 + \omega_0^2} d\tau - 2\varepsilon\omega_0 v_0 \int_0^\infty I_0(\tau r) \frac{(v_0\tau^2 + \omega_0^2)\sin t'\sqrt{\frac{\omega_0^2}{\mu_1^2} + \frac{v_T^2}{\mu_1^2}\tau^2}}{(v_0^4\tau^4 + 2\tau^2\omega_0^2 v_0^2 + \omega_0^4)\sqrt{\frac{\omega_0^2}{\mu_1^2} + \frac{v_T^2}{\mu_1^2}\tau^2}} \tau\, d\tau, \qquad \left(\text{if} \quad t' = t - \frac{z}{v_0} > 0\right). \tag{53.14}$$

Since it is the retardation that interests us, we pass to $\frac{\partial\Phi_n}{\partial z}$. Differentiating with respect to z and setting $t' = t - \frac{z}{v_0} = 0$, we have:

$$-\frac{\partial\Phi_{\text{п}}}{\partial z}\Big|_{t'=-0} = E_z^{(1)} = -\varepsilon\omega_0^2\int\limits_0^\infty \frac{I_0(\tau r)\,\tau\,d\tau}{v_0^2\tau^2+\omega_0^2},$$

$$-\frac{\partial\Phi_{\text{п}}}{\partial z}\Big|_{t'=+0} = E_z^{(2)} = -2\varepsilon\omega_0^2\int\limits_0^\infty I_0(\tau r)\frac{(v_0\tau^2+\omega_0^2)\,\tau\,d\tau}{v_0^4\tau^4+2\tau^2v_0^2\omega_0^2+\omega_0^4}+$$

$$+\varepsilon\omega_0^3\int\limits_0^\infty I_0(\tau r)\frac{\tau\cdot d\tau}{v_0^2\tau^2+\omega_0^2} = -\varepsilon\omega_0\int\limits_0^\infty I_0(\tau r)\frac{\tau\,d\tau}{v_0^2\tau^2+\omega_0^2},\tag{53.15}$$

from which we see that the field strength is discontinuous upon passing through the plane $t-\frac{z}{v_0}=0$.

Thus the strength of the electric field for $v_0 > v_T$ in the plane t' = 0, is equal to:

$$E_z = E_z^{(1)} = E_z^{(2)} = -\varepsilon\omega_0\int\limits_0^\infty I_0(\tau r)\frac{\tau\,d\tau}{v_0^2\tau^2+\omega_0^2}.\tag{53.16}$$

Calculation of this integral is performed as follows: we set

$$\frac{1}{\tau^2+\frac{\omega_0^2}{v_0^2}} = \int\limits_0^\infty e^{-\left(\tau^2+\frac{\omega_0^2}{v_0^2}\right)\eta}\,d\eta;\tag{53.17}$$

in which case

$$E_z = -\frac{\varepsilon\omega_0}{v_0^2}\iint\limits_0^\infty I_0(\tau r)\,e^{-\tau^2\eta}\,d\eta\,e^{-\frac{\omega_0^2}{v_0^2}\eta}\,d\tau.\tag{53.18}$$

Integrating with respect to τ can be performed by expanding in a series $I_0(\tau r)$. We easily obtain:

$$\int_0^\infty I_0(\tau r)\,\tau e^{-\tau^2\eta}\,d\tau = \frac{1}{2\eta} e^{-\frac{r^2}{4\eta}}. \tag{53.19}$$

Substituting this in E_z, we find:

$$E_z = -\frac{1}{2}\frac{\varepsilon\omega_0^2}{v_0^2}\int_0^\infty \frac{e^{-\left(\frac{\omega_0^2}{v_0^2}\cdot\eta + \frac{r^2}{4\eta}\right)}}{\eta}\,d\eta. \tag{53.20}$$

The last integral is Macdonald's function K_0 of zero order

$$E_z = -\frac{\varepsilon\omega_0^2}{v_0^2} K_0\left(\frac{\omega_0}{v_0} r\right). \tag{53.21}$$

Since $F_{\text{тор}} = \varepsilon E_z|_{r\to 0}$ and taking into consideration that

$$K_0(x) \underset{x\to 0}{\longrightarrow} \ln\frac{2}{\gamma x} \tag{53.22}$$

(γ being Euler's constant), we have:

$$F_{\text{тор}} = \frac{\varepsilon^2\omega_0^2}{v_0^2}\ln\frac{2v_0}{\gamma\omega_0 r}\left(\text{для } x \ll 1, \text{ т. е. } \frac{v_0}{\omega_0} r \ll 1\right). \tag{53.23}$$

The retarding force is easily expressed in terms of the energy loss over unit length $F_{\text{тор}} = -\frac{dE}{dz}$, and, thus we finally have for the loss of energy in the case $v_0 > v_T$:

$$-\frac{dE}{dz} = \frac{\varepsilon^2\omega_0^2}{v_0^2}\cdot\frac{1}{2}\ln\left(1 + \frac{r^2}{q^2}\right). \tag{53.24}$$

In the theory of paired collisions, the formula for the energy loss over unit path has the following form:

$$-\frac{dE_{st}}{dz} = \frac{\varepsilon^2\omega_0^2}{v_0^2}\int_0^r \frac{p\cdot dp}{p^2+a^2} = \frac{\varepsilon^2\omega_0^2}{v_0^2}\cdot\frac{1}{2}\ln\left(1+\frac{r^2}{a^2}\right), \qquad (53.25)$$

where $a = \frac{\varepsilon e(M+m)}{Mmv_0^2}$ and r is the maximum distance, which remains remains indeterminate in the theory of paired collisions.

Comparing (53.24) and (53.25), we see that both theories lead to diverging expressions, but whereas the theory of paired collisions gives divergency for $r\to\infty$, formula (53.24) diverges for $r \to 0$.

Since the quantity a is very small:

$$a \sim \frac{e^2}{mv_0^2} = \frac{e^2}{mc^2}\frac{c^2}{v_0^2} = 10^{-13}\frac{c^2}{v_0^2},$$

we see that for velocities that are not too small, we may consider $r_{max} \gg a$, in formula (53.24), and, consequently, formula (53.24) can be rewritten as:

$$-\frac{dE_{st}}{dz} = \frac{\varepsilon^2\omega_0^2}{v_0^2}\cdot\ln\frac{r}{a} \quad (\text{для } r\gg a). \qquad (53.26)$$

The sum of (53.24) and (53.26) gives as its result the magnitude of retardation independent of the indefiniteness in establishing the boundary (r has been shortened!). It should be taken into account that the condition exists

$$\frac{e}{mv_0^2} \ll r \ll \frac{v_0}{\omega_0}. \qquad (53.27)$$

It is this condition that defines the limit of applicability of the formula for the total energy as a sum of (53.23) and (53.26). For a sufficiently high velocity, condition (53.27) is always satisfied.

For

$$\frac{v_0}{\omega_0} = \frac{v_T v_0}{\omega_0 v_T} = r_0 \frac{v_0}{v_T} > r_0 \text{ и } \frac{e^2}{mv_0^2} < N^{-\frac{1}{3}}$$

(where r_0 is the Debye distance). Since for most cases $r_0 > N^{-\frac{1}{3}}$ then for $v_0 > v_T$ condition (53.27) always applies in practice.

2. Formulation of problem of anomalous dispersion of electrons. Some time ago, Langmuir experimentally established the anomaly in the dispersion of a beam of electrons entering a plasma. It was found that even at a small distance from the boundary of the plasma (from the cathode), the electrons of the beam are subject precisely to the Maxwell law of distribution by velocities. It was mysterious that the relaxation length, corresponding to the redistribution of velocities of electrons in the beam, should be as long as it was. By Langmuir's data, the observed relaxation length was approximately 1/6000 of the calculated length.

This phenomenon was repeatedly studied, experimentally and theoretically. Ditmer[+] and Pennig[++] found in their experiments that the dispersion of the electrons is accompanied by oscillations of the plasma. Druveystein and Warmoltz[+++] observed that the changes in velocity distribution, space potential and electron density take place in different regions. The most careful experimental investigation was made by Merrill and Webb[++++]; they studied in detail the relationship between the oscillations of the plasma and the dispersion of the beam of primary electrons. This research also disclosed new details of the phenomenon: it was found that the regions in which oscillations arise in the plasma as the effect of the beam, and the regions of dispersion have a special periodic structure. Their publication makes it possible to make a quantitative evaluation of the fundamental qualities that play a part in the phenomenon of the anomalous dispersion of electrons in plasma. The theoretical problem arising in connection with this phenomenon was formulated by Langmuir[+++++] in work dating from 1926 - 1928, and consisted in the need for explaining the sharp discrepancy between the experimental and theoretical values for the relaxation lengths.

+ A. Ditmer, Phys. Rev. 28, 507 (1926).
++ F. Pennig, Physica 6, 241 (1926).
+++ M. Druveystein and N. Warmoltz, Physica 4, 54 (1937).
++++ I. Merrill and Webb, Phys. Rev. 55, 1191 (1939).
+++++ I. Langmuir, Proc. Nat. Acad. 14, 627 (1928).

The difficulty of the problem lies in the fact that none of the possible hypotheses as to the dispersion mechanism is of the order of magnitude that is required by experiment. In all the work referred to above, the calculated lengths are ten thousand times and more the observed lengths.

Obviously, the following possible causes for the redistribution of electron velocities in the beam must be taken into account:

1) The interaction of the electrons of the beam with neutral atoms;

2) The interaction with ions;

3) The interaction with electrons of the plasma;

4) Excitation of oscillations in the plasma by the electrons of the beams;

5) Dispersion of the electrons along already existing oscillations of the plasma of thermal origin;

6) Dispersion by microinhomogeneities of the plasma (microfield);

7) Interaction between electrons of the beam itself;

8) Interaction with the field of radiation.

The influences of the first and last factors may be eliminated at once. In all cases of anomalous dispersion, the density of the neutral particles is such that the length of the free path between atoms exceeds even the dimensions of the vessel. In the experiments of Merrill and Webb, the pressure was three microns, $\lambda = \frac{1}{\pi\sigma N_A} \sim 10^2$ cm, while the observed distance from the cathode to the regions of dispersion is of the order of 0.5 cm, σ is the effective cross-section, which may be regarded as that of kinetic theory of gases for the velocities under consideration ($\sim$ 20 electron volts). A similar situation exists in the experiments of Langmuir and Mott-Smith, et al. Thus the presence of neutral atoms has no direct relationship to the phenomenon observed. The interaction with the field of radiation, comparable with the interaction between electrons, appears, as is known, only in the field of relativity velocities and, therefore, may be left out of consideration in the present case.

The interaction of electrons with ions is less than the interaction with electrons, because of the greater mass of the ions, and therefore it, too, may be ignored. The fundamental effect must be due to interaction with the plasma electrons, and the question is the correct calculation of this interaction.

Calculating the interaction of the electron beam with the plasma electrons by means of the theory of paired collisions have been made by a number of authors (Langmuir, Landau, Druveystein, Davydov, Gvozdover). All the treatments reduce in practice to a single formula, the individual instances differing only in the maximum range.

Actually, for the amount of energy lost per unit of path, we have:

$$-\frac{dE_{st}}{dz}=\frac{\varepsilon^2\omega_0^2}{v_0^2}\ln\frac{r_{\max}}{a},\quad a=\frac{e^2}{mv_0^2}. \tag{53.28}$$

The value of the relaxation length is determined by the formula

$$\frac{1}{\lambda_{st}}=\frac{1}{E}\left|\frac{dE^{st}}{dz}\right|, \tag{53.29}$$

from which

$$\lambda_{st}=\frac{v_0}{\omega_0}\,\frac{mv_0^2}{\frac{\omega_0}{v_0}\ln L},\quad \left(L=\frac{r_{\max}}{a}\right). \tag{53.30}$$

Under the conditions of the experiments of Merrill and Webb,

$v_0\approx 3\cdot 10^8$ cm/sec, $N=3\cdot 10^{10}$ el/cm^3, $\lambda_{st}\approx 10^6$ cm, the observed

magnitude is only several millimeters...

Under Langmuir's experimental conditions, the magnitude of the relaxation length as calculated from formula (53.30), is 6000 times the observed amount. It is thus obvious that the theory of paired collisions has no direct relationship to the observed relaxation lengths. Romp and Steenbek considered the action on the moving electron of chaotically distributed field in the space between the electrons of the plasma (action of microfield); this led to the appearance of a retarding effect, but the length of the relaxation line that they obtained is of the same order of magnitude as in the theory of paired collisions.

The experimental fact of the formation of oscillations in the dispersion of a beam led Ditmer to the hypothesis that the anomalous dispersion is due to the existence of oscillations in the electron plasma. We considered above the question of retardation as due to the excitation of plasma oscillations by a uniformly moving charge. The formula for retardation (53.24) obtained indicates that this effect as well, calculated for a single electron (see below), differs from the result of the theory of paired collisions only in the expression under the logarithm sign, that is, we obtain for the relaxation length the same order of magnitude as for the expression in the theory of paired impacts. However, the question still remains as to the influence on the dispersion of the beam of oscillations of the plasma of fluctuational origin. It will be natural, by analogy with the retardation of bound electrons due to the radiation fields, to call this an induced retardation due to thermal excitation of the plasma oscillations. Then the retardation effect analyzed above acts as a spontaneous emission of radiation. Langmuir had in view only the induced retardation. He made a rough estimate of the magnitude of this effect, which led once again to the same order of magnitude for the relaxation length.

A quantum-mechanical treatment makes it possible to take into account both types of retardation at once. This problem is discussed in detail in the monograph cited above. The expression obtained there for induced emission shows that under the conditions of a gaseous electron plasma, it is less than the spontaneous emission. We thus find that both the spontaneous and reduced retardation are unable to account for the observed phenomena for the relaxation lengths. If we take into account that the spontaneous and induced retardations exhaust the interaction between the moving electron and the oscillations of the plasma, it appears that in general the role of oscillations in the phenomena of dispersion is practically negligible.

However, in contrast to this, the experimental data indicate that the oscillations have an especial role in the dispersion phenomena. First, as Langmuir observed, the amplitudes of the observed oscillations considerably exceed those that correspond to the case of thermal equilibrium. In the second case, the experiments of Merrill and Webb established that the cause of dispersion is intense plasma oscillations occurring in confirmed regions. The cause for their excitation, and their spatially-periodic structure are unknown.

The problem of a rational theory for the phenomenon reduces not only to obtaining a relaxation length of the order of magnitude of the observed length, but also to explaining the entire complex of relationships occurring therewith.

The fundamental qualitative characteristics of the phenomenon are:

1) The regions of dispersion are small as compared to the distance from the cathode; these regions have a spatially-periodic structure;

2) To the regions of dispersion, there correspond regions of strong plasma oscillations with frequency approximately ω_0; these, too, have a periodic structure;

3) The distance from the cathode to the regions of dispersion (the magnitude playing the role of the relaxation length λ) is considerably less than the length given by the theory.

3. Role of fluctuational double layer at the plasma boundary. We desire to show that if we take into account, in addition to the space oscillations, to which we have confined ourselves hitherto, the surface oscillations of plasma as well, the contradictions to which reference has been made between theory and experience in connection with the relaxation lengths are eliminated, the properties of the phenomena, to which reference has also been made, are explained, and quantitative evaluation of the fundamental quantities, where this is possible, leads to a satisfactory agreement with experiment.

It seems at first sight that surface oscillations cannot have any essential influence on the dispersion of the electrons in the beam. This is not so, however, since it must be taken into consideration that small perturbations in the velocity of the electrons of the beam entail major consequences: the distances between the electrons of the beam will change proportionally to the time, and, consequently, this change can no longer be considered as small for large periods of time. The notion of the electrons is in this sense not stable. It is necessary, accordingly, to consider the role of superficial oscillations in detail.

We saw earlier that surface oscillations lead, on the boundaries of the plasma, to the appearance of a double electric layer of fluctuational origin. The characteristics of this layer are as follows: we have seen that the power of the double layer fluctuates with the characteristic frequency of the plasma, the thickness of the wave is small, on the order of the Debye distance, and the potential difference upon passing through the double layer is of the order of KT.

As the beam of electrons passes through such a double layer, it undergoes a velocity modulation. The role of this modulation essentially determines the density distribution of the beam beyong the layer. A similar problem in the modulation of a beam of electrons, only in that case in a vacuum, has already been analyzed in the literature in connection with the theory of radiation generators. The role of the double layer with variable potentials is played there by a pair of screens, with varying potential difference, placed perpendicularly to the beam.

The treatment of this problem for the case of a plasma is practically the same. For, as is shown by the result of subsequent calculations, the retarding action of the plasma leads to relaxation lengths considerably greater than the distances at which the modulation effect is of importance. This makes it possible to analyze the velocity modulation without taking into account the retardation action.

We explain the properties of the distribution of current density in the beam beyond the double layer. We follow a group of electrons, $j_0 dt_0$, passing through the double layer during time interval dt_0. Because of the change in velocities, the magnitude of this interval will be different at some distance x from the double layer.

If $eV(t_0)$ is the difference in the potential energy before and after the layer, then the change in the kinetic energy upon passing through the layer, in the case $\frac{mv_0^2}{2} \gg eV$ is

$$eV \approx v_0 v_1 m,$$

from which the variation of the velocity is determined $v_1 = v - v_0$.

The transit time to point x is equal to:

$$t = t_0 + \frac{x}{v_0 + v_1} = t + \frac{x}{v_0}\left(1 - \frac{v_1}{v_0}\right) \quad (v_1 \ll v_0), \tag{53.31}$$

from which

$$dt = dt_0\left(1 - \frac{x}{mv_0^3} \cdot \frac{deV}{dt_0}\right).$$

Let $V(t_0) = V_0 \sin \omega_0 t_0$: then

$$dt = dt_0\left(1 - \frac{xeV_0\omega_0}{mv_0^3} \cos \omega_0 t_0\right). \tag{53.32}$$

We seek those points at which dt = 0, that is, the points through which the electrons leaving in the course of the interval dt_0, pass through at the same time. It will be seen from (53.32) that such points exist. A number of authors have called these points <u>foci</u>. We have for them

$$x = \frac{v_0}{\omega_0} \cdot \frac{mv_0^2}{eV_0} \cdot \frac{1}{\cos \omega_0 t}. \tag{53.33}$$

The least value of x corresponding to the first focus is

$$x_{\min} = \frac{v_0}{\omega_0} \cdot \frac{mv_0^2}{eV_0}, \tag{53.34}$$

since by assumption $mv_0^2 \gg eV_0$, then $x_{\min} \gg \frac{v_0}{\omega_0}$.

It follows from the condition of the conservation of the number of particles $j_0 \, dt_0 = j \, dt$, using (53.32) for the current density j, that

$$j = j_0 \frac{1}{1 - \frac{x}{x_{\min}} \cos \omega_0 t_0}. \tag{53.35}$$

We have, for particle density N:

$$N = N_0 \frac{1}{1 - \frac{x}{x_{\min}} \cos \omega_0 t_0}. \tag{53.36}$$

These expressions, together with

$$t = t_0 + \frac{x}{v_0} \left(1 - \frac{eV_0}{mv_0^2} \cdot \sin \omega_0 t_0\right) \tag{53.37}$$

define j and N as functions of t and x.

As will be seen from (53.35) and (53.36), at point $x = x_{\min}$ B j and N become infinite at certain moments of time. The cause for the divergence is the fact that dt is equal to zero at the focus. It follows from this that the electrons are concentrated in a point at the focus.

In the case we consider, that of a fluctuational double layer $eV_0 \sim kT$. In this way, the first focus is located at a distance λ_Φ from the cathode:

$$\lambda_\Phi = \frac{v_0}{\omega_0} \cdot \frac{mv_0^2}{kT}. \qquad (53.38)$$

As was pointed out above, the discussion given is correct on condition that the relaxation lengths λ_{st} are greater in any case than λ_Φ, but

$$\lambda_{st} = \frac{v_0}{\omega_0} \cdot \frac{mv_0^2}{2e^2 \frac{\omega_0}{v_0} \ln L}, \qquad (53.39)$$

from which

$$\frac{\lambda_{st}}{\lambda_\Phi} = \frac{kT}{2e^2 \frac{\omega_0}{v_0} \ln L}. \qquad (53.40)$$

In all the cases of practical interest, this magnitude greatly exceeds unity. For example, in Merrill and Webb's experiments

$$\left.\begin{aligned} &v_0 \sim 3 \cdot 10^8 \text{ cm/sec} \\ &\omega_0 = 2\pi\nu_0 \sim 2\pi\, 10^9 \frac{1}{\text{sec.}}, \\ &T \sim 30000\ \text{K}, \\ &N \sim 3 \cdot 10^{10}\ \text{el/cm}^3 \end{aligned}\right\}$$

from which

$$\frac{\lambda_{st}}{\lambda_\Phi} \approx 10^6.$$

Thus, it is clear that phase focusing occurs much earlier than all the effects listed above can appear.

4. Amplitudes of oscillations and periodic structure of polarization. We now consider the extent to which the phase focusing effect of a beam of electrons influences the excitation of plasma oscillations.

We considered above, the polarization state of electron plasma around a charge moving in uniform rectilinear motion. It is a characteristic feature that at velocities exceeding the velocity of sound in the electron plasma, polarization is concentrated behind the charge (in the cone), and has an alternating periodic structure. In a motion of system of coordinates, this spatial periodicity is perceived as a fluctuation of the field with a frequency close to ω_0.

In the case of moving electrons in an unfocused current, this effect practically does not appear because the chaotically distributed electrons on the average mutually extinguish the alternating polarization. However, when there is phase focusing, the situation is quite different. At the focus (electrons situated in the focal plane), the polarization of the individual electron is not only mutually compensated, but is summated. In the case of sufficient density, when the polarization cones intersect, we should obtain a reinforcement of the polarization state of the plasma, but even if the electron density is so small that the cones practically do not overlap, because of the synchronous nature of the action of the electrons, we should still have reinforcement of the oscillations.

First, it is necessary to distinguish the cases of high and low densities of the electron in the beam, according to whether the polarization cones overlap or not. In the neighborhood of the focal plane, this overlapping is of the greatest importance, since at great distances, the electron density in the beam will decrease.

The criterion of greater density should be the condition that overlapping begins even with the first spatial period of the cone. For this it is necessary that the mean distance between the electrons in the focal plane be less than the radius of the cross-section of the cone at a distance of the first period from the vertex. If n_0 is the density of the electrons in an unfocused beam, then the surface density of electrons in focal plane N_ϕ is equal to the number of electrons going through the double layer in the time equal to half the period of the oscillations of the field in the double layer:

$$N_\phi = n_0 v_0 \frac{T_0}{2} = \pi \frac{v_0}{\omega_0} n_0 . \qquad (53.41)$$

The mean distance of electrons in the focal plane is

$$S = N_\phi^{-\frac{1}{2}}. \qquad (53.42)$$

A change in the plasma density because of the action of one moving electron has the following form (see Sec. 50):

$$\sigma = -m\frac{\varepsilon}{2\pi e}\frac{\omega_1^2}{v_T^2}\cdot\frac{1}{\beta_1}\frac{\cos a_1\sqrt{\frac{1}{\beta_1^2}[z - v_0(t-t_0)]^2 - r^2}}{\sqrt{\frac{1}{\beta_1^2}[z - v_0(t-t_0)]^2 - r^2}}\times$$

$$\times e^{-\frac{\alpha v_0}{2v_T^2\beta_1^2}[z - v_0(t-t_0)]} \tag{53.43}$$

for

$$r < \frac{1}{\beta_1}[z - v_0(t-t_0)] \tag{53.44}$$

and equal to zero in the remainder of the region, where

$$\beta_1^2 = \frac{v_0^2}{v_T^2} - 1,\quad a_1^2 = \frac{\omega_0^2}{v_T^2}\left[1 - \left(\frac{\alpha}{2\omega_0\mu_1}\right)^2\right],\quad \mu_1^2 = 1 - \frac{v_T^2}{v_0^2}. \tag{53.45}$$

We designate the position of the electron by the coordinates x_k, y_k, z_k; then

$$\left.\begin{aligned} r^2 &= (x - x_k)^2 + (y - y_k)^2,\\ z_k &= v_0(t - t_0). \end{aligned}\right\} \tag{53.46}$$

The radius of the cross-section of the cone is

$$r = (z_k - z)\frac{1}{\beta_1} = (z_k - z)\frac{1}{\sqrt{\frac{v_0^2}{v_T^2} - 1}}. \tag{53.47}$$

We set the distance from the vertex of the cone equal to the first period $(z_k - z) = 1$; we then have

$$a_1 \frac{1}{\beta_1}(z_k - z) = 2\pi; \; l = \frac{2\pi v_T}{\omega_0} \cdot \sqrt{\left(\frac{v_0^2}{v_T^2} - 1\right)\left[1 - \left(\frac{\alpha}{2\omega_0\mu_1}\right)^2\right]}. \tag{53.48}$$

The radius of the cross-section at a distance $\frac{l}{2}$ from the vertex will therefore be:

$$r = \frac{\pi}{2}\frac{v_T}{\omega_0}\frac{1}{\sqrt{.1 - \left(\frac{\alpha}{2\omega_0\mu_1}\right)^2}}. \tag{53.49}$$

Hence, the condition for high densities will be formulated as:

$$N_\Phi^{-\frac{1}{3}} \ll \frac{l}{2} \quad . \tag{53.50}$$

or

$$\frac{l^3}{4} \cdot N_\Phi = n_0 \frac{\pi^3}{4}\left(\frac{v_T}{\omega_0}\right)^3 \frac{v_0}{\omega_0} \cdot \left[1 - \left(\frac{\alpha}{2\omega_0\mu_1}\right)^2\right]^{-1} \gg 1. \tag{53.51}$$

For usual conditions in plasma (see below), the criterion obtained is completely justified.

To obtain the summated polarization Σ, we must integrate with respect to the coordinates of all the electrons, making use of the density n (Z_k) of the electrons in the beam (53.36)

$$\Sigma = -m \frac{\varepsilon}{2\pi c}\frac{\omega_0^2}{v_T^2} \times \frac{1}{\beta_1} \int\int\int \frac{\cos a_1 \sqrt{\frac{1}{\beta_1}(z - z_k)^2 - (x - x_k)^2 - (y - y_k)^2}}{\sqrt{\frac{1}{\beta_0^3}(z - z_k)^2 - (x - x_k)^2 - (y - y_k)}} \times$$

$$\times e^{-\eta (z_k - z)} n(z_k)\, dx_k\, dy_k\, dz_k \tag{53.52}$$

$$\left(\eta = \frac{\alpha v_0}{2v_T^3\beta_1^2}\right).$$

The integration reduces to the calculation of an integral I, where

$$I = \int n(z_k) e^{-\eta(z_k - z)} dz_k \int_0^{\frac{1}{\beta_1}(z_k - z)} \frac{\cos a_1 \sqrt{\frac{1}{\beta_1^2}(z - z_k)^2 - r^2}}{\sqrt{\frac{1}{\beta_1^2}(z - z_k)^2 - r_k^2}}\, dx_k dy_k. \tag{53.53}$$

Setting

$$\xi^2 = \frac{1}{\beta_1^2}(z_k - z)^2 - r^2, \text{ т. е. } r_k\,dr_k = -\sqrt{\frac{1}{\beta_1^2}(z - z_k)^2 - r_k^2}\,d\xi$$

We have:

$$I = \frac{2\pi}{a_1}\int_0^z e^{-\eta(z_k - z)} n(z_k) \sin a_1 \cdot \frac{1}{\beta_1}(z_k - z)\,dz_k. \tag{53.54}$$

The integral obtained is not taken directly, by virtue of the complexity of n (z_k). However, we can evaluate its fundamental value. Since n (z_k) has its greatest value at $z_k = z_\phi$, then

$$I = \frac{2\pi}{a_1} e^{-\eta(z_\phi - z)} \sin a_1 \cdot \frac{1}{\beta_1}(z_\phi - z)\int_0^z n(z_k)\,dz_k. \tag{53.55}$$

The number of electrons per unit of the focal plane is equal to:

$$\pi n_0 \frac{v_0}{\omega_0}.$$

Consequently,

$$\Sigma \cong -me\frac{\omega_0}{v_m}\cdot\frac{1}{\beta_1}\sin a_1\cdot\frac{1}{\beta_1}(z_\phi - z)\,\pi n_0\frac{v_0}{\omega_0}e^{-\eta(z_\phi - z)}. \tag{53.56}$$

It must be kept in mind that the formula obtained is, strictly speaking, correct in the case where the electron concentration in the beam differs from zero only on the focal plane. Therefore, the result obtained is valid only at a distance of a few periods from the focal plane. At greater distances, where no noticeable focusing can be observed any longer, the summated effect of polarization of the individual electrons on the average, because of the alternating nature of polarization, may be considered as compensated. Because of the complexity of the integral, it is not possible to say anything further as to the behavior of Σ at a distance from the focal plane.

The expression obtained for the summated polarization shows first that in the neighborhood of the focal plane, there is a periodic structure of polarization with the half period l:

$$l = \frac{2\pi}{\omega_0} v_0 \sqrt{1 - \left(\frac{v_T}{v_0}\right)^2} \qquad (53.57)$$

(for $\alpha \ll \omega_0$, $v_0 > v_T$), and secondly, that the polarized state of the medium is reinforced in the neighborhood of the focal plane. We compare the summated expression for the polarization with the amplitude of polarization set up by a single electron.

We have:

$$\frac{\Sigma}{\sigma} \cong 2\pi^2 \left(\frac{v_T}{v_0}\right)\left(\frac{v_0}{\omega_0}\right) n_0 z. \qquad (53.58)$$

At a distance $z = \frac{l}{2}$ (in the first half period)

$$\mu = \frac{\Sigma}{\sigma} \cong 2\pi^3 \left(\frac{v_0}{\omega_0}\right)^2 \cdot \frac{v_T}{\omega_0} \sqrt{1 - \frac{v_T^2}{v_0^2}}\, n_0. \qquad (53.59)$$

For $T = 30{,}000°$ K, $v_T \sim 0{,}6 \cdot 10^8$ cm/sec, $v_0 \sim 3 \cdot 10^8$ cm/sec,

$\frac{\omega_0}{2\pi} \sim 10^9$ 1/sec (the condition in Merrill and Webb's experiments)

$$\frac{v_T}{\omega_0} \sim 10^{-2}\ \text{см}, \quad \frac{v_0}{\omega_0} \sim 5 \cdot 10^{-2}\ \text{см},$$

from which

$$\mu \sim n_0 10^{-3}. \qquad (53.60)$$

Thus, for ordinary concentrations in a beam $\sim 10^7 - 10^9$, we have an increase of ten thousand times or more.

The intensity of the field set up by the polarization cloud of a similar electron is [see (53.21), (53.23)]

$$E_\sigma = \frac{\varepsilon\omega_0^2}{v_0^2} \ln Lg; \quad Lg = \frac{2v_0}{\gamma\omega_0 r_{min}}. \qquad (53.61)$$

Consequently, the summated intensity of the field is

$$E_z = \frac{\varepsilon \omega_0^2}{v_0^2} \mu \ln Lg. \tag{53.62}$$

Fields like this, not changing their sign, act over a half period. The energy change upon passing through such a region is equal to

$$eV = \frac{\varepsilon^3 \omega_0^2}{v_0^2} \mu \frac{l}{2} \ln Lg. \tag{53.63}$$

For velocities v_0 not very close to v_T (ignoring $\left(\frac{v_T}{v_0}\right)^2$ in comparison with 1, quantity $l = \frac{2\pi}{\omega_0} v_0$. Therefore

$$eV = \mu\pi \frac{\varepsilon^2 \omega_0}{v_0} \ln Lg. \tag{53.64}$$

Under the experimental conditions given above

$$\frac{\varepsilon^2 \omega_0}{v_0} \sim 5 \cdot 10^{-8} \text{ CGSE} \sim 3 \cdot 10^{-4} \text{ эл-в}$$

the change of energy amounts to

$$eV \sim 3 \cdot 10^{-4} \cdot 2\pi\mu \ln Lg,$$

but since μ is of the order of tens of thousands and more, the change in the energy constitutes several volts, that is, the amount required in the Merrill-Webb experiments. The ratio of e V and k T is, on the basis of (53.59) and (53.64)

$$\frac{eV}{kT} \approx 2\pi^3 \frac{n_0}{N} \frac{v_0}{v_T} \ln Lg. \tag{53.65}$$

It will be seen from this formula under what conditions eV may even considerably go beyond the magnitude of KT.

It may be said, on the basis of what has gone before, that the excitation of oscillations due to the electron-focusing effect, gives enough energy to account for the anomalously strong redistribution in velocities at distances greatly smaller than those given by the theory of paired collisions.

5. Comparison with experiments of Merrill and Webb. In Merrill and Webb's work, a detailed experimental investigation was made of the relationship between plasma oscillations and the dispersion of a beam

of primary electrons in an electron plasma.

In brief, the conditions of the experiment were as follows: In mercury vapors at pressures of 3 microns and 7 microns, for currents from 0.020 to 0.100, with oxidized cathode (with heating), an electron plasma was set up that was almost homogeneous in density. The voltage drop at the cathode was from 16 tO 21 v, and at the anode a negative drop from 5 to 8 v.

In the region between the electrodes (in the electron plasma) the potential was practically constant. The electron density was of the order of $3 \cdot 10^{10}$ el/cm^2 and T = 30,000° K. In this region was placed a tungsten wire 0.05 mm in diameter and 6 mm in length, serving as a probe. The volt-ampere characteristics of the probe were taken for motions of 0.1 mm. The probe circuit is usually closed through the anode, and included a circuit for detecting and measuring oscillations of high frequency caught by the probe. Definition of the distribution of velocities, densities and temperature is made by analyzing the volt-ampere characteristics according to Langmuir. The frequency of the oscillations was determined by means of a Lecher frame.

The results of the investigation are as follows:

1. In the region of an electron plasma located close to the cathode, there was a beam of primary electrons, which had not yet noticeably lost velocity, and secondary electrons of the plasma. At a distance of the order of several millimeters, the velocity distribution of primary electrons in the beam was practically monochromatic, with velocities ranging in various experiments from 16 to 21 v. At distances from the cathode of the order of several millimeters, no changes were noted in the distribution of velocities of dispersion or retardation.

2. Beginning with a certain distance (for example, 4.3 mm for electron concentration of $1.77 . 10^{10}$ el/cm^3 and velocities of primary electrons at 19.8 v), a first region was observed with correspondence to a redistribution of velocities in the beam, involving the formation of oscillations. The breadth of the region was small as compared with the distance to the cathode (of the order of 0.2 mm). The redistribution of velocities corresponded to a break in the distribution function. The change in velocity, both increase and decrease, came to approximately 6 el-v. From 4.5 mm to 6.1 mm, the distribution of velocities again became practically unchanged. At 6.1 mm, a second region of dispersion and formation of oscillations was found. Between this region and the anode, there was little dispersion and retardation.

Thus, there was observed a step-by-step structure of regions of dispersion and formation of plasma oscillations (up to three periods).

It was noted that the maximums of plasma oscillations were shifted from the corresponding points of dispersion by approximately half the distance between them.

The problem of the theory for these experiments includes:

1. Discovering the mechanism of the phenomenon and qualitatively explaining all the details observed in it:

a) The absence of any noteworthy dispersion and oscillations over the entire volume of the beam and the presence of strong oscillations and of dispersion only in a relatively small region,

b) The periodic structure of the regions of oscillation and dispersion,

c) The presence both of retardation and of acceleration of electrons in the field of strong redistribution of velocities.

2. Giving a quantitative evaluation of the characteristic quantities that appear:

a) The length from a cathode to the region of maximum dispersion and formation of oscillations,

b) The magnitude of the spatial period,

c) The frequency of the observed oscillations,

d) The change in energy per electron upon passing through the region of dispersion.

The results of the preceding two sections show that the mechanism of the phenomenon as explained there gives a complete explanation of the qualitative picture of the whole group of observed details. For:

1. The existence of the focusing plane guarantees the unity of phase of the plasma oscillations excited by each electron separately. Without the focal plane, the electrons are in various phases, and, hence, the alternating structure of polarization from the various electrons is compensated in the mean.

2. The periodic structure of the regions of oscillations involves a periodic structure for the summation polarization in the neighborhood of the focal plane.

3. The redistribution for velocities is a secondary effect, due to the presence of strong regions of oscillations.

4. The fact that dispersion is linked with the existence of a potential, changing of time, at the focal plane, likewise determines retardation and acceleration of electrons through this medium, depending on the phase of oscillations at which the passage takes place.

5. It follows from the mechanism adduced that the maximums of oscillation should coincide with maximums of dispersion. The presence of the observed small shift is explained, as mentioned by the authors themselves, by the peculiar nature of the operation of the

Quantative correspondence is given by the formulas obtained in the previous sections:

1. The focusing distance λ_ϕ (distance from the cathode to the first maximum of dispersion from the anode side) equals:

$$\lambda_\phi = \frac{v_0}{\omega_0} = \frac{mv_0^2}{kT},$$

where v_0 is the velocity of the electrons in the beam.

2. The spatial period (difference between maximum and minimum oscillation $l = \frac{1}{2}$) [see (53.57)]

$$l = 2\pi \frac{v_0}{\omega_0} \cdot \sqrt{1 - \frac{v_T^2}{v_0^2}},$$

where

$$v_T^2 = \frac{kT}{m}.$$

3. The frequency of oscillations

$$\omega = \frac{2\pi}{l} \cdot v_0 = \frac{\omega_0}{\sqrt{1 - \left(\frac{v_T}{v_0}\right)^2}}.$$

4. The energy change upon passing through the region of oscillations, calculated for a single electron [see (53.64)]

$$eV = \mu\pi \frac{\varepsilon^2 \omega_0}{v_0} \ln Lg,$$

$$\mu = 2\pi^3 n_0 \left(\frac{v_0}{\omega_0}\right)^2 \frac{v_T}{\omega_0} \cdot \sqrt{1 - \left(\frac{v_T}{v_0}\right)^2}.$$

The results of comparing theoretical and experimental values are given in the following table.

Temperature and pressure	Velocity of electrons in beam, in cm/sec	Electron concentration in el/cm^2	Frequency of vibrations $\times 10^9$ $\frac{1}{sec}$ Theory	Exp.	Focusing distance in mm Theor.	Exp.	Spatial period in mm Theor.	Exp.	Relaxation length according to theory of paired impacts
30,000° 7 micr.	$2.45 \cdot 10^8$	$1.93 \cdot 10^{10}$	1.23	1.17	4.5	5.0	0.96	0.75	
	2.42	3.03	1.58	1.34	3.3	4	0.77	0.5	
	2.45	3.85	1.76	-	2.9	3	0.55	0.4	
	2.39	3.65	2.14	-	2.3	-	-	-	
30,000° 3 micr.	1.65	1.77	1.20	1.18	6.0	6.0	1.1	1.0	st 10^6 cm
	2.72	2.56	1.44	1.14	5.0	4.5	0.9	0.5	
	2.72	3.38	1.64	1.50	4.4	4.1	0.8	0.5	
	2.73	3.72	1.73	-	4.4	-	-	-	
	2.78	3.94	1.79		4.4	2.0	-	-	

The value of eV is determined by the concentration of the electrons of the beam n_0. No data are given in Merrill and Webb's experiments. However, for probable values of these concentrations $n_0 \sim \frac{1}{10} N_0$ we obtain 10 el.-v for eV. The experimental value is $\sim$ 5 el.-v

$$\lambda_{st} = \lambda_{\Phi} \frac{kT}{2e^2 \frac{\omega_0^2}{v_0} \ln L} \sim 10^8 \text{ cm.}$$

Note to Table. The values of the experimental data in the first three columns are taken from a table attached to Merrill and Webb's publication. The experimental values for λ_ϕ and $\frac{1}{2}$ are determined from the graph in the same publication. The theoretical values for the magnitudes λ_ϕ, l, ω are determined from the formulas in points 1, 2, and 3 (see above).

The last column gives the values for the relaxation length determined from the theory of paired impact.

The comparison adduced showed that the quantitative agreement of the theoretical and experimental values may be considered as in general good.